Quantitative and Dynamic Plant Ecology

Third edition

Kenneth A. Kershaw F.R.S.C.
Professor of Biology

and

John Henry H. Looney
Research Associate
McMaster University, Ontario

Edward Arnold

First published as *Quantitative and Dynamic Ecology* 1964
by Edward Arnold (Publishers) Limited,
41 Bedford Square, London WC1B 3DQ

Reprinted 1966
Second Edition 1973
Reprinted 1974
Reprinted 1975
Reprinted 1978
Third Edition 1985

Edward Arnold (Australia) Pty Ltd,
80 Waverley Road, Caulfield East,
Victoria 3145, Australia

Edward Arnold,
300 North Charles Street,
Baltimore, Maryland 21201, U.S.A.

British Library Cataloguing in Publication Data
Kershaw, Kenneth A.
Quantitative and dynamic plant ecology.—3rd ed.
1. Botany—Ecology—Statistical methods
I. Title II. Looney, John Henry H.
581.5′072 QK901

ISBN 0 7131 2908 5

Printed in Great Britain by
Thomson Litho Ltd, East Kilbride, Scotland

Preface to the First Edition

The study of Ecology now necessitates a knowledge of many sciences in addition to Botany and Zoology and with the rapid development of statistics the student of Ecology is often faced with the problem of using and understanding the simpler statistical procedures which are in common use.

The aim of this book is to present the general theoretical concepts that have been formulated in relation to plant ecology together with the approaches and statistical methods that have been used in developing these concepts. The early work in plant ecology was, of necessity, largely concerned with the description of vegetation in relation to the environment. This branch of ecology is still very actively pursued and will continue to be of importance until all vegetation is described on a world-wide basis. From this early descriptive ecology came a realization of the difficulties of sampling vegetation and its dynamic nature, and consequently sampling measures were developed and tested and different measures of the abundance of a species devised. The concept of plant succession was also developed from this early descriptive work and pointed the way to many of the concepts in use today. With the basic idea of a dynamic vegetation accepted, further examination in still greater detail outlined some of the processes involved and demonstrated the close interrelationships that exist between species and between individuals of the same species. On reaching this level of development the necessity for sensitive sampling measures and statistical procedures to extract from a mass of data the relevant information became of paramount importance.

This book follows the general sequences of development outlined above. The early chapters are concerned with the methods of describing vegetation and the problem of sampling. The subsequent chapters deal with the fundamental concepts in use today. These include succession, cyclic change, plant interactions and the non-random distribution of individuals of a species in vegetation. The final chapters outline the controversy over the validity of the plant community as a distinct and recognizable entity and the methods that have been employed in this branch of ecology.

No detailed description of vegetation types or account of environmental factors and their measurement is attempted – both these aspects of ecology are now so extensive as to merit a book to themselves.

Finally, it gives me considerable pleasure to thank Dr N. Waloff of the Department of Zoology, Imperial College, for her reading of the draft manuscripts and her many valuable criticisms and suggestions. I also owe a considerable debt to Dr A.J. Rutter of this Department for the many hours he has spent in criticizing and discussing various sections of this book.

Imperial College, London, 1964 K.A.K.

Preface to the Second Edition

Since the publication of the first edition of this book there has been a tremendous increase of interest in classification and ordination of vegetational samples or units. As a result the original chapter on classification and ordination methods has been extensively revised and developed as two separate chapters. The emphasis is towards an understanding of the principles involved rather than an attempt at explaining the mathematical complexities. Two other developments in plant ecology, population studies and computer methods, are also well established areas of ecology necessitating a numerical approach. The former has been a gradual development over the past ten years and forms a new chapter (Chapter 6). The latter development has been more explosive and has also necessitated a chapter on computer methods (Chapter 11). This again is not intended as a detailed introduction of programming method but an attempt at a general outline of some of the more powerful features of the present generation computer. Chapter 12 covers the principles of computer simulation but avoids the mathematical treatments often mistakenly identified with simulation studies. Finally, new data have been incorporated into Chapters 4 and 8.

I am indebted to Dr G.P. Harris for his criticisms of the drafts of the new chapters and his helpful suggestions. I am grateful to him also together with Dr J.N.A. Lott for their permission to use some of their data on hysteresis effects in algal populations prior to its publication. I am most grateful to Mrs Janet Edmonds for her help in tracing references, and her invaluable assistance in the preparation of the manuscript for the press, and to Miss Judy Street, who has typed successive drafts.

Thanks are due to the Literary Executor of the late Sir Ronald A. Fisher, F.R.S., Dr Frank Yates, F.R.S. and to Longman Group Ltd., London, for permission to reprint the tables on pages 239–50 which are from *Statistical Tables for Biological, Agricultural and Medical Research* (formerly published by Oliver and Boyd, Edinburgh).

McMaster Univerity, Ontario, 1973 K.A.K.

Preface to the Third Edition

Over the last decade there has been a continued interest in ordination techniques and as a result we have included two additional chapters which cover the most recent advances. There has also been some debate as to what form simulated data must take if it is to be used to examine the level of distortion that may be present in an ordination method, if artificial data is acceptable at all in this context. Accordingly we have included as appendixes, two *field* data sets that are ideally suited for this purpose and which may perhaps be of use as a teaching aid at the undergraduate level. We have similarly included as an appendix a simple pattern/covariance data set together with its analysis in response to a number of requests that we have received for such material.

Encouragingly there has been some thoughtful re-examination of plant succession with some successful synthesis of these new ideas with mechanisms of species replacement and both the chapter on succession and the section on allelopathy have been updated. The extensive work on plant demography and plant population changes has produced disappointingly few concepts and largely has stubbornly remained at a descriptive level. Some of this new material has been included in Chapter 6. Finally, the explosion of computer applications and simulation studies has forced redundancy of the original chapters on these topics, and clearly the material is now so extensive and embraces such diverse forms it is no longer appropriate to attempt inclusion here.

We are grateful to a number of external reviewers of our proposed additions to this edition, who confirmed and amplified our original ideas, and particularly for the patience and cooperation of the word-processing staff at McMaster for their conscientious work with the manuscript production.

McMaster University Ontario, 1985 K.A.K. and J.H.H.L.

Contents

1

The Description of Vegetation

A considerable proportion of all ecological work in the past and to a considerable extent the present time, has been directed towards the description of vegetation. The object of such description is to enable people other than the observer to build a mental picture of an area and its vegetation and to allow the comparison and ultimate classification of different units of vegetation. Before any serious or detailed work can be commenced in an area it is necessary to know what species are present, their distribution and the relative degree of abundance of each species. Additional information on the morphology of the species characteristic of the area and a general account of the environment are also essential. Thus the ***floristic*** composition, simply expressed as a list of species, ***life form*** composition and ***structure*** of vegetation are a necessary basis of all ecological work; this approach to ecology has dominated the subject until quite recently and will continue to be prominent until an adequate description of all vegetation both on a regional and on a world-wide basis is completed.

Life form

The best-known description and classification of life forms, and the use of life form to construct a ***biological spectrum*** are due to Raunkiaer (1909, 1928, 1934). He arranged the life forms of species in a natural series in which the main criterion was the height of perennating buds, and he showed that this criterion appeared to reflect adaptation to climate. He based this series on the assumption that the original environment was considerably more moist and uniformly hot than it is now, and accordingly the most primitive life form is that which dominates tropical vegetation at the present time. Conversely the more highly evolved species are found in the colder areas of the world. Raunkiaer established the following types (see also Fig. 1.1):

Phanerophytes

The perennating buds or shoot apices borne on aerial shoots.

(*a*) EVERGREEN PHANEROPHYTES without bud scales (tropical tree species)

(*b*) EVERGREEN PHANEROPHYTES with bud scales

(*c*) DECIDUOUS PHANEROPHYTES with bud scales

(Each of the above groups can be sub-divided into Nanophanerophytes (<2 m in height), Microphanerophytes (2–8 m), Mesophanerophytes (8–30 m) and Megaphanerophytes (>30 m in height). Two further groups are usually included – epiphytic Phanerophytes and stem-succulent Phanerophytes.)

Chamaephytes

Perennating buds or shoot apices borne very close to the ground.

(*a*) SUFFRUTICOSE CHAMAEPHYTES Characterized by more or less erect aerial shoots which die away in part at the onset of an unfavourable season of the year. Perennating buds arise on the lower portions of the erect stems where they are less exposed to the environment. Very characteristic of many parts of the Mediterranean.

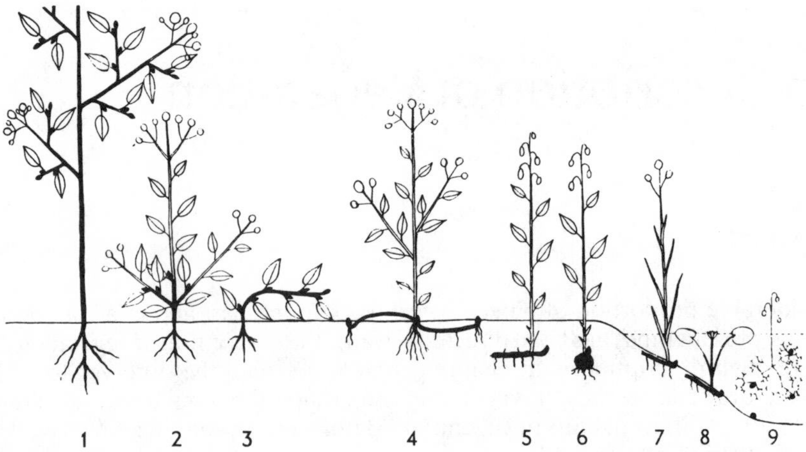

Fig. 1.1 The relative positions of the perennating parts of four life forms. 1 Phanerophytes, 2–3 Chamaephytes, 4 Hemicryptophytes and 5–9 Cryptophytes. The persistent axes and surviving buds are shown in black. (From Raunkiaer, 1934; courtesy of Clarendon Press, Oxford.)

(*b*) PASSIVE CHAMAEPHYTES Similar to the last group but at the onset of adverse conditions the weakened erect axes fall over and buds arise along the horizontal stems where at ground level they obtain some protection from the environment; e.g. *Stellaria holostea*.

(*c*) ACTIVE CHAMAEPHYTES The vegetative shoots are persistently orientated along the ground usually rooting along their length; e.g. *Lysimachia nummularia, Empetrum nigrum*.

(*d*) CUSHION PLANTS Basically a reduced and very compacted form of the last group.

Hemicryptophytes

Perennating buds at ground level, all above ground parts dying back at the onset of unfavourable conditions. Stolons may or may not be present.

(*a*) PROTO HEMICRYPTOPHYTES The lowermost leaves on the stem are less perfectly developed than the upper ones, the perennating buds arise at ground level; e.g. *Rubus idaeus*.

(*b*) PARTIAL ROSETTE PLANTS The best developed leaves form a rosette at the base of the aerial shoot, but some leaves are present on the aerial stems; e.g. *Ajuga reptans, Achillea millefolium*, etc.

(*c*) ROSETTE PLANTS The leaves are restricted to a rosette at the base of the aerial shoot; e.g. *Bellis perennis, Taraxacum officinale, Hypochaeris radicata*, etc.

Cryptophytes

Perennating buds below ground level or submerged in water.

(*a*) GEOPHYTES Rhizome, bulb or tuber geophytes, over-wintering by food stores under ground from which arise the buds to produce the next season's aerial shoots.

(*b*) HELOPHYTES Those plants which have their perennating organs in soil or much below water-level with aerial shoots above water-level; e.g. *Typha, Alisma*, etc.

(*c*) HYDROPHYTES Those plants with their perennating buds under water, and with their leaves submerged or floating. The buds may occur on rhizomes as in *Nuphar, Nymphaea*, etc, or winter buds may become detached from the plant and sink to the bottom of the water and there survive the unfavourable season (*Potamogeton obtusifolius, P. pusillus*, etc.).

Therophytes
Annual species which complete a life history from seed to seed during the favourable season of the year. Their life span can be as short as a few weeks, and they are characteristic of desert regions and cultivated soil where the interference of man protects them from their (slower growing) natural competitors.

Using the life form as a basis of comparison it is possible to compare areas which are widely separated geographically and do not contain species common to both (the normal basis of comparison of areas). The life form is assumed to have evolved in direct response to the climatic environment and accordingly the proportion of life forms in an area will give a good indication of its climatic zone. Thus the biological spectrum is a useful measure for comparison on a geographical scale and shows the number of species within each life form group (Table 1.1).

Table 1.1 The percentage representation of different life forms in the flora of Denmark (1084 species) and St. Thomas and St. Jan, West Indies (904 species). (From Raunkiaer 1934; courtesy of Clarendon Press, Oxford.)

	Denmark	**St. Thomas and St. Jan**
Mega and Mesophanerophytes (MM)	1	5
Microphanerophytes (M)	3	23
Nanophanerophytes (N)	3	30
Epiphytes (E)	0	1
Stem succulents (S)	0	2
Chamaephytes (Ch)	3	12
Hemicryptophytes (H)	50	9
Geophytes (G)	11	3
Hydrophytes and Helophytes (HH)	11	1
Therophytes (Th)	18	14

Raunkiaer gives several interesting examples from which he concludes that the zonation of vegetation progresses from a phanerophytic type in the tropical zone to therophytic in the sub-tropical desert areas, hemicryptophytic in the cold temperature and finally a chamaephytic type in the cold zone (Table 1.2). Similar trends can be seen with increase in altitude which reflects the increasing severity

Table 1.2 The biological spectrum for different geographical zones. (From Raunkiaer 1934; courtesy of Clarendon Press, Oxford.)

Area	**Percentage distribution of the species among the life forms** (key to symbols in Table 1.1)										
	S	E	MM	M	N	Ch	H	G	HH	Th	
Seychelles	1	3	10	23	24	6	12	3	2	16	Tropical forest
St. Thomas and St. Jan.	2	1	5	23	30	12	9	3	1	14	Tropical forest
Ellesmereland	—	—	—	—	—	23.5	65.5	8	3	—	Tundra
Baffinland	—	—	—	—	1	30	51	13	3	2	Tundra
Death Valley, California	3	—	—	2	21	7	18	2	5	42	Desert
Libyan Desert	—	—	—	3	9	21	20	4	1	42	Desert

of the climate. Thus there are more Chamaephytes and correspondingly fewer Hemicryptophytes, with increase in altitude in the Alps (Table 1.3). In general the life form spectrum is not of great significance in comparing adjacent areas of vegetation but is of value in the realm of 'Plant Geography'.

Table 1.3 The biological spectrum at different altitudes in the Alps. (From Raunkiaer 1934; courtesy of Clarendon Press, Oxford.)

The nival region of the Alps	**Percentage distribution of the species among the life forms** (key to symbols in Table 1.1)									
	S	E	MM	M	N	Ch	H	G	HH	Th
Above 3600 m						67	33			
3300–3600 m						58	42			
3150–3300 m						58	42			
3000–3150 m						52.5	45			2.5
2850–3000 m						33	62	2		3
2700–2850 m						33	61	3		3
2550–2700 m					0.5	28	65.5	2		4
2400–2550 m					1	24	67	4	0.3	4

Structure

The structure of vegetation is defined by three components: the vertical arrangement of species, i.e. the stratification of the vegetation; the horizontal arrangement of species, i.e. the spatial distribution of individuals; and, finally, the abundance of each species. The latter component of vegetation can be expressed in several ways, ranging from a direct count of the number of individuals in an area (density) to the dry weight of vegetable material produced in a given area (yield).

Stratification of vegetation

A casual inspection of a woodland will demonstrate the marked stratification of the vegetation ranging from two layers (a herb layer at ground level and the canopy of the tree layer) to a more complicated arrangement with a herb layer, several shrub and sapling layers, and the overall canopy of the tree layer. In tropical forest the stratification of the vegetation has been long recognized as one of its characteristic features, though the variation of height within any one stratum has led to some argument as to the distinctness of the various strata recognized by different workers. The same criteria of layering can be applied to herbaceous vegetation and detailed description of the stratification of vegetation forms an important part in the understanding and recording of its structure. The method of description of the stratification of forest vegetation is due to Davis and Richards (1933–4) who constructed, to scale, profile diagrams taken from narrow sample strips of tropical forest. (The concept of stratification is of obvious importance in any consideration of tropical forest and accordingly most of the published work on this topic relates to this type of vegetation.) The method consists of laying out a narrow rectangular sample plot, of required length and width (usually not less than 60 m in length, and 8 m has been found to be a convenient width). All vegetation lower than a chosen arbitrary height is cleared away, the remaining trees and shrubs being carefully mapped and the diameters of the trunks noted. The total height, height to first large branch, lower limit of crown and width of crown are measured. To obtain these measures it is often necessary to fell all the trees in the strip, commencing with the smallest trees so as to avoid the crushing of these smaller trees by the larger. From the measurements obtained an accurate scaled profile diagram can be obtained showing the spatial relation of the different species to each other both in a horizontal sense as well as a vertical sense.

Tropical forest is usually divided into five layers *A, B, C, D* and *E* (see Richards, 1952). The *A* layer is composed of the tallest trees present in the area (usually termed 'emergents'), the *B* and *C* layers form two strata of tree species, *D* the shrub layer and *E*, the lowest stratum, the ground or field layer. Two examples of profile diagrams are given below (Figs 1.2 and 1.3) taken from Richards (1936) and Beard (1949). They illustrate the differences between primary mixed Dipterocarp forest in

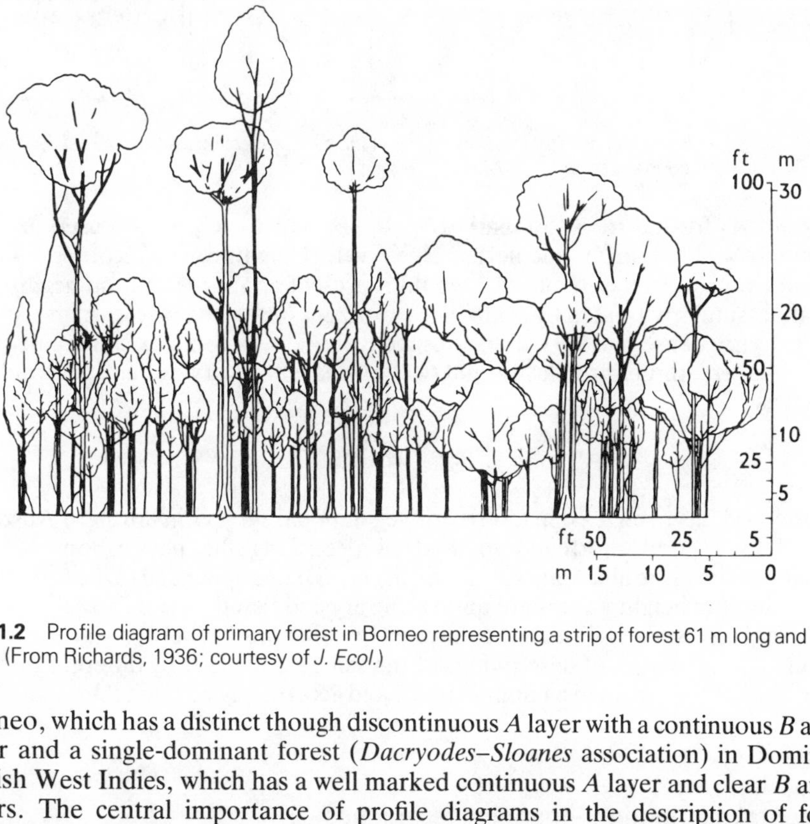

Fig. 1.2 Profile diagram of primary forest in Borneo representing a strip of forest 61 m long and 7.6 m wide. (From Richards, 1936; courtesy of *J. Ecol.*)

Borneo, which has a distinct though discontinuous *A* layer with a continuous *B* and *C* layer and a single-dominant forest (*Dacryodes–Sloanes* association) in Dominica. British West Indies, which has a well marked continuous *A* layer and clear *B* and *C* layers. The central importance of profile diagrams in the description of forest structure, particularly in the tropics, is evidenced by its long-established usage (see for example Proctor *et al.*, 1983).

Profile diagrams can also be usefully employed in vegetation of lower stature to illustrate the relationship between topography and the distribution of individuals of a species, or the distribution of root systems in relation to the topography and drainage of an area. Thus, the distribution of vegetation in relation to hummocks in calcareous mire in Teesdale (Pigott, 1956) is very conveniently expressed as a diagram (Fig. 1.4).

A simultaneous approach to stratification of vegetation, abundance and spatial distribution of each species can be made in a way suggested by Christian and Perry (1953) who developed a method especially suitable for large-scale surveys. Letters and figures are assigned to the tree, the shrub and ground stories, to their component layers and to the density of each. Thus, A_1 is used for low trees, A_2 for medium sized

Fig. 1.3 Profile diagram of primary forest in the West Indies representing a strip of forest 7.6 m wide. (From Beard 1949; courtesy of *Oxf. For. Mem.*)

trees and A_3 for tall trees. Similarly, B_1, B_2, B_3 and C_1, C_2, C_3 are used for low, mid-height and tall shrubs and herbs respectively. Heights can be recorded for each stratum as mean values. Thus, A_3^{15} would represent a tall tree 15 m in height, etc. Density is similarly treated by prefixing *x*, *y* or *z* for dense, average or sparse and *xx* or *zz* for very dense and very sparse respectively. Thus a complex community with two tree layers, three shrub layers and two grass layers at varying densities would be written as

$$A_3^{60}z,\ A_2^{30}y,\ B_3^{8}z,\ B_2^{3}x,\ B_1^{1}x,\ C_2^{1}zz,\ C_1^{3}zz$$

This method has been used in a series of vegetational surveys in northern Australia and offers a good shorthand method of describing the physiognomy of the vegetation, the more abundant species of any layer being appended to the formula. It is clear that the height groups are quite arbitrary and that the errors associated with such visual assessments should be controlled by collaboration between different field workers. Comparisons of descriptions of the same stand made by different workers would reduce the errors to a minimal level, and accordingly increase the value of the method.

Horizontal distribution – mapping techniques

The detailed mapping of vegetation is obviously a very laborious and time-consuming operation and its use is normally reserved for the detailed study of small areas of vegetation made over a period of years. The mapped areas are in the form of 'permanent quadrats'. Permanent quadrats are established and carefully marked out in such a manner as to enable the exact position to be located again in subsequent years. Each individual plant, or shoot, or clump of shoots is recorded accurately. This enables a composite picture of change in vegetation with time to be built up. It is advisable to sub-divide the permanent quadrat to facilitate the positioning of individuals within the area, and similarly it is more convenient to record the actual positions of the individual plants on 'ruled square' paper. The relationship between *Festuca ovina* and *Hieracium pilosella* in grassland after the removal of rabbit grazing has been investigated by Watt (1962) who gives a series of permanent quadrats

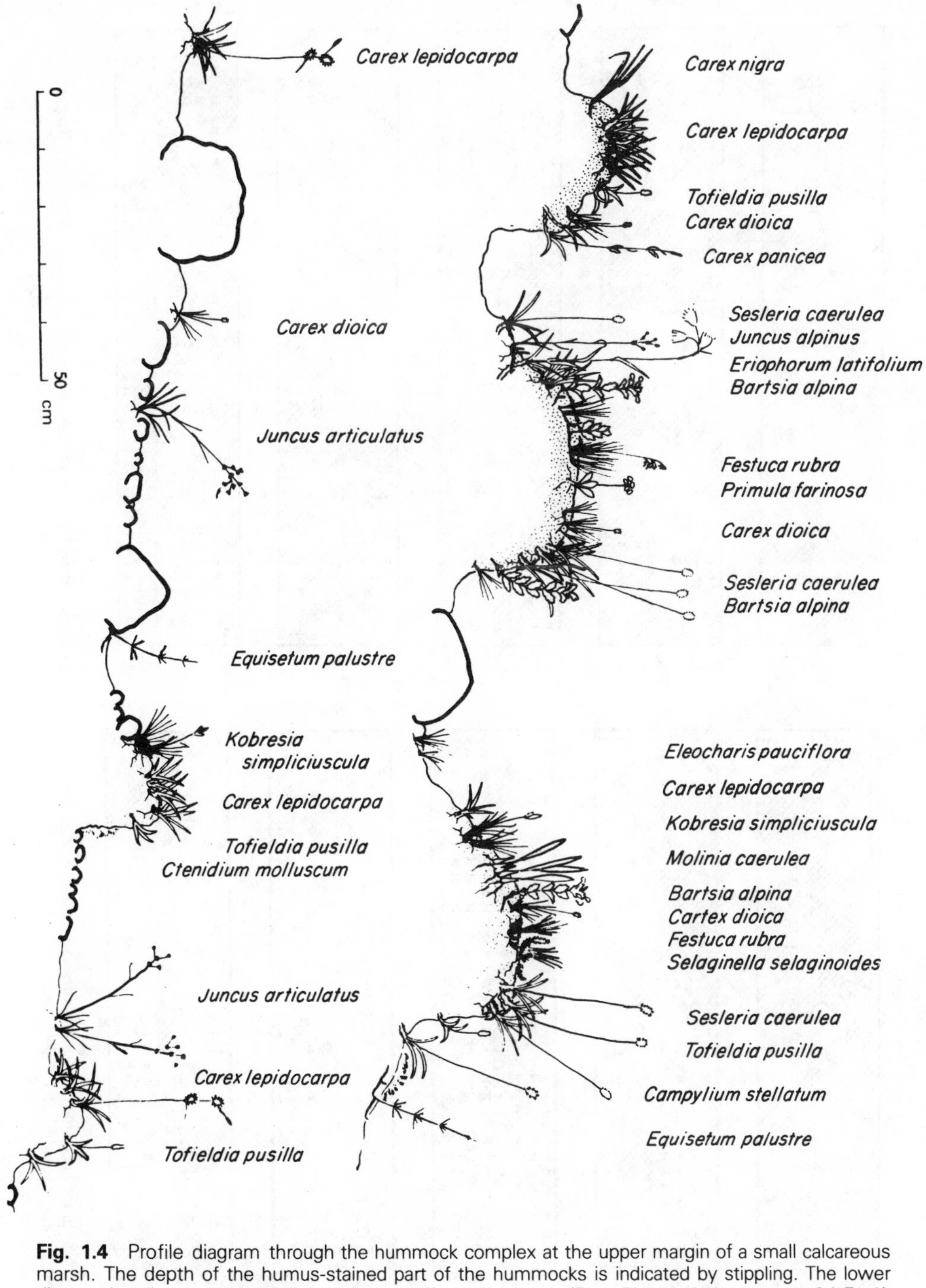

Fig. 1.4 Profile diagram through the hummock complex at the upper margin of a small calcareous marsh. The depth of the humus-stained part of the hummocks is indicated by stippling. The lower diagram is a direct continuation along the same line as the upper. (From Pigott, 1956; courtesy of *J. Ecol.*)

recorded at intervals from 1946 onwards (Fig. 1.5). The permanent quadrats show quite clearly the change over the years of the relative abundance of the two species. Austin (1981) has usefully summarized the continued importance of permanent quadrat studies and has outlined the required experimental designs which allow the

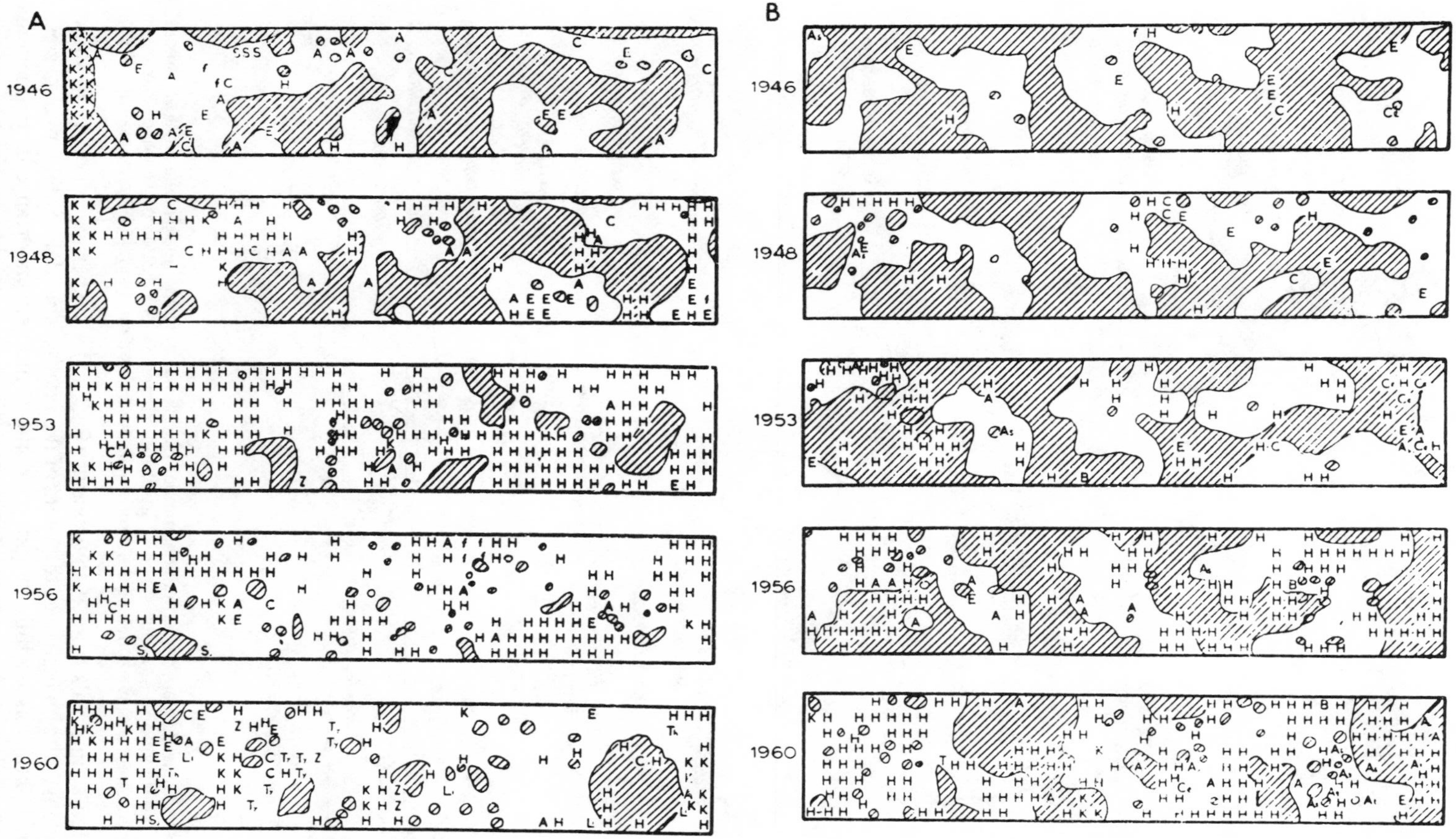

Fig. 1.5 The use of permanent quadrats to show the change of abundance of *Festuca* (shaded) and *Hieracium* (H) in enclosed (A) and control (B) plots. (From Watt, 1962; courtesy of *J. Ecol.*)

roles of location, contiguity, perturbation or synthetic experimentation, and demographic measurement, to be examined.

In general, the use of both permanent quadrats and the mapping of vegetation, as descriptive techniques, should be limited to those problems where the vegetational change is likely to be fairly rapid and at the same time very marked. Where vegetational change is less marked more refined techniques are necessary (see p. 11).

The subjective assessment of abundance

Frequency symbols

The simplest and most rapid way of describing vegetation is to list the species present within a sample area and to attach to each species a subjective assessment of its abundance. This method of approach has been used by numerous workers and several convenient classes of plant abundance have been employed. One of the most widespread of the subjective ratings used by British ecologists employs five classes of abundance: Dominant, Abundant, Frequent, Occasional and Rare with prefixes 'very' and 'locally' to qualify distribution and abundance where necessary. Thus a species of clumped habit, scattered over the area, would be referred to as locally frequent. Further classes can be used if thought necessary but it becomes increasingly difficult to distinguish between adjacent abundance classes as the total number of such classes increases. This in fact is one of the great weaknesses of this method of assessing the relative abundance of a species, and it is clear that other factors will often unknowingly influence the judgement of an observer. Small species tend to be rated lower than conspicuous ones, and similarly a species in full flower receives a higher rating than one which was in a vegetative state at the time of sampling.

Again, species that form obvious clumps are more conspicuous than species that are more evenly scattered and would be preferentially classed in an abundance scale. Thus such visual assessment of species abundance in an area includes a large unconscious error of judgement reflecting size, pattern of distribution and state of flowering of the species under consideration. Further the assessment is not constant from person to person and a species that is referred to as 'Frequent' by one field worker may well be classed as 'Abundant' or 'Occasional' by others. This problem has been discussed by Hope-Simpson (1940) who investigated the causes of error in subjective assessment of abundance, including such effects as annual fluctuation, seasonal change and choice of sampling site. These latter errors are common to most measures of plant abundance and do not concern us here. However, comparisons of the data, derived from 'ordinary' and 'specially careful' recordings, illustrate the inconsistency of subjective methods. It is possible to compare the data if one establishes 'agreement classes' rated as 'Exact', 'Approximate', 'Intermediate' agreement or complete 'Disagreement'.

Table 1.4 Comparison by agreement classes expressed as a percentage of the total number of species. (From Hope-Simpson, 1940; courtesy of *J. Ecol.*)

	Comparison of two independent samples from the same site	**Comparison of ordinary and specially careful recording**
Exact agreement	44	36
Approximate agreement	8	12
Intermediate	36	38
Disagreement	12	14

The data show clearly that subjective assessment of plant abundance contains a very large error and will, at the most, only give an approximate indication of abundance. Accordingly the use of frequency symbols should be strictly limited to initial description, and the degree of error present always borne in mind. A further drawback to the method lies in the rather unfortunate choice of the word 'Dominant' as the class representing the most abundant species, since in addition to the meaning here (the most abundant in a physiognomic sense) it is often confused with dominant in a sociological sense, meaning that species which exerts the most influence on the other species of the community. A species may be dominant in both senses of the word, but this need not necessarily be the case. For example Pearson (1942) has shown that regeneration of *Pinus ponderosa* depends on the type of ground vegetation. No regeneration occurs where the ground vegetation is *Festuca arizonica*, a species which grows during both the dry and the wet season, and can successfully compete with pine seedlings and prevent their establishment. On the other hand there is active regeneration when the ground vegetation is *Muhlenbergia montana* which only grows during the 'wet' season when water is plentiful for both species. In a sociological sense the dominant species is *Festuca arizonica*, whereas the immediate visual impression would put *Pinus* as the dominant. In many instances it is legitimate to equate the two definitions of 'dominant' but this is by no means always justified as the above example demonstrates. A further confusion that may arise is by the use of the word dominant in forestry literature as applied to a forest canopy, where those trees whose crowns are more than half exposed to full illumination are often referred to as 'dominants' (see Richards *et al.*, 1940). It is usual that a dominant species of tree in the latter sense is one which is dominant in a sociological sense (one exception is given above), and thus it is recommended that 'dominant' as a frequency symbol be rejected – it can be very readily replaced by the term 'very abundant', which to all intents and purposes is synonymous.

Braun-Blanquet's system of rating

A more elaborate and better defined approach has been suggested by Braun-Blanquet (1927) who proposed what is now a recognized procedure for describing a stand of vegetation (see also Poore, 1955). Two scales are used, one combining the number and cover of a species, and the second giving a measure of the grouping:

	+	= sparsely or very sparsely present; cover very small
	1	= plentiful but of small cover value
	2	= very numerous, or covering at least 5% of the area
	3	= any number of individuals covering 25–50% of the area
	4	= any number of individuals covering 50–75% of the area
	5	= covering more than 75% of the area
Soc.	1	= growing singly, isolated individuals
Soc.	2	= grouped or tufted
Soc.	3	= in small patches or cushions
Soc.	4	= in small colonies, in extensive patches, or forming carpets
Soc.	5	= in pure populations

The assessment of plant abundance and distribution is obtained from a series of adjacent quadrats of increasing size which are laid down so as to continuously increase the sample area (Fig. 1.6). An exact list of all plant species including lichens and bryophytes, with an attendant visual estimate of their sociability and abundance, is taken from the sample area. The measure of plant abundance used here is a definite improvement on the more traditional approach outlined previously.

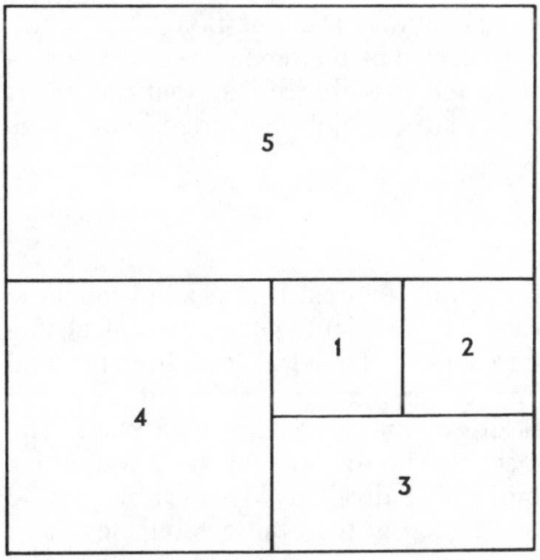

Fig. 1.6 The systematic arrangement of a series of quadrats to increase successively the size of the sample area. (From Poore, 1955; courtesy of *J. Ecol.*)

The scale combines, in part, abundance (number) with coverage of the area. It separates these to some extent from pattern of distribution and plant size which bias to an undefined extent the earlier abundance scale used. Accordingly the Braun-Blanquet scale has much to recommend it for a rapid survey of large areas of vegetation, since with practice it is possible for an individual recorder to minimize error from site to site. However, unconscious errors due to bias of different recorders will always be present and the extent to which they detract from the value of the method will remain debatable.

The Braun-Blanquet scale has several modifications and one of the more widely-used scales, i.e. the Domin scale, is set out in Table 1.5.

Table 1.5 The Domin and Braun-Blanquet scales.

	Domin scale	Braun-Blanquet scale
Cover about 100%	10	5
Cover > 75%	9	
Cover 50–75%	8	4
Cover 33–50%	7	3
Cover 25–33%	6	
Abundant, cover about 20%	5	2
Abundant, cover about 5%	4	
Scattered, cover small	3	1
Very scattered, cover small	2	
Scarce, cover small	1	
Isolated, cover small	x	x

The increased number of divisions enables a more detailed assessment to be made of plant coverage and field workers claim a high degree of accuracy can be obtained with sufficient experience. The Domin scale has the additional advantage in that it can be directly converted into the Braun-Blanquet scale wherever necessary and enable direct comparison of different sets of data.

Quantitative assessment of abundance

Since the realization that a large degree of error is inherent in subjective evaluation of abundance, ecologists have become increasingly conscious of the necessity of using quantitative measures to describe vegetation. The question as to which abundance measure should be used is often affected by the time it takes to obtain a sample of the population of adequate size, and it is true to say that the better quantitative measures available to ecologists are certainly but unavoidably time consuming.

Density

This is a count of the number of individuals within an area. It is usual to count the number of individuals within a series of randomly distributed quadrats, calculating the average number of individuals relative to the size of quadrat used, from the total sample. The method is accurate, allows direct comparison of different areas and different species and is an absolute measure of the abundance of a plant. The disadvantage of the method is usually associated with the time involved in counting what, potentially, can be very large numbers of individuals. This can in part be circumvented by counting within small quadrats, but there always remains those cases where the 'individual' cannot be picked out with any great degree of certainty, e.g. *Vaccinium myrtillus*, many grasses, sedges, etc. In some instances a convenient unit of the vegetation can be used instead of the complete individual; thus, individual tillers, flowering culms, erect shoots, or, indeed, whole clumps could be a suitable reflection of the density of a species. When these approaches fail, as for example with a species such as *Festuca rubra*, it is necessary to find an alternative approach.

Where density measurements are to be made in forest areas, a number of authors have emphasized the advantages of 'plotless' sampling by using a range of nearest neighbour measures. However, the methods that are most frequently used, because of their simplicity, all assume a random distribution of individuals. This then allows the distribution of distances to be converted into an estimate of density using estimators that have been appropriately derived for the range of methods. Greig-Smith (1983) has reviewed the range of methods and has emphasized that the assumptions of random distribution of individuals is not necessarily always correct. Thus Cottam and Curtis (1956) compared estimates from plotless sampling with complete enumerations and somewhat surprisingly found little discrepancy. However, from our now detailed knowledge of the extensive non-random nature of plant distribution (Chapters 4, 7 and 8, for example), it is strongly recommended that the advantages of direct random counts from plots far outweight the modest disadvantage of the time requirement for the establishment of the plot boundaries.

Cover

This is defined as the proportion of the ground occupied by perpendicular projection on to it of the aerial parts of individuals of the species under consideration (Greig-Smith, 1957, 1983). It is in fact an estimation of the area covered by given species, usually expressed as a percentage of the total area and estimated from a number of sample points. The cover value can be obtained either by a visual estimate, a certain percentage of the total area of a quadrat being covered by a given species, or measured by taking a number of points from the sample area and determining at those points which species if any is covering the surface of the ground. Visual estimates are obviously subject to the same personal bias as has been discussed already in relation to frequency symbols (p. 9) and should be avoided where possible, despite their relative speed and ease of use. Measures of cover are conveniently obtained using a frame of pins that can be adjusted to the height of the vegetation. The pins are lowered one at a time, and the species touched by each pin in turn recorded. The final number of 'hits' from a number of sample 'frames' is then expressed as a percentage of the total number of pins. It should be noted that the total percentage cover for all species in an area will nearly always exceed 100% since in closed vegetation leaves of several different species frequently overlap one another at a particular point.

Table 1.6 The dependence of estimated cover on the diameter of pin. (From Goodall, 1952*a*, courtesy of *Aust. J. Sci. Res.*)

Locality	Species	Number of points	Estimated cover using pin different diameters (mm)		
			0	1.84	4.75
Seaford	*Ammophila arenaria*	200	39.0	66.5	71.0
	Ammophila arenaria	200	60.5	74.0	82.0
Black Rock	*Ehrharta erecta*	200	74.5	87.0	93.5
Sorrento	*Lepidosperma concavum*	200	19.5	22.0	27.5
	Spinifex hirsutus	200	35.0	48.5	61.0
Carlton	*Fumaria officinalis*	200	20.5	31.5	30.0
	Ehrharta longiflora	200	24.5	25.5	37.5
	No contact	200	53.0	42.5	38.5
	Lolium perenne	200	65.0	85.5	82.5

The main error involved in this measure of plant abundance (apart from small personal errors which are inevitably present and are only of any consequence in tall vegetation) is the exaggeration of the estimate of percentage cover when a large diameter pin is used (Goodall, 1952*a*). Data on the effect of pin diameter are given in Table 1.6.

For the majority of the species there is a considerable difference which is statistically significant, between the percentage cover estimate obtained by using an optical cross-wire apparatus sighted on the vegetation (0 mm pin diameter) and the percentage cover reading given by the pins. This is simply due to the area of the tip of the pin being finite instead of zero. If absolute measures of percentage cover are required it is important to use an optical method, which will give a reliable measure of the true cover of a species. Conversely if data are being collected for purposes of comparison, providing the pin diameter is kept constant from site to site the exaggeration of the percentage cover is not of great importance except where the species tends to have a very high cover value. In the latter instance all areas will show a 100% cover value whereas in fact the true value may range from 70 to 80% in the different sites. If such very abundant species are to be included in a composite study it is necessary to use an optical method.

Percentage cover is a good measure of plant abundance and is very widely used in grassland where it is impossible to define where one individual starts and another ends. The main disadvantage of the method is the rather tedious and slow sampling involved. This can make the measure impractical in a large-scale survey for example. The difficulty in accurately observing the point of the needle in tall vegetation does lead to some error and there is a definite limit to the practicability of the method. The measure can be modified to take into account the stratified nature of the ground layer. The number of times a given species is encountered by a pin is recorded, rather than just the one hit noted in the normal measure of cover. Such a measure, often termed 'cover repetition', gives an estimate of the 'bulk' of plant material and can thus be related to yield (see p. 16). Personal error in the measure of cover repetition tends to be rather high, and if an investigation calls for such a detailed approach it is probably more satisfactory to measure directly the actual weight of plant material in a sample area.

Frequency

The frequency of a species is a measure of the chance of finding it with any one throw of a quadrat in a given area. Thus, if a species has a frequency of 10% then it should occur once in every ten quadrats examined. The measure is obtained very simply by noting whether a species is present or not in a series of randomly placed quadrats. For example, if 86 quadrats out of a total of 200 contain a given species the frequency is $86/200 \times 100 = 43\%$. If a value for the percentage frequency of a species within a single quadrat is required it can be simply obtained by sub-dividing the quadrat into a number of smaller units, then counting the number of sub-divisions within which the species occurs and expressing this as a percentage of the total number of sub-divisions. The results from a series of random quadrats can be finally averaged giving a value often termed local frequency.

Two convenient forms of frequency are used, ***shoot frequency*** and ***rooted frequency*** (Greig-Smith, 1957; 1983). The former is obtained by including as present all parts of the foliage that overlaps into a quadrat; the latter measure only includes a species as present when it is actually rooted within the area of the quadrat. The only advantage of frequency as a measure of abundance lies in the ease and rapidity with which an area can be sampled. Accordingly it has been very widely used as a measure of abundance, often without a realization of the grave disadvantages of the method. The error involved in estimating frequency compared with cover or density is negligible, but the final frequency figure obtained is dependent on several factors, which considerably offset the advantages of speed and facility.

(*a*) THE EFFECT OF QUADRAT SIZE It is clear from Fig. 1.7 that quadrats *A* and *B* of different size will give very widely differing frequency values for the species

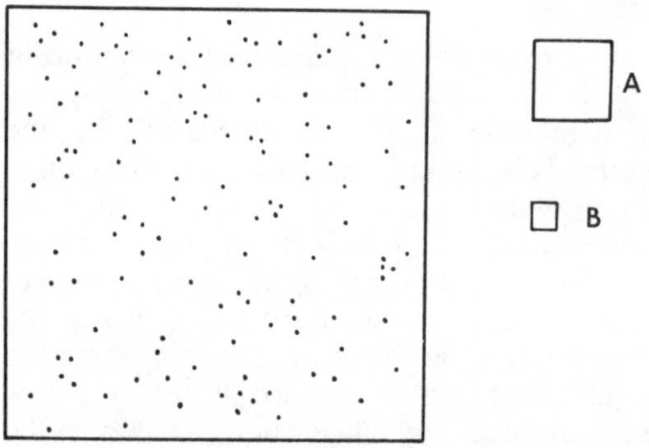

Fig. 1.7 The dependence of percentage frequency on quadrat size. The two sizes of quadrat A and B will give widely differing percentage frequency values from the same diagrammatic community.

represented in the layout. Quadrat A would give a frequency value of about 100 per cent whereas quadrat B would give a frequency value in the low intermediate range. Thus, frequency is dependent on quadrat size and it is important to state the size of quadrat used in an estimate of percentage frequency. Consequently if data are to be compared from different sample plots it is essential to use the same quadrat size throughout.

(*b*) THE EFFECT OF PLANT SIZE Again it is obvious that shoot frequency will vary markedly for the two species in Fig. 1.8. Both species have the same density but, using the size of quadrat shown, species B will have a high and species A a low percentage frequency. On the other hand if rooted frequency is used both species will have the same frequency value. Again it is essential that the type of frequency measure used is clearly stated.

(*c*) THE EFFECT OF SPATIAL DISTRIBUTION OF INDIVIDUALS Three distributions of individuals are shown in Fig. 1.9 all having the same density but a different

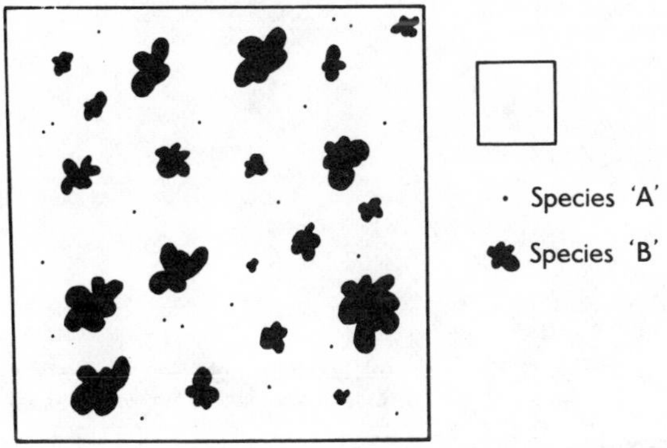

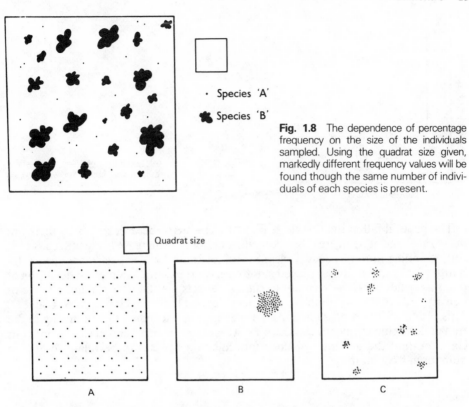

Fig. 1.8 The dependence of percentage frequency on the size of the individuals sampled. Using the quadrat size given, markedly different frequency values will be found though the same number of individuals of each species is present.

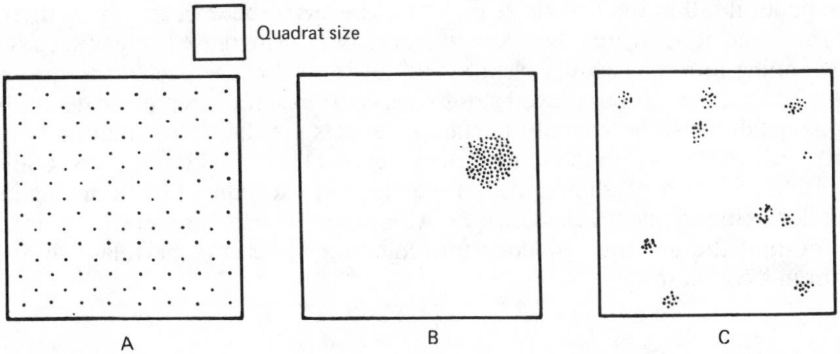

Fig. 1.9 The dependence of percentage frequency on the pattern of the distribution of individuals. With the same number of individuals present in each case and using the same size of quadrat, widely differing values will be obtained from the three communities.

pattern. With the quadrat size shown the estimated frequency of the species in A will be very high, in B very low and in C intermediate. In A the uniform distribution of individuals makes it almost impossible to take a random sample with zero presence, whereas the marked clumped arrangement in B gives the other extreme.

It is clear from the above considerations that frequency is not only dependent on the density of individuals but also on several other factors whose effects in any given sample area are usually unknown. Thus, despite the desirability of finding a simple conversion factor for transforming frequency data, which are readily obtained in the field, into density data, the situation is so complex as to make it impossible. Under random conditions it is possible to derive a satisfactory relationship, but random distributions of individuals in vegetation are infrequent and such a transformation is of academic interest rather than practical use. In general, frequency is a useful measure of abundance where comparisons are to be made on a large scale, as long as the limitations are borne in mind but it would seem worth while (despite the time necessary for the sampling) using some other method wherever possible.

A further point over which there has been considerable controversy is Raunkiaer's Law of Frequencies often referred to (in graphical form) as Raunkiaer's *J*-shaped distribution curves (Raunkiaer, 1928). The law of frequency simply states that the numbers of species of a community in the 5 percentage frequency classes, i.e. those of 0–20, 21–40, 41–60, 61–80 and 81–100 per cent frequency are distributed as in Fig. 1.10,

$$\text{i.e. } A > B > C \gtreqless D < E$$

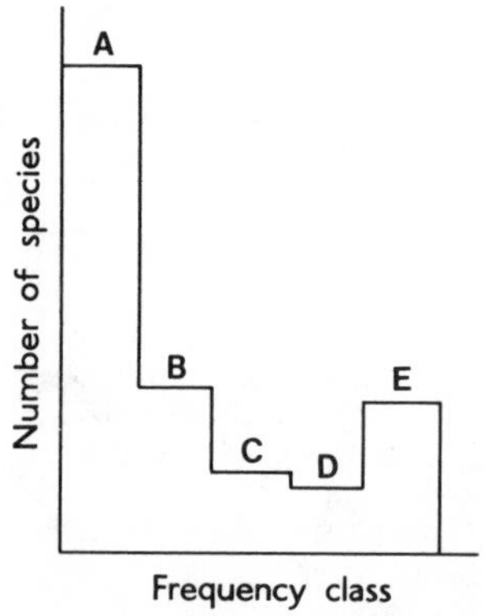

Fig. 1.10 The relationship between frequency class and the number of species falling within each of the classes, showing the typical *J*-shaped form of the curve.

The general fall in the first three or four classes is to be expected since there are more rare species in an area than common ones but the increase in the fifth class (*E*), corresponding to the frequency class 81–100%, is not to be expected and, as Greig-Smith (1957) states, 'It has played a notorious part in some attempts at definition of plant associations'. The increase in class *E* reflects the theoretical infinite range of density and contrasts with the more strictly defined limits for classes *A, B, C* and *D*. This fifth class has a possible density range greatly exceeding that occurring in the remainder of the frequency classes *A–D*. Accordingly many more species fall into this class, despite the general tendency for 'common' species to be relatively few in number in a community.

Yield

As a measure of abundance, yield is being used increasingly widely in ecological work. It is determined by clipping a series of sample quadrats, sorting the clippings, drying and weighing them, so that for each species present a figure for dry weight is available. Under field conditions it is very time consuming to sort out clippings into the component species and accordingly its use as a measure of abundance is limited. Where a series of culture experiments or a series of field trials are being undertaken, then yield offers a reliable and absolute measure of both abundance and vigour of a species.

Performance

In some instances it is necessary to know how well a plant is growing in a particular area and a measure of some suitable character can often be made, which reflects the relative vigour or performance, as it is termed, of an individual. Suitable measures include leaf length, leaf width or a ratio of the two, these measures being related to leaf area (the area available for photosynthesis), or flower number, length of flowering spike, number of seeds per capsule, etc., reflecting the reproductive capacity of an individual. The choice of character depends entirely on the species under investigation, but in general leaf length or width is often suitable and gives a good measure of performance. Such a measure was used by Kershaw (1960*b*, 1962*a*) on *Alchemilla alpina* and *Carex bigelowii* (pp. 70–1). Phillips (1954*b*) has shown that the ratio of the length of triangulate tip in a leaf of *Eriophorum angustifolium*, to the channelled portion of the leaf, is controlled by the relative growth rates of shoot apex and leaf primordia. As a result this ratio offers a good measure of performance of *Eriophorum*, and its general vigour under different environmental conditions can be readily determined in the field (Phillips, 1954*a*).

2

Sampling, Tests of Comparison, Application of Quadrat Measures

The importance of sampling a plant community so as to obtain the maximum amount of information from one set of samples, has been emphasized by numerous workers and Greig-Smith (1957; 1983) gives a comprehensive and detailed account of the problems involved. However, it is necessary to outline the more important aspects since a series of conclusions will be based on a single sample, and if the sampling procedure is incorrect, then the derived conclusions may be invalid. The importance of sampling has been stressed to the point where an impression is gained that the procedure is very complicated and in the hands of the uninitiated a dangerous tool. This is only partly true, since the approaches to this problem are all based on common sense. Often the correct procedure and also the one that will give the maximum amount of information on the problem in hand, can be readily devised by careful thought beforehand and with only an elementary knowledge of statistical methods. On the other hand since complex sampling methods are widely used advice from a statistician at an early stage may save wasted labour in the field. The necessity to carefully plan the sampling beforehand is important in yet another sense; recording the contents of a quadrat is nearly always a lengthy and a tedious operation and it is very much better to obtain all the required information from one set of quadrats rather than to have to go back and repeat the sample so as to obtain additional information. Thus no hard and fast rules can be laid down; each sampling procedure should be thought out in relation to a specific problem, the information required being borne in mind.

Statistical methods

In all ecological work there is a certain error involved in any field observations or experimental results. This error is partly due to the inadequacies of the equipment and, more important, to an unknown amount of variation always found in biological material. Thus it is necessary to know whether the difference between, for example, the number of buttercups in two plots is a real difference reflecting some variation in the environment of the two plots, or whether in fact this difference is more than covered by the error attached to the data. In order to assess the reality of the results it is necessary to compare the difference between them with an estimate of the error, or in other words to test the significance of the result.

Normally it is impractical to count every individual present within an area and accordingly a sample of the density is taken from the total population by using a number of randomly placed quadrats.

Variance and standard error

The following data represent the number of individuals of a species in ten quadrats placed at random in two sample areas:

Plot I	6	3	0	1	8	2	7	0	1	4
Plot II	7	11	15	8	14	9	12	14	15	10

The average density is obtained by summing the numbers and dividing by the number of quadrats in each plot:

$$\text{Plot I} \quad \bar{x} = \frac{\Sigma(x)}{n} = \frac{32}{10} = 3.2$$

$$\text{Plot II} \quad \bar{x} = \frac{\Sigma(x)}{n} = \frac{115}{10} = 11.5$$

The problem with which the observer is now faced is to decide whether the difference between these two mean values has an ecological significance or whether the sampling error is large enough to explain it. Both sets of data show a range of values reflecting the biological variation of the data and the range (8) is almost equal to the difference between the mean values. It would thus be reasonable to account for the difference in the means by this obvious biological variation. The variation of the density reading is expressed statistically as the ***variance*** of the sample and is calculated as follows:

$$\text{Variance} = \frac{\Sigma(x - \bar{x})^2}{n-1}$$

This is algebraically identical with

$$\text{Variance} = \frac{\Sigma(x^2) - \frac{(\Sigma x)^2}{n}}{n-1}$$

The latter offers an easier alternative to the rather laborious subtraction of the individual values from the mean value and squaring the results. The divisor $(n-1)$ is known as the degrees of freedom of the sample, and is used instead of n to compensate for using the sample mean instead of the true population mean. That is to say, if the whole population had been included in the sample, the mean density would be an accurate value. As it is, only ten quadrats were taken and the mean value thus obtained is only an estimate of the true mean, and accordingly the use of this sample mean will slightly bias the estimate of sample variance. By using $(n-1)$, some compensation for this bias is achieved. Usually the degrees of freedom of a sample are one less than the number of samples taken, though there are exceptions to this. The significance of the difference between the two mean values is assessed by comparing it with the standard error of the difference which itself is related to the variance. The calculation is as follows:

I $\quad \Sigma(x) = 32, \; n = 10, \; \bar{x} = 3.2$

$$\Sigma(x^2) = 6^2 + 3^2 + \cdots + 4^2 + 1^2 = 180$$

$$\Sigma(x)^2 = 1024$$

$$\Sigma(x^2) - \frac{(\Sigma x)^2}{n} = 180 - 102.4 = 77.6$$

$$\text{Variance} = \frac{77.6}{n-1} = 8.6$$

$$\text{II} \quad \Sigma(x) = 115, \; n = 10, \; \bar{x} = 11.5$$
$$\Sigma(x^2) = 1401$$
$$\Sigma(x)^2 = 13\,225$$
$$\Sigma(x^2) - \frac{(\Sigma x)^2}{n} = 78.5$$

$$\text{Variance} = \frac{78.5}{n-1} = 8.7$$

The estimate of the error of the difference between the two means is obtained from the variance of the total data.

Estimate of variance of the total data is given by

$$V = \frac{\left(\Sigma x_1^2 - \frac{(\Sigma x_1)^2}{n_1}\right) + \left(\Sigma x_2^2 - \frac{(\Sigma x_2)^2}{n_2}\right)}{n_1 + n_2 - 2} = \frac{77.6 + 78.5}{18} = 8.67$$

This is the variance of the total data, and the variance of the ***difference of the means*** accordingly is $V\left(\frac{I}{n_1} + \frac{I}{n_2}\right)$ and the standard error is

$$\sqrt{V\left(\frac{I}{n_1} + \frac{I}{n_2}\right)} = \sqrt{\frac{8.67}{5}} = \sqrt{1.734} = 1.317$$

t-Tests

The standard error of the difference is now used to assess the difference between the means by a *t*-test:

$$t = \frac{\text{Difference between the means}}{\text{Standard error of the difference}}$$

i.e. $\dfrac{11.5 - 3.2}{1.317} = 6.302$ with 18 degrees of freedom

i.e. the difference between the means is more than six times as great as the standard error of the difference and this difference is highly significant. The level of significance is readily obtained from a table of t (Fisher and Yates, 1957). With $n = 18$ the probability of exceeding a value of t of 3.92 by chance is very small ($p = 0.001$). Thus with $t = 6.3$ the difference between the two mean values arising by chance is very much less than 1000:1 and this difference is highly significant. [A 5% level of significance (1 in 20) is usually taken as the level of statistical significance.]

It is important to realize that a significant value of t does not necessarily mean a real difference between the two values. A large value of t can arise either by the difference between two mean values being significant, or by the variance of the two populations being significantly different, or both. Thus it may be necessary in comparing data to use a modified version of the *t*-test when the variance of the two samples is significantly different. This is readily confirmed by a variance ratio test. In the above example the two sample variances were 8.6 and 8.7 respectively and the ratio of the larger variance to the smaller is approximately 1. From variance ratio tables entered with 9×9 degrees of freedom this difference in variance is not significant (a variance ratio of 3.2 has to be exceeded before the difference can be regarded as statistically significant) and the use of a t-test is quite legitimate. (For details of the modified t-test, see Snedecor and Cochran, 1980.)

Confidence limits or fiducial limits

In a normal population with a mean of $\bar{x}$ the shape of the curve simply relates individual sizes or numbers to this mean value. Thus most of the population is roughly equal to the mean value but a proportion of the individuals are much larger or smaller. Just as the variance of the population is a measure of the variability then the ***standard deviation*** (standard deviation $= \sqrt{\text{variance}}$) is a linear standard of measurement along the x-axis (Fig. 2.1). The total population of individuals in fact is

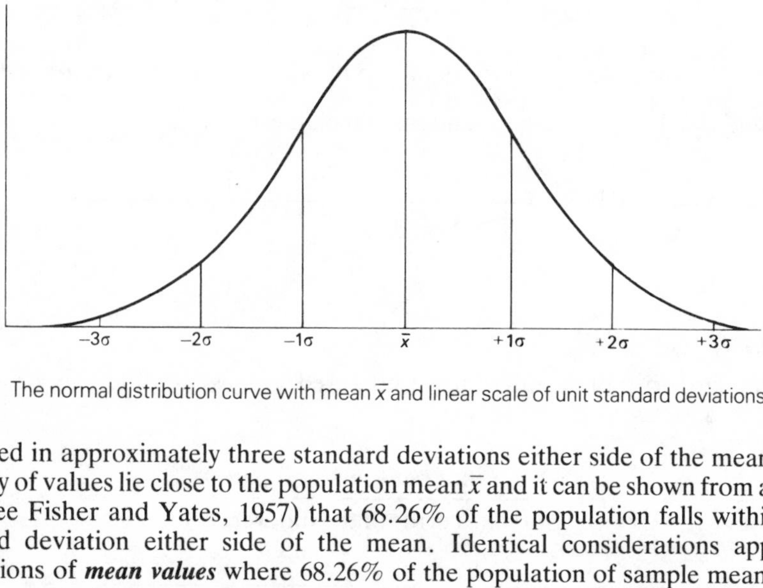

Fig. 2.1 The normal distribution curve with mean $\bar{x}$ and linear scale of unit standard deviations σ.

contained in approximately three standard deviations either side of the mean. The majority of values lie close to the population mean $\bar{x}$ and it can be shown from a table of C (see Fisher and Yates, 1957) that 68.26% of the population falls within one standard deviation either side of the mean. Identical considerations apply to populations of ***mean values*** where 68.26% of the population of sample means falls within one standard error of the true population mean value (S.E.M.). The 68.26% ***confidence limits*** are then simply:

$$\text{Upper limit} = \bar{x} + 1\ \text{S.E.M.}$$
$$\text{Lower limit} = \bar{x} - 1\ \text{S.E.M.}$$

Similarly for a 5% level of significance, the 95% confidence limits are:

$$\text{Upper limit} = \bar{x} + 1.96\ \text{S.E.M.}$$
$$\text{Lower limit} = \bar{x} - 1.96\ \text{S.E.M.}$$

and the 99% confidence limits

$$\bar{x} + 2.58\ \text{S.E.M.}$$

From the data above

$$\text{I} \qquad \text{Variance} = 8.6 \qquad n = 10$$

$$\text{S.E.M.} = \sqrt{\frac{V}{n}} = \sqrt{\frac{8.6}{10}} = 0.93$$

and the 95% confidence limits are:

$$\bar{x} \pm 1.96 \,.\, \text{S.E.M.} = 3.2 \pm 1.96 \times 0.93$$
$$= 3.2 \pm 1.82$$

The probability is 0.95 that 1.4 and 5.0 include the true population mean.

Similarly the 99% confidence limits are 3.2 ± 2.40. A statement of the confidence limits, as a scale mark in relation to graphed data, enables a rapid visual assessment of both the significance of any differences as well as the quality of the data.

χ^2-Test for association between species, scatter diagrams

Useful information can be obtained from field data on the possible inter-relationships between different species by counting the number of quadrats out of the total in which one or both of a pair of species are found. For example in n quadrats, a contained species A and B, b species B alone, c species A alone and d quadrats contained neither species A or B. A 2×2 contingency table is then made up as follows:

		Species A +	Species A −	
Species B	+	a	b	$a+b$
	−	c	d	$c+d$
		$a+c$	$b+d$	n

It can be calculated that the number of quadrats containing both A and B, provided they are not associated in any way, should be $[(a+b)(a+c)]/n$. Thus we have a, the observed number of quadrats containing both species and we can calculate the expected number assuming an independent relationship between the two species. The other expected frequencies of the different A and B quadrat combinations can readily be obtained simply by subtraction from the marginal totals. Thus the expected number of quadrats $+A$, $-B$ would be

$$(a+c) - \frac{(a+b)(a+c)}{n} \quad \text{and so on.}$$

The number of observed quadrat combinations of A and B can now be compared with the expected number by a χ^2-test.

The calculation of χ^2 is conveniently done from the formula

$$\chi^2 = \frac{(ad - bc)^2 \,.\, n}{(a+b)(c+d)(a+c)(b+d)}$$

The χ^2 tables are entered with one degree of freedom. The χ^2 value shows whether the difference between the observed and expected numbers of quadrats is significant or merely due to a chance fluctuation. The trend of the association, i.e. whether the association is positive or negative, is given by a comparison of the expected and observed values for the joint occurrence of *A* and *B*. Thus if *A* and *B* occur together *more* frequently than the expected number of joint occurrences, then *A* and *B* are positively associated. Similarly if *A* and *B* occur together *less* than expected they are negatively associated.

The level of association between species is very quickly calculated but since the data are basically a measure of frequency the result obtained is dependent on quadrat size. Consider the relationship of species *A*, *B* and *C* as shown below (Fig. 2.2). Clearly by inspection *A* and *B* are markedly positively associated; *A* and *C* and similarly *B* and *C*, negatively associated. Now consider the effect of sampling in a random fashion such a population with quadrat size 1, roughly equal in size to the individuals *A*, *B* and *C*. Most quadrats would contain either *A* or *B* or *C* or none of them, but hardly any quadrats with *AB*, *AC* or *BC*. Accordingly the χ^2-test would

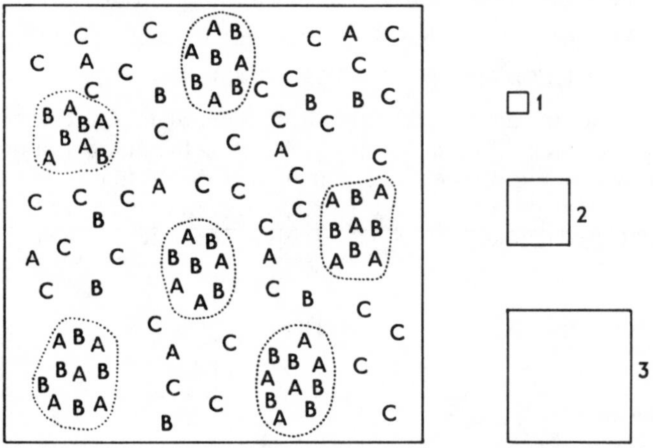

Fig. 2.2 The relationship between quadrat size and trend of association between three species *A*, *B* and *C* (see text).

show strong negative association between all species. With quadrat size 2, A and B would on the average be found frequently together and species C on its own. A χ^2-test would confirm the obvious trends of association between *A*, *B* and *C*. However consider finally the sampling of the layout with quadrat size 3. Nearly all quadrats would contain species *A*, *B* and *C* and, in this instance, the number of quadrats containing neither of the pairs of species would be zero and χ^2 cannot then be calculated. However, assume the same pattern but with a much lower density of individuals, then the final size of quadrat would show nil association or possibly a positive association between all paired combinations of the three species. It is possible to get three trends of association merely by changing the size of sample quadrat and considerable care must be taken in interpreting a set of positive and negative associations.

Obviously in this instance the pattern of distribution relative to the size of quadrat is important and, since most if not all vegetation contains pattern (see Chapter 8), the trend of association will always vary considerably with different sizes of quadrats. As Greig-Smith (1957, 1983) points out, very little attention has previously been directed to this important point and yet it has far reaching effects in the considerations of statistical analysis of community groupings (p. 156). If however the scale or scales of pattern are known, then the variation of trend of association between two species with increasing quadrat size gives valuable information as to the detailed interrelationships of the species in the area examined.

When data is in the form of a quantitative measure as opposed to mere presence and absence, then a correlation coefficient is a more appropriate measure of the degree of association; details of the calculation are obtainable in any book of elementary statistics (but see below).

An indication of a possible relationship between two factors can be very easily obtained by means of a scatter diagram. An example is given in Fig. 2.3 showing the

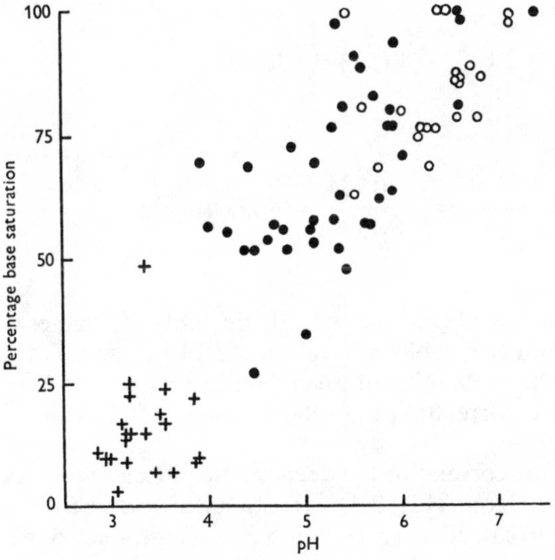

Fig. 2.3 The relationship between pH and base saturation expressed as a scatter diagram. (From Gorham, 1953; courtesy of *J. Ecol.*)

marked relationship between pH and percentage base saturation. The relationship is so clear-cut it is probably unnecessary for a correlation coefficient to be calculated; where the scatter of points becomes greater and the degree of correlation correspondingly lower, the visual confirmation of a suspected correlation becomes impossible and it is necessary to calculate the actual coefficient. A scatter diagram showing the relationship between percentage cover of *Festuca* and *Agrostis* in chalk grassland (see also Fig. 2.7) is given in Fig. 2.4, and illustrates a more typical degree of scatter. In most cases it is worthwhile to calculate the correlation coefficient since it can be very tempting to place too much weight on an apparent correlation but one that only has a few degrees of freedom. The raw data from Fig. 2.4 are given in Table 2.1 and the correlation coefficient is calculated as follows:

$n = 128$

$\Sigma x_1 = 2379, \qquad \Sigma x_1^2 = 63\,803, \qquad \frac{(\Sigma x_1)^2}{n} = 44\,216$

$$\Sigma(x_1 - \bar{x}_1)^2 = \Sigma x_1^2 - \frac{(\Sigma x_1)^2}{n}$$

$$= 63\,803 - 44\,216 = 19\,587$$

$\Sigma x_2 = 2183, \qquad \Sigma x_2^2 = 50\,413, \qquad \frac{(\Sigma x_2)^2}{n} = 37\,230$

$$\Sigma(x_2 - \bar{x}_2)^2 = \Sigma x_2^2 - \frac{(\Sigma x_2)^2}{n}$$

$$= 50\,413 - 27\,230 = 13\,183$$

$\Sigma(x_1 \,.\, x_2) = 29\,829, \qquad \frac{\Sigma x_1 \,.\, \Sigma x_2}{n} = 40\,573$

$$\Sigma[(x_1 - \bar{x}_1)\,.\,(x_2 - \bar{x}_2)] = \Sigma x_1 x_2 - \frac{\Sigma x_1 \,.\, \Sigma x_2}{n}$$

$$= 29\,829 - 40\,573 = -10\,744$$

Correlation coefficient

$$r = \frac{\Sigma[(x_1 - \bar{x}_1)\,(x_2 - \bar{x}_2)]}{\sqrt{[\Sigma(x_1 - \bar{x}_1)^2 \,.\, \Sigma(x_2 - \bar{x}_2)^2]}} = \frac{-10\,744}{16\,069} = -0.669$$

The number of pairs of observations is 128 and accordingly the table of r-values is entered with 126 degrees of freedom (for table of r-values see Fisher and Yates 1948). For 100 degrees of freedom the probability of exceeding an r-value of 0.3211 by chance is 0.001, i.e. the negative correlation between *Festuca* and *Agrostis* is a highly significant one.

This result is not unexpected and the correlation between the two species is in fact readily appreciated from the scatter diagram (Fig. 2.4). It is often worthwhile to plot a scatter diagram before attempting to calculate a correlation coefficient, since those cases which are clearly not related in any way can be rejected without resorting to what can be a lengthy computation. Valuable information can thus be readily obtained of the level of association between pairs of species or of correlation between

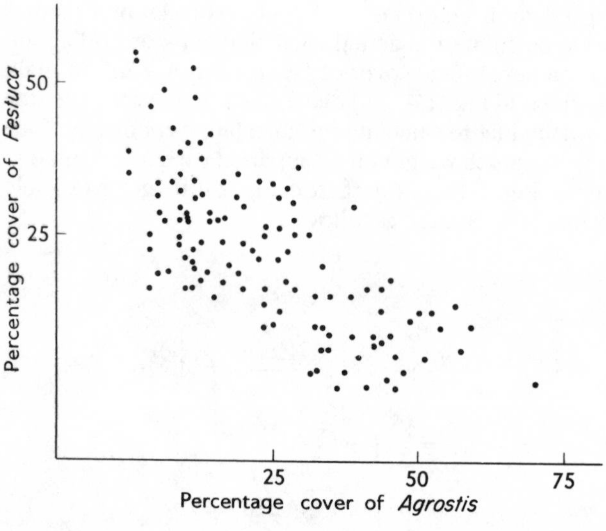

Fig. 2.4 A scatter diagram of the relation between the percentage covers of *Festuca* and *Agrostis* in chalk grassland.

Table 2.1 (see text)

***Agrostis tenuis* cover values**

	7	11	11	3	23	7	11	10	7	13	21	13	18	8	1					
	12	9	4	3	3	6	1	9	8	5	4	4	3	8	6					
x_1	7	5	15	13	21	42	36	40	31	34	19	37	35	47	20					
	31	36	13	16	22	46	35	31	18	9	27	19	16	22	19					
				12	21	7	10	9	9	8	–	10	5	–	7	33	9	3	19	15
				11	10	25	27	15	7	6	8	14	8	22	17	4	23	7	10	11
				23	19	7	7	30	23	16	19	39	35	23	26	32	26	20	21	34
				36	27	20	26	29	28	45	43	28	38	25	56	41	28	33	26	37

***Festuca rubra* cover values**

	22	22	33	37	24	22	1	30	19	22	21	19	17	32	43						
	22	38	25	20	18	34	43	27	13	30	27	15	13	22	15						
x_2	31	22	15	14	17	10	7	10	24	7	11	4	6	8	8						
	7	14	27	19	14	5	13	12	27	25	16	8	13	18	–						
					12	25	26	25	18	16	23	28	32	29	31	28	13	42	28	20	28
					15	19	20	8	25	19	27	17	16	22	26	18	23	13	6	14	23
					29	13	24	7	2	24	24	21	9	10	20	8	4	12	24	10	6
					1	5	1	2	–	7	11	8	12	2	2	1	4	5	–	2	–

a species and the environmental factors, but such analyses can be under-taken only when the initial collection of data is adequately made. When designing a sampling method it is always advisable to bear in mind the possibility that there will be some degree of association or correlation between the species themselves and also with the environment, and wherever possible one should obtain the field data in a quantitative form. Where data exist that are partly (or wholly) qualitative, the analysis can present some difficulty and the resulting information is often limited. (For detailed consideration of such methods the student is referred to Greig-Smith (1957).) *It is important to realize that trends in correlation will depend on the sizes of the quadrats which are used in the sampling* and results should always be considered with this fact in mind.

When one of the variables is dependent on the value of the other, an *alternative* approach is to express the relationship between them as a ***linear regression***, effectively a linear equation that will *predict* the value of the dependent variate from the independent variate. The reader is referred to Clarke (1980), Parker (1979) for suitable treatments of regression analysis. Where the relationship in the scatter diagram is clearly not linear it is more appropriate to apply a linear regression to the ***logarithm*** of the independent variate or, in more complex situations, to utilize a ***curvilinear*** regression. In other situations where information is available perhaps on *a range* of environmental parameters (for example, soil pH, together with a number of elemental soil analyses) a ***multiple*** regression analysis is a suitable approach for examining *the relative importance* and significance of each of the variables together with their ***combinations***. The reader is referred to Hunt (1978, 1982), Clarke (1980), and Snedecor and Cochran (1980) for an introductory treatment of curvilinear and multiple regression analysis, and to Snedecor and Cochran (1980), and Draper and Smith (1981) for a more advanced treatment.

Sampling methods

Random sampling

It is usually necessary to distribute a set of quadrats in an area in such a way that the position of each quadrat is independent of all the other quadrats and is also independent of any prominent feature of the area. This may be made clear by considering the hypothetical case where a series of quadrats are being taken from a sample plot containing a few large and conspicuous thistles. Subconsciously, if not consciously, the very size of the thistles will lead to a feeling that they *really ought* to be included in the sample. If the randomization of the quadrats is being attempted by the closing of eyes or throwing over the left shoulder, then inevitably the tendency is to try to include at least *one* thistle in a quadrat throw. This leads to over-sampling of the area closely adjacent to the thistles. This error is surprisingly common and very often leads to completely biased sampling since the effect of *trying* to make a random throw often produces a final distribution of quadrats which are too regularly distributed. Thus conscious or unconscious prejudgement has to be eliminated, otherwise the final non-random sample will be of little use for further comparison. The use of a *t*-test, or the calculation of any standard error, demands that the original sample should be taken at random. If there has been any introduction of non-random sampling then the calculated standard error is a fictitious figure and a statistical comparison is not valid.

Size of sample necessary

It is impossible to make a general rule as to the number of sample quadrats or measurements to take from an area. Each case has to be decided independently and considered against the amount of sampling involved to reach a certain level of accuracy. It is possible to obtain a subjective assessment at least of the size of sample necessary for any given problem. The method consists of taking a sample of five quadrats or readings and calculating a mean value. The number of samples is then increased to 10, 15, 20 . . . and a fresh mean is calculated each time. A graph relating mean value to the sample size is constructed and the point (i.e. the sample size) at which the mean value ceases to fluctuate is easily determined. An example is given in Fig. 2.5 for performance data of *Calamagrostis neglecta* expressed as leaf length. The curve fluctuates considerably up to a sample size of about 50 after which there is little variation and a sample size of 80 would give a reasonably accurate measurement of mean performance in that particular site. (For additional methods of estimating the

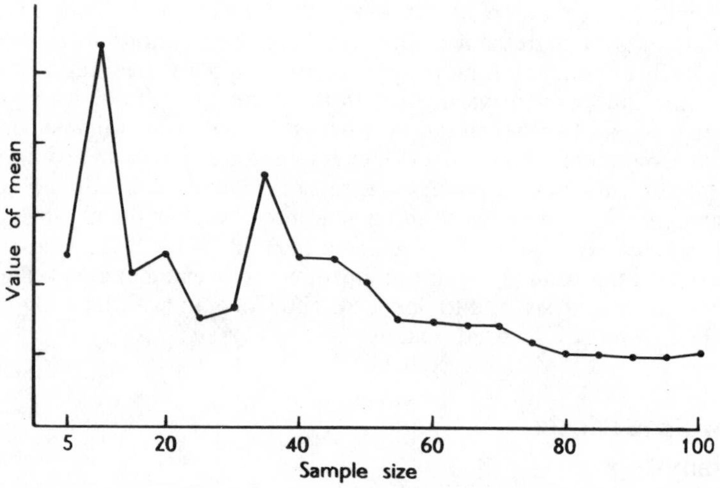

Fig. 2.5 The reduction of the variations in the value of the mean, as the size of sample is increased (see text).

required sizes of sample see Greig-Smith 1957; 1983; Snedecor and Cochran 1980, etc.)

It is obvious that as the sample size is increased a better measure of the mean of the population is obtained, but to reduce the standard error the number of samples has to be increased greatly and in general it is recommended to take as large a sample as time will permit. The calculation of a series of mean values as the sample size is increased merely acts as a guide and even if the fluctuations are reduced markedly at a given sample size it does not mean that the fluctuations have ceased altogether. In the case quoted above additional samples would continue to show a slight fluctuation but certainly less than 0.1 mm and since the measurements were taken to the nearest millimetre it would have been pointless to increase the sample beyond 100. Similar considerations can be applied to quadrat density, frequency or percentage cover. However, if very slight differences are suspected to be of ecological significance then a large sample will be unavoidable.

Size and shape of quadrat

If the distribution of individuals in a population is completely random then the size of quadrat used is immaterial except from the standpoint of convenience. Any size convenient for the particular vegetation which is investigated can be used – a small quadrat if the species are small and numerous so that the counting of individuals is made easier, or a large quadrat if the species are large or thinly scattered. However most individuals are not distributed at random and the size of the quadrat has a considerable effect on the variance of the data obtained. This is discussed more fully later (p. 128) but, briefly, if there is a tendency for individuals to be grouped together, the measure of the variance of the data is at a maximum when the size of quadrat is approximately equal to the mean area of the groups of individuals. Since it is usually impossible to predetermine the scale of this pattern (as it is termed) and since such patterns are often repeated on larger scales, the size of quadrat has to be chosen quite arbitrarily. The most suitable quadrat on theoretical grounds being the smallest possible, relative to the type of vegetation and to the practicability of the enumeration of such a quadrat size.

The shape of a quadrat has been a square almost by tradition but some advantages are obtained by the use of rectangular quadrats. Clapham (1932) showed that the variance between rectangular strips was less than between squares of the same area, in one instance at least. Presumably this resulted from a rectangle crossing a high density patch and an adjacent low density patch simultaneously and thus levelling out the variance over the whole area. However, this would not always be true and the only consistent advantage for the use of rectangular quadrats is the increased facility with which the quadrat can be studied. Thus in large quadrats there is a tendency to crush part of the vegetation by trampling or leaning and this can be a serious disadvantage when several types of measurements are being taken from the same quadrat. One common error associated with quadrat work which is often not appreciated is the 'edge effect'. Frequently a decision has to be taken as to whether a species on the edge is 'in' or 'out' of a quadrat. Where the edge of a quadrat is large relative to the area inside, this error can be quite marked and thus very long thin quadrats should not be used.

Regularly spaced samples

From the considerations outlined above it is quite clear that in the majority of cases samples of a population taken at random with their resultant standard error are of considerably more use than samples taken at regular intervals. However, there are some exceptions to this general rule. Where data about the abundance of a species along an environmental gradient are required, it is worthwhile to sample systematically along a transect (p. 30) in order to record the abundance of a species in relation to position. Similar considerations apply to isonomes (p. 32) and to the grid analysis of pattern (Chapter 7,) both of which relate the abundance of a species to location of the quadrats, the former more as a descriptive method, the latter as a statistical analysis of non-randomness. On the other hand regular sampling can lead to a considerable bias of the mean value as would occur in the case in Fig. 2.6. This is

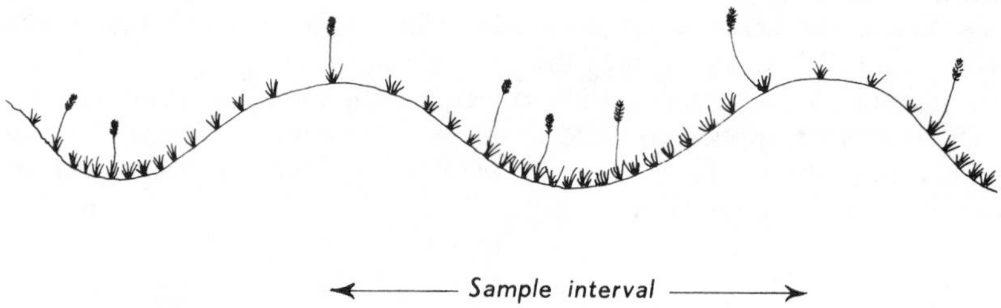

Fig. 2.6 A hypothetical ridge and furrow distribution of a species which when sampled at regular intervals would produce a very biased estimate of abundance.

a rather extreme example of the hypothetical distribution of a species on furrows and ridges which developed in relation to the drainage of a field. The regularly spaced sample points will coincide exactly with the equally regular furrow and ridge, and the mean value for the abundance will be considerably different from that obtained by random sampling.

A similar false impression can be obtained for the optimum level of pH of a species if the sample is not randomized. Thus numerous studies have been made relating the distribution of a given species to pH by taking a soil sample at each site examined where the species was growing. This in fact is a sample of the variation of soil pH in a given area and the resultant data relates to the frequency distribution of pH in that particular area. Such data offers no real proof of an interrelationship between a species and pH since samples are not randomized, although in fact such a relationship may exist. Emmett and Ashby (1934) present a set of data for *Pteridium aquilinum* and *Vaccinium myrtilis* (Table 2.1) obtained from a random sample. The

Table 2.1 Frequency distribution of occurrence of *Pteridium* and *Vaccinium* in classes of pH. (From Emmett and Ashby, 1934; courtesy of *Ann. Bot., Lond.*)

pH class means	Number of samples	Frequency of *Pteridium*	Frequency of *Vaccinium*	Percentage frequency of *Pteridium*	Percentage frequency of *Vaccinium*
4.8	2	0	2	0	100.00
4.9	2	0	2	0	100.00
5.0	2	0	2	0	100.00
5.1	5	1	4	20.00	80.00
5.2	13	6	8	46.15	61.54
5.3	17	8	11	47.07	64.71
5.4	7	4	3	57.14	42.86
5.5	44	27	28	61.36	63.64
5.6	78	41	54	52.56	69.23
5.7	16	5	12	31.25	75.00
5.8	7	5	1	71.43	14.29
5.9	9	4	3	44.44	33.33
6.0	7	5	1	71.43	14.29
6.1	3	2	0	66.67	0

most frequently occurring pH value in the samples is pH 5.6 and it is not surprising that the frequency of occurrence of *Pteridium* and *Vaccinium* are apparently correlated with this particular soil pH as an optimum for both species. In fact the data merely demonstrates three frequency distributions present in an area which may or, equally well, may not be correlated. The two final columns of the table give the percentage occurrence of the two species, and are derived from the second, third and fourth columns, for example at pH 5.1 one sample out of five contained *Pteridium* which is in fact a 20 per cent occurrence.

The data presented in this form show immediately the dissimilar relationship of each species with pH. It is probably worthy of note that the most efficient way of approaching this particular problem is to take a series of random samples noting the pH and abundance of the species in question. A correlation coefficient can then be readily calculated.

Partial random sampling

If an area on inspection appears to be heterogeneous, it is pointless to sample at random and to obtain what may be merely an arbitrary set of figures of species abundance. In such a circumstance the recognized method is to sub-divide the area into a convenient number of equal sized areas and to take random samples within each sub-division. If the area proves to be uniform after all then the data can all be lumped together, but on the other hand if the area is variable then information is available which is relative to each sub-division of the plot and is accordingly of considerable interest. The use of partial random samples would be equally applicable to the hypothetical relationship outlined above between the distribution of a species and topography (Fig. 2.6) and the approach in general is ideally suited for sample plots which contain a slight variation of surface topography. An excellent example of this is given in Chapter 11 (p. 208) where the appropriate division of a sequence of sample plots located on a beach ridge allowed the characterization of a lower, upper and crestal zones in terms of a number of dominant species. In this example the ecology of these species strongly reflects the contrasting soil moisture status and exposure of each of the ridge zones (see p. 218, for further discussion and Figs 11.11, 11.12, 12.5, and 12.6).

Transect and isonome studies

Transects are of considerable importance in the description of vegetative change along an environmental gradient, or in relation to some marked feature of topography. The method simply consists of laying out a line running across the zones to be sampled and then placing quadrats at known intervals along the line. Various measures can be employed in the description of a transect, density or frequency determinations being made at intervals along the line, and cover determination most conveniently made by using a frame of pins orientated at right angles to the transect at fixed intervals. The data are suitably represented, either by means of a graph or a histogram, of abundance of a species plotted against position on the transect.

In Fig. 2.7 the transect data illustrating the change of composition of grassland on the North Downs are presented; the transect runs from the zone of acidic clay with flints down into neutral or slightly alkaline chalk grassland. The data are in the form of percentage cover, 80 points being taken from each 80 cm strip of the transect and expressed as a percentage. The general distribution of the species in relation to each other and to the slope of the area can be readily appreciated.

A further use of transect studies is in the description of zonation representing stages of a succession and Alvin (1960) gives data on the frequency of different lichen species occurring in different zones of sand-dune vegetation (Fig. 2.8). The sampling in

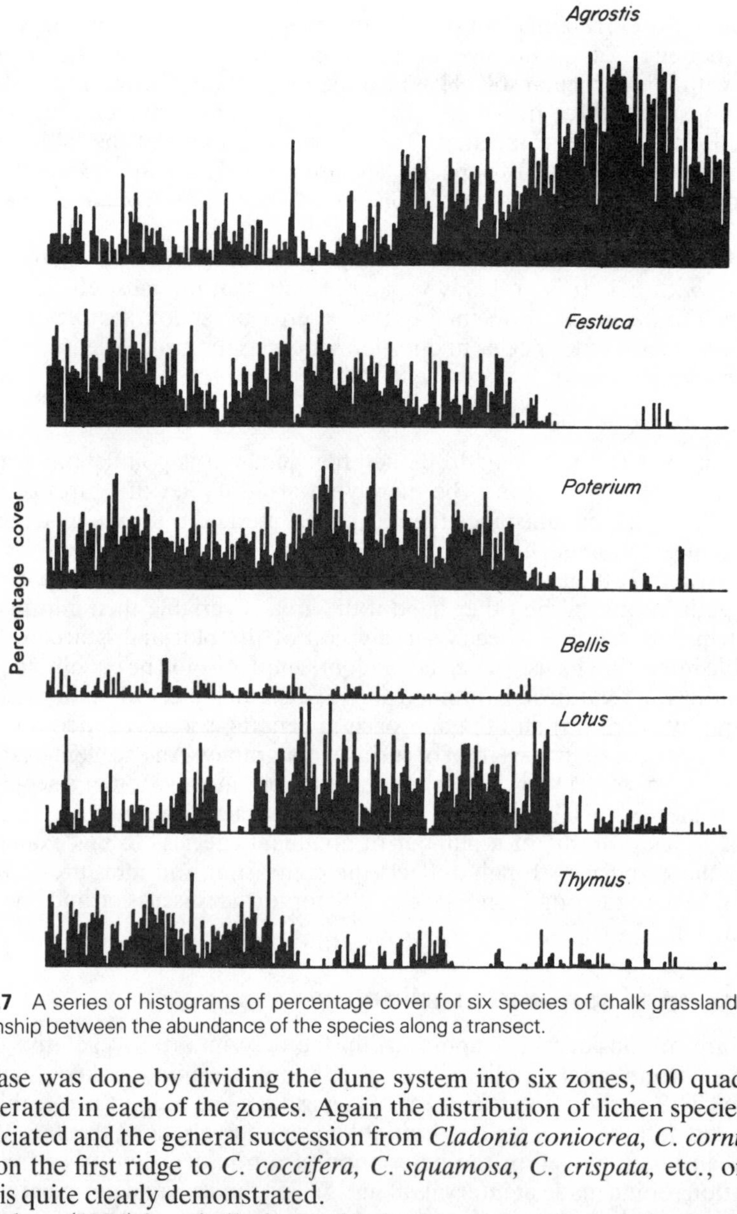

Fig. 2.7 A series of histograms of percentage cover for six species of chalk grassland showing the relationship between the abundance of the species along a transect.

this case was done by dividing the dune system into six zones, 100 quadrats being enumerated in each of the zones. Again the distribution of lichen species is readily appreciated and the general succession from *Cladonia coniocrea, C. cornutoradiata*, etc., on the first ridge to *C. coccifera, C. squamosa, C. crispata,* etc., on the third ridge is quite clearly demonstrated.

Kershaw (1976) has similarly provided a description of a young salt-marsh with a classical linear species sequence, utilizing a number of replicate transect and corresponding histograms (Fig. 11.7, p. 214). These data have been used in an additional context in this book but do illustrate well the effectiveness transect studies can have when used on well-defined gradients.

Thus the use of a transect is of greater importance where the variation of vegetation in response to a changing environmental factor is fairly well marked. If the change is only slight it is then necessary to lay out a number of sample plots at intervals along a transect and make a detailed analysis by means of random quadrats, within each plot. By this means slight trends in the density of a species down a slope may be detected and if necessary correlated with environmental factors which are measured at the same time.

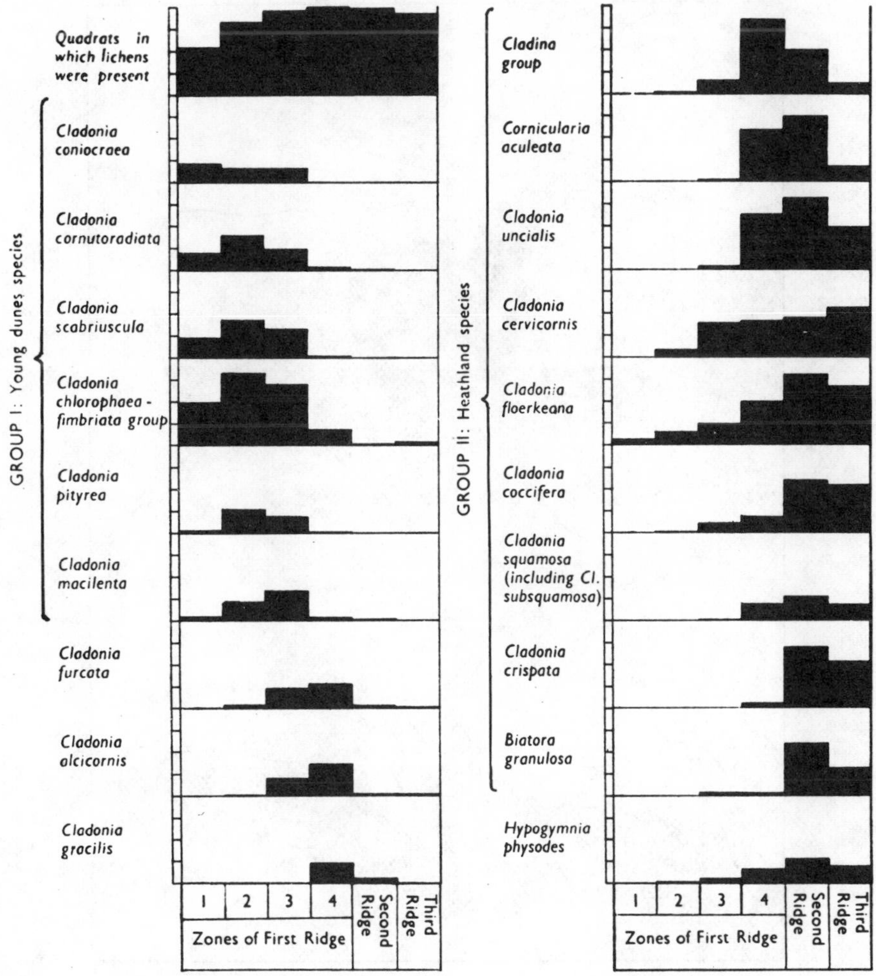

Fig. 2.8 Histograms representing the distribution of the main lichen species in six zones of sand-dune system. One division of the vertical scale represents 25% frequency. (From Alvin, 1960; courtesy of *J. Ecol.*)

An analogous approach to variation in the spatial distribution of a species in a sample plot can be made by the use of the isonome method outlined by Ashby and Pidgeon (1942). The method consists of laying out a grid of contiguous quadrats over a sample area and recording for each quadrat the abundance of the species to be considered. For each species separately the density distribution is then reproduced on squared paper and those squares with approximately equal abundance values joined by a series of lines, giving a 'contoured' distribution of species abundance. This method can be extended to suitable measurements of the environment so that the distribution of a species in an area can be related to the distribution of other species or to environmental factors. The method may be made clearer by considering an actual example.

The microtopography of an area of hummocks in limestone grassland at Malham Tarn, Yorkshire, is presented in the form of an isonome (Fig. 2.9) constructed from a series of height measurements taken from a levelled datum line. At the same time the densities of *Carex flacca* and *C. panicea* were recorded for each unit of the grid and the

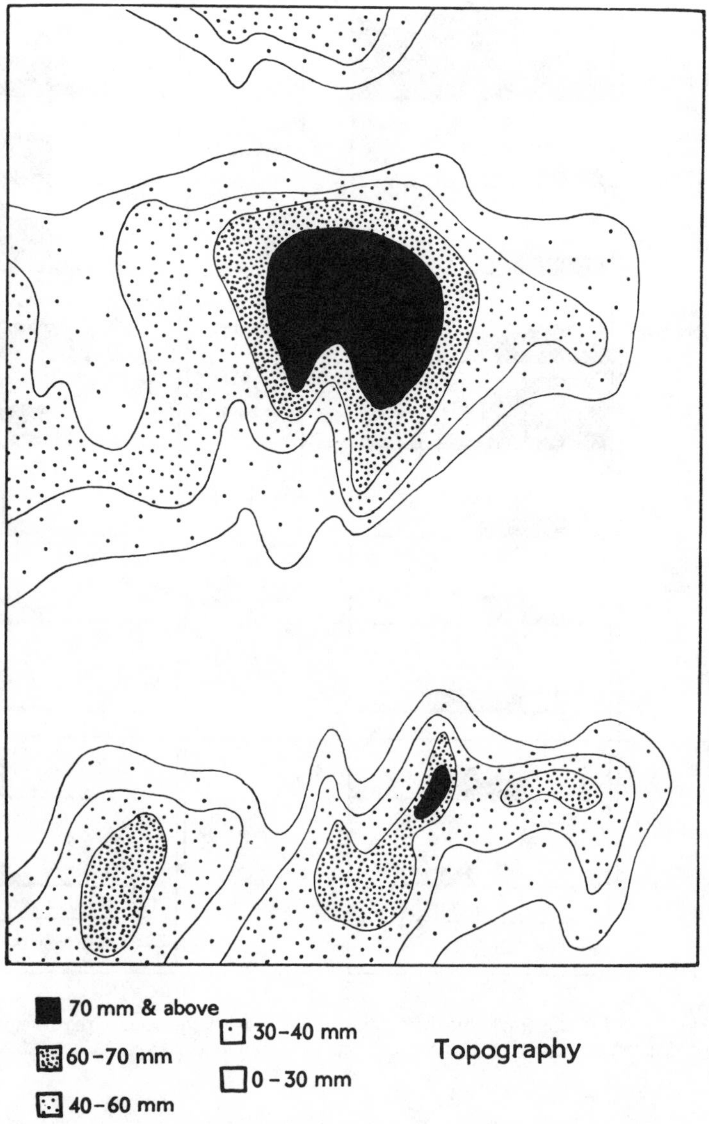

corresponding isonomes constructed from the density data for both species. The marked relationship of both *Carex* species with the hollows, is readily appreciated and the isonome method is in general a very convenient way of depicting relationships between species or between a species and some environmental factor. Clearly the method is only applicable where the distribution of species varies quite markedly; where the distribution is not readily appreciated by the visual inspection of an isonome it is necessary to use a more elaborate method (see Chapter 7).

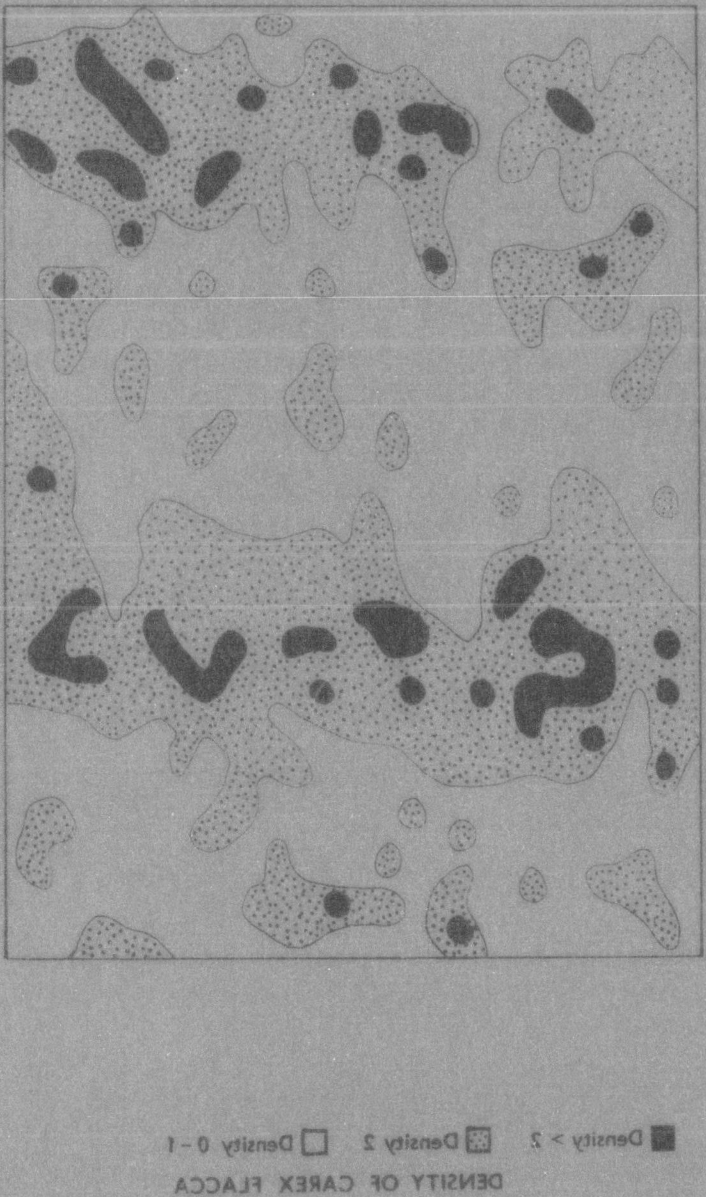

Fig. 2.9(a) Isonomes of the density of *Carex flacca* showing the marked relationship between their relative abundance and micro-topography.

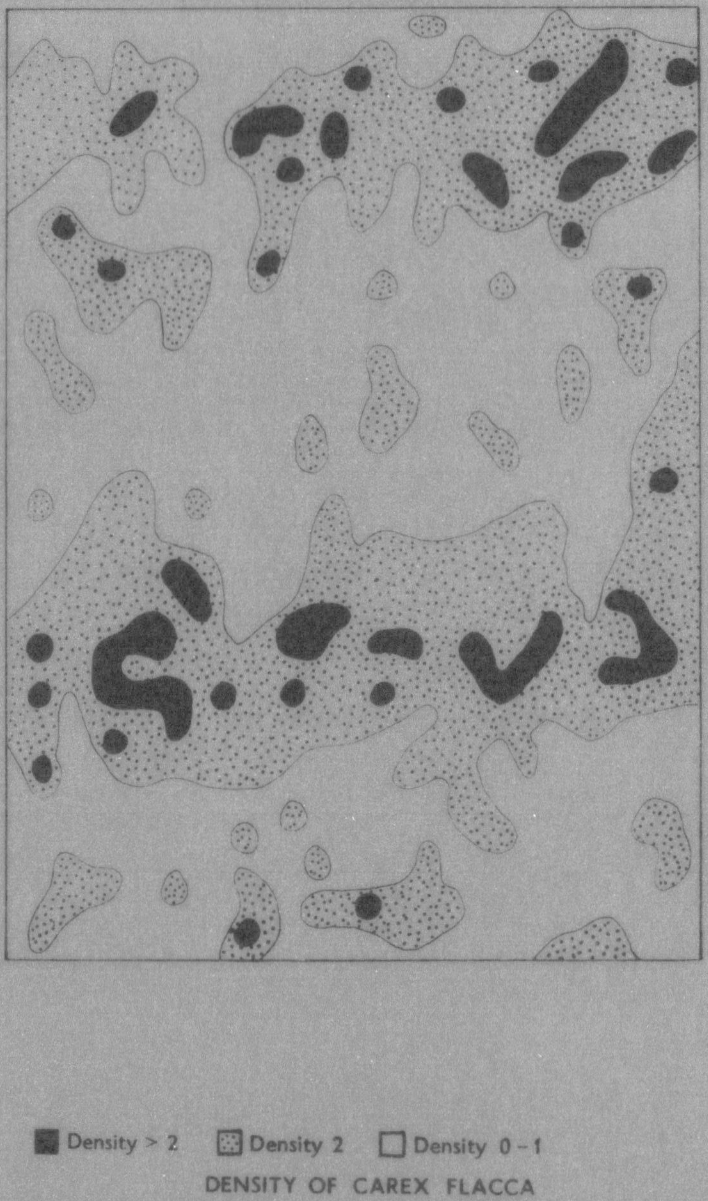

Fig. 2.9(a) Isonomes of the density of *Carex flacca* showing the marked relationship between their relative abundance and micro-topography.

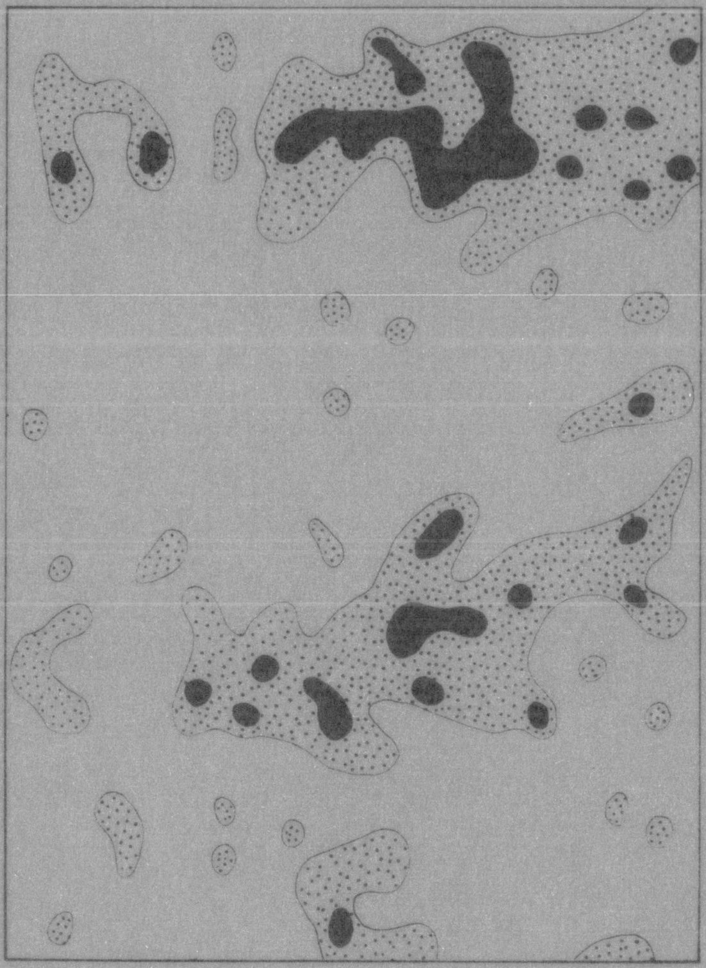

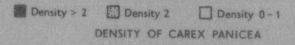

Fig. 2.9(b) Isonomes of the density of *Carex panicea* showing the marked relationship between their relative abundance and micro-topography.

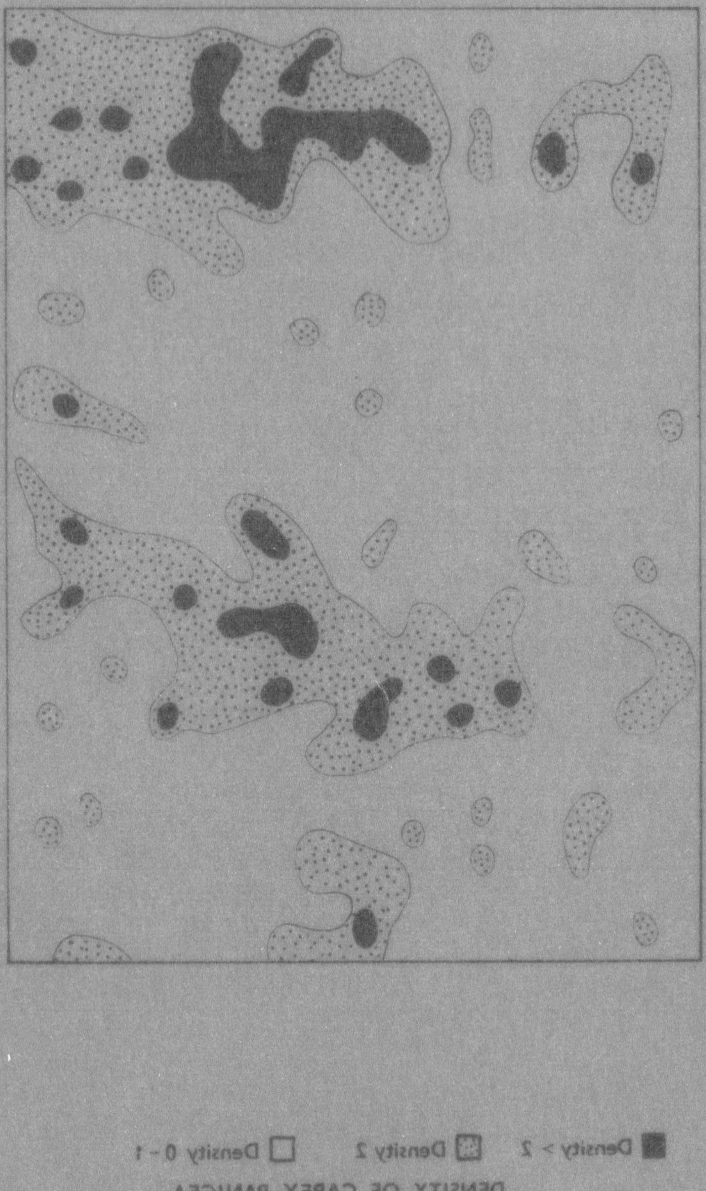

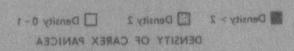

Fig. 2.9(b) Isonomes of the density of *Carex panicea* showing the marked relationship between their relative abundance and micro-topography.

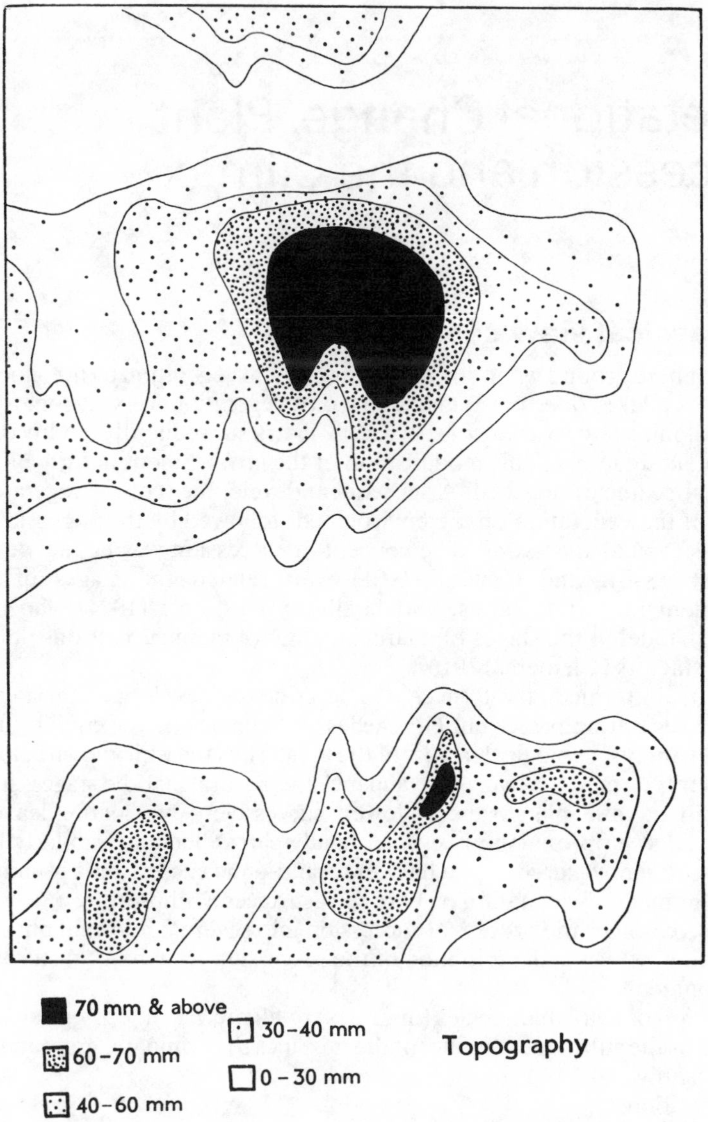

Both transect and isonome methods illustrate the use of regularly spaced samples in describing an area and it should be again emphasized that the sampling procedure most suitable for a given problem is usually chosen or designed for that individual problem. The choice of measure, the size of quadrat, the size of sample and the random or regular position of the quadrats can all be decided more by common sense than resorting to complex statistical theory.

3

Vegetational Change, Plant Succession and the Climax

The classical view of succession

An area of bare ground when stripped of its original vegetation by fire, flood or by the drainage of a lake, does not remain devoid of vegetation for very long. The area is rapidly colonized by a variety of species which will subsequently modify one or more environmental factors. This modification of the environment in turn allows further species to become established. A subsequent development of the vegetation, by this reaction of the vegetation on the environment followed by the appearance of fresh species, is termed ***succession***. The concept of succession was largely developed by Warming (1896), and Cowles (1901) who related the stages of sand-dune development in a time series, and finally by Clements (1904) who outlined in considerable detail the stages of a large number of plant successions, initiated by a variety of factors (Clements, 1916).

Clements introduced the term *sere* to describe the developmental stages through which the vegetation passes until it reaches an ultimate state of equilibrium with the climate and major geological factors of the area. Thus the well known colonization of open water is termed the hydrosere and includes several marked stages, starting with submerged aquatic plants and followed successively by floating-leaved aquatic plants, reed swamp, marsh or fen and finally carr. Clements similarly termed the stages of salt marsh succession a halosere, sand-dune succession a psammosere and the development of vegetation on bare rock surfaces a lithosere.

Analysis resolves the process of succession into several important phases:

1. Nudation, which is the initiation of the succession by a major disturbance in the environment.
2. Migration of available species (migrules) to fill the vacant ecological niches.
3. Ecesis or the subsequent ability of the migrules to germinate, grow and reproduce successfully.
4. Competition.
5. Reaction.
6. Final stabilization. (Clements, 1916.)

The reaction of a species on the environment is one of the outstanding features of succession and constitutes the major mechanism of environmental change which allows further immigrants to enter. A very obvious example of such a reaction is found in the formation of sand-dunes. The establishment of a rhizome fragment of *Ammophila arenaria* and the subsequent production of aerial shoots constitutes a disturbance of the environment at that point. The aerial parts offer an increased impedance to the wind, any sand particles carried by the wind being deposited around the plant and gradually burying it. As the sand accumulates, continued growth of the *Ammophila* keeps it at a generally higher level than the low mound that is formed, and accumulation of wind-borne sand continues. Production of axillary buds and the development of the typical clumped habit of *Ammophila* accelerates the rate of 'dune' formation and gradually over a period of years the dune gets progressively larger (Fig. 3.1). This is a very marked example of reaction of a plant on the

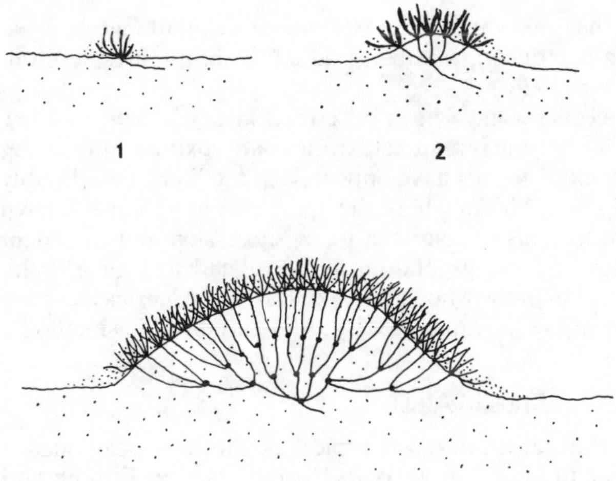

Fig. 3.1 Dune formation by the gradual deposition of wind-carried sand particles around the aerial shoots of *Ammophila arenaria*.

environment and is comparable to the increase in the rate of silting of a lake by the early colonisers, which constitute the first stage of a hydrosere. The same mechanism operates, that is increased impedance to water-borne particles. These are deposited around the vegetative parts of the plants. The depth of water decreases over a period of years until other species with their vegetative parts above water begin to invade. The rate of silt accumulation (plus vegetable remains from the plants themselves) increases and opens the way to the gradual development of reed swamp and then fen vegetation.

These two examples of reaction are fairly obvious, but frequently the reactions which are contributing to the impetus of what is clearly succession are far from clear. Vegetational change was divided by Clements into two kinds: ***primary succession*** initiated on a bare area where reactions are, initially at least, often fairly clear, and ***secondary succession*** initiated by a major environmental disturbance and disrupting a previously initiated succession or producing a marked modification in stable vegetation. For example removal of grazing from grassland, or a forest fire, both initiate a secondary succession and it is in such examples of succession that reactions are often very obscure. It may be that ecesis and competition play a more decisive role in such instances.

Other classification systems have sub-divided secondary successions into a varying number of divisions, but in general little is gained by a complex classification of successions.

The direction in which a succession will develop has been argued at considerable length in ecological literature. Phillips (1934) states 'Succession is *progressive* only; all examples of apparent retrogression are explicable in terms of some disturbing agency'. Similarly *true* succession is defined by Michelmore (1934) as inherent succession of the vegetation which takes place through the plants altering their environment. This is to be distinguished from physiographic succession due to an initial change in the environment which should apparently not be included. Similarly Tansley (1916; 1920) employs the terms ***autogenic*** and ***allogenic*** to distinguish between these two distinct successional causes. For an understanding of succession at a level which serves as a *useful* concept, such a degree of fine definition is not essential. Any *directional change* in vegetation, be it due to intrinsic properties of the plants, or to changing extrinsic factors (the environment), gives a useful and workable definition of succession as a dynamic process. This was admirably expressed by Cooper (1926) who stated that 'Succession is the universal process of change which is embodied in the great vegetational stream; all vegetational changes, whether internally or externally induced, whether gradual or abrupt, are therefore successional.' (See also Gleason, 1926.)

Numerous examples of succession have been described in detail, but very few cases are available where the succession is related to even an approximate time scale. One or two outstanding exceptions to this have appeared in the literature, notably colonization of glacial moraines in Alaska where the age of the moraines is known with a fair degree of exactness. This enables the plant succession and attendant modification of the substratum by the vegetation to be related to a time scale. Similarly the sand-dune succession in the Michigan Lake Basin in America has been studied in detail and related to a time scale obtained by radio-carbon dating methods.

Plant succession on glacial moraines in Alaska

The Glacier Bay region in south-east Alaska is typical of the large-scale glacial retreat that has occurred over the last 200 years in North America, Europe and Scandinavia. The position of the snout of the glacier in historic times can be estimated by growth ring counts of the present spruce forest. Thus ring counts of trees growing beyond the maximum extent of glacier activity (easily recognized from aerial photographs) show the oldest trees to be over 600 years of age and growing on a 'raft' of fallen submerged trunks. The oldest trees on the last morainic ridge are approximately 200 years old, whilst the maximum age of trees on moraines nearer the present site of the glacier progressively decreases. Cooper (1937), Field (1947) and Lawrence (1953) all contributed to a detailed study of the recession of the glacier and a clear picture is now available (Fig. 3.2). Their findings are confirmed by comparison of written records extending back to the 1890s and photographic surveys commenced around 1916.

The stages of succession have been described by Cooper (1923, 1931, 1939) and by Lawrence (1953). The pioneer stage is characterized by *Rhacomitrium canescens*

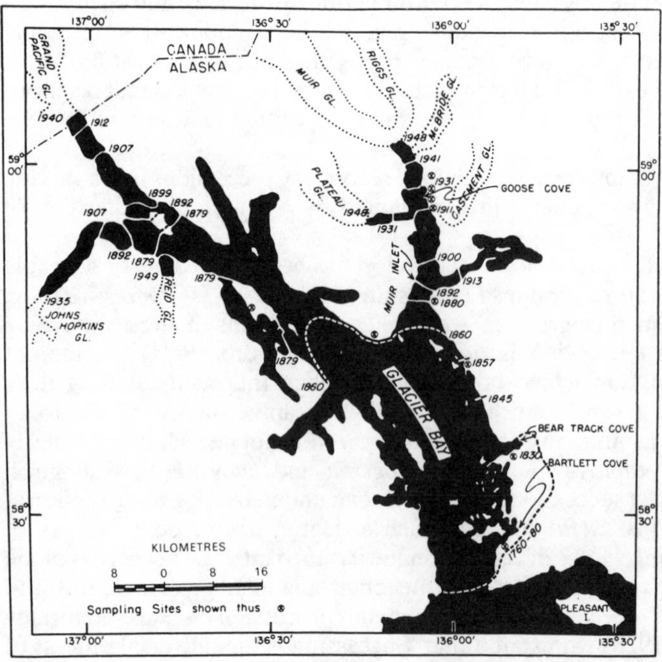

Fig. 3.2 The Glacier Bay fiord complex showing the rate of ice recession. (From Crocker and Major, 1955; courtesy of *J. Ecol.*)

Brid., *R. lanuginosum* (Hedw.) Brid., *Epilobium latifolium* L., *Equisetum variegatum* Schleich., *Dryas drummondii* Rich. and *Salix arctica* Pall. The next stage is the appearance of *Salix barclayi* And., *S. sitchensis* Sans. and *S. alexensis* (And.) Colville, which start as prostrate forms eventually developing an erect habit and forming dense scrub. *Alnus crispa* Ait. Pursh. dominates the next stage of the succession and eventually forms almost pure thickets (with some scattered individuals of *Populus trichocarpa* T. and G.), which are finally invaded by *Picea sitchensis* Carr. *Picea* initially forms pure stands, but over a period of time *Tsuga heterophylla* Sarg. and *T. mertensiana* Carr. enter the community. On well drained slopes this stage represents the climax vegetation of the area. However, where the ground is gently sloping or flat *Sphagnum* species subsequently invade the moss mat of the forest floor and through the growth of *Sphagnum* and its ability to retain large amounts of water, the forest floor becomes soggy and root aeration decreases. This leads to the elimination of the tree species and to the establishment of muskeg dominated by *Sphagnum* species, with occasional scattered individuals of *Pinus contorta* Laws, which can apparently tolerate the wet conditions which prevail.

Crocker and Major (1955) have studied the time sequence of this succession and the reaction of each stage on the properties of the soil. In the earliest stages of soil development following glacial recession the processes involved appear to be greatly affected by the type of vegetational cover.

The reaction of the soil parent material is quite marked with a pH of 8.0–8.4 due to the occurrence of marble in the geological formations of the area. The soil pH falls rapidly with the establishment of the vegetation (Fig. 3.3) and furthermore the rate

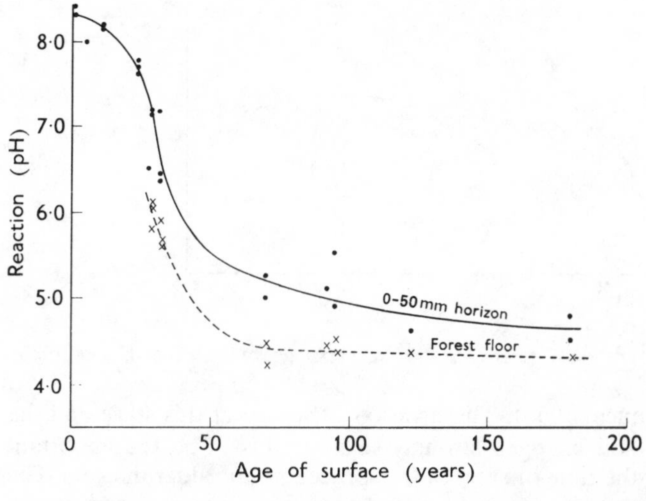

Fig. 3.3 pH of litter residues and surface horizons of the mineral soil with increasing age of surface. (From Crocker and Major, 1955; courtesy of *J. Ecol.*)

of increase of acidity is markedly affected by the type of vegetational cover. Bare surfaces show almost no change of pH over a 30-year period, while *Populus, Dryas* and *Salix* sites show a more marked decrease of pH, and *Alnus* produces a striking acidifying effect, lowering the pH of the surface soil from about 8.0 to 5.0 in 30–50 years (Fig. 3.4). This rapid decrease of pH ceases with the invasion of *Picea*, the leaf litter of this species appears to be approximately of equal acidity to *Alnus* and has no significant effect on the mineral soil.

The organic carbon and total nitrogen concentration in the soil shows equally marked change with time (Figs 3.5 and 3.6). The vast increase of soil nitrogen is related to the presence of bacterial nodules on the roots of *Alnus* which actively fix atmospheric nitrogen and contribute largely to the reserves of nitrogen in the soil via

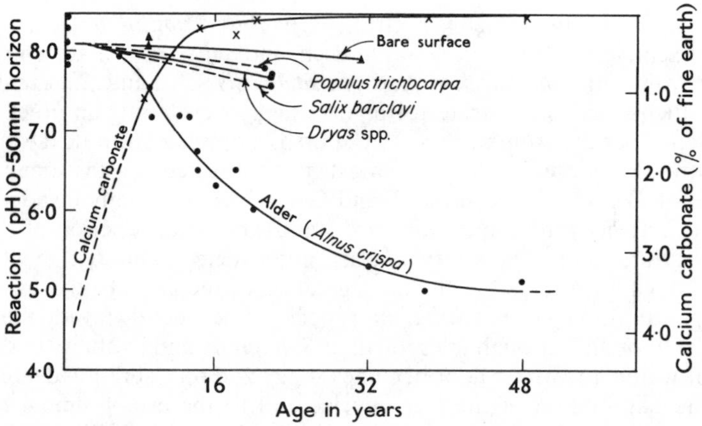

Fig. 3.4 Rate of change of pH in 0 –50 mm horizon relative to different vegetational cover and rate of calcium carbonate change under *Alnus*. (From Crocker and Major 1955; courtesy of *J. Ecol.*)

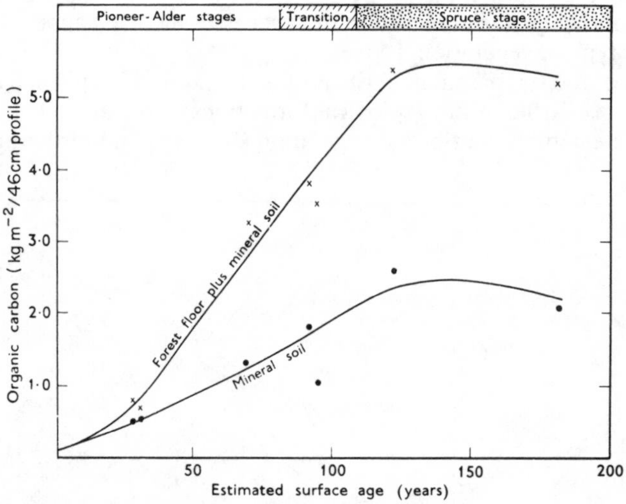

Fig. 3.5 Organic carbon accumulation in mineral soil and forest floor. (From Crocker and Major, 1955; courtesy of *J. Ecol.*)

leaf fall (Fig. 3.7) (Lawrence, 1958). The invasion of spruce at this stage and the subsequent reduction of soil nitrogen strongly suggests this to be the important consideration controlling the time of entry of the spruce into the alder thickets. The elimination of the alder and its root nodules decrease the net annual addition of nitrogen as leaf litter, for spruce has no such root structures and accordingly the soil nitrogen is reduced.

The complete succession from bare morainic detritus to mature spruce forest takes approximately 250 years during which time the most significant reaction of the vegetation appears to be the building up of soil nitrogen. The reduction of soil pH is probably also closely related to this. The building up of organic carbon will influence to a considerable extent the development of soil structure and will lead to a soil with 'crumb-structure' which is markedly different from the rather amorphous glacial till. In turn this will affect soil aeration and movement of water in the soil, and both of these factors probably influence the succession. However, very little data is available which would enable a general assessment of the effect of such factors on vegetation

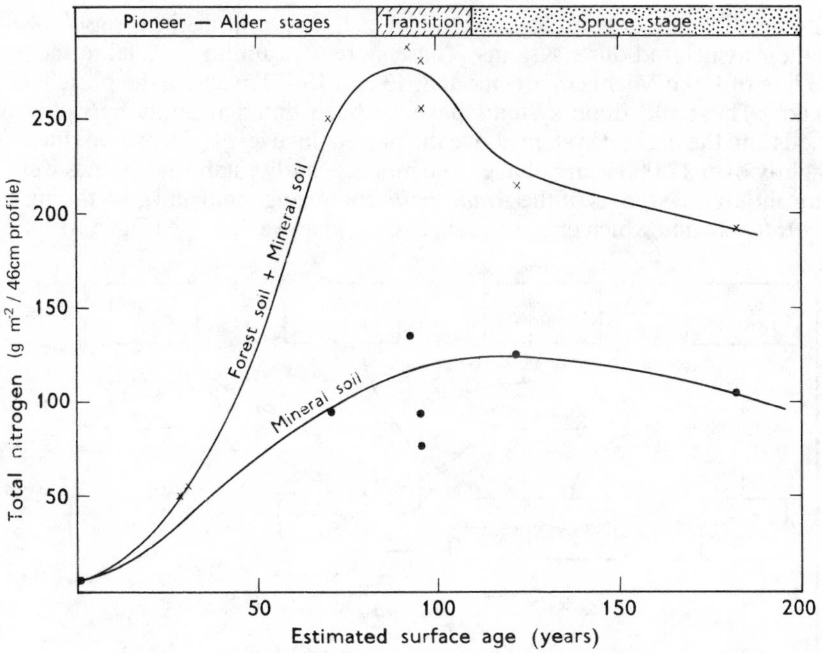

Fig. 3.6 Change of total nitrogen content of soils on surfaces of varying age. (From Crocker and Major, 1955; courtesy of *J. Ecol.*)

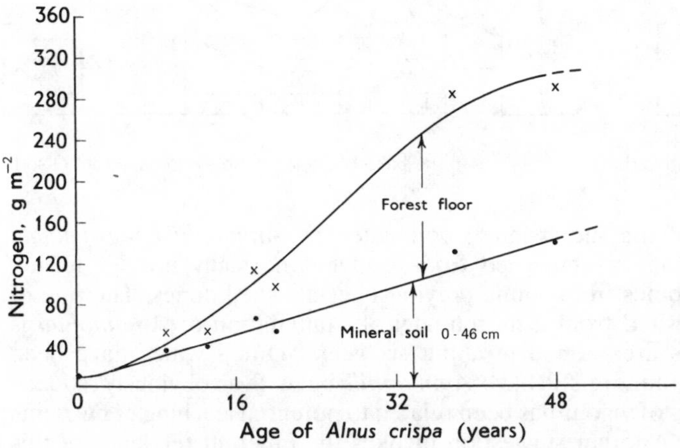

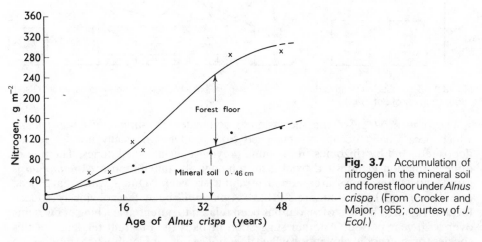

Fig. 3.7 Accumulation of nitrogen in the mineral soil and forest floor under *Alnus crispa*. (From Crocker and Major, 1955; courtesy of *J. Ecol.*)

apart from the rather more obvious relationships that exist. Clearly a detailed study is required of the exact environmental requirements of each species contributing to the succession before the complete mechanism of the succession can be fully appreciated.

Lake Michigan sand-dune succession

Cowles (1899) classic description of the sand-dune vegetation of Lake Michigan was one of the early contributions to the concept of plant successional stages in this area in relation to both the development of the soil and to the age of each dune system.

During and after the retreat of the ice-age glaciers from the Great Lakes region the resultant fall in the general level of the lakes left several distinct 'raised beaches', with their associated dune systems. These systems, running parallel to the present shore line of Lake Michigan are about 8, 13 and 18–20 m above the present level of the lake. These old dune systems have all been dated recently by radio-carbon methods and the highest system above the present lake level (Glenwood) turns out to be slightly over 12 000 years of age. The more recently established dunes were aged by morphological studies of the *Ammophila* colonizing them and growth ring counts of *Pinus banksiana* which enters the succession at an early stage (Fig. 3.8).

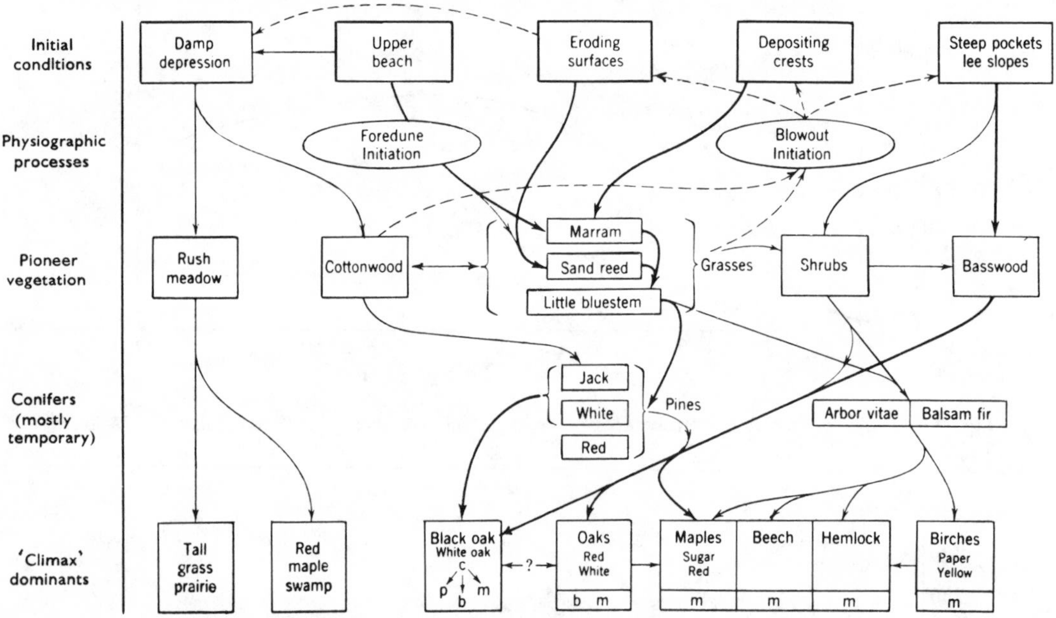

Fig. 3.8 Schematic alternative dune successions in the Lake Michigan dune systems. (From Olson, 1958; courtesy of *Bot. Gaz.*)

The pioneer stage of the succession is dominated by *Ammophila breviligulata* which occasionally becomes established from seed but normally invades freshly deposited sand by rhizomes from young previously established dunes. The rate of dune formation established from a morphological examination of *Ammophila* is rapid and stable dunes are formed in about six years. Dunes which have been established for 20 years or more still have *Ammophila* but with an obviously reduced level of vigour. This loss of vigour has been related to potential leaching of nutrients over a 20-year period. Another suggestion invokes the inherent tendency of this species for internodal elongation which does not cease completely when stabilization has been reached. This results in the growing apex being pushed into the dry surface layers of the sand which are unfavourable for plant growth, with the consequent reduction of vigour of the plant as a whole. A further possibility may be found in the average age of the plants themselves, the loss of vigour being due to the slowing down of physiolgical mechanisms, resulting directly from old age (see Chapter 4). The next stage of the main line of succession is the arrival of *Pinus banksiana* and *P. strobus* which enter immediately after dune stabilization is reached. These species eventually are replaced by the final stage dominated by *Quercus velutina*. This main succession is summarized in Fig. 3.8 together with the modifications present in those dune systems where local variation of the environment can change the course of the succession markedly. Thus damp depressions caused by impeded drainage develop into a grassland community and, on the other hand, sheltered pockets on the lee slopes progress to a *Tilia americana* dominated intermediate stage. This different course of succession is related, in part at least, to the microclimate of these pockets which are largely protected from excessive solar radiation and wind effects. Accordingly moisture availability will be higher, which in addition to the obvious relation to plant growth would contribute some protection from fires which occur frequently in the area, the relatively moist pockets tending to be by-passed. In addition these dune pockets tend to accumulate fallen leaves brought in from the more exposed parts of the dune system and the general nutrient status of the soil increases at the expense of the rest of the area. The reaction of the vegetation on the properties of the soil is broadly similar to the example discussed above but the rate of change is considerably slower. The leaching of carbonate from the soil is very rapid (Fig. 3.9) and the surface layers (10 cm) are more or less devoid of carbonate from the soil after a few hundred years. There is a slow building up of organic carbon (Fig. 2.10) and soil nitrogen (Fig. 3.11). These three environmental changes are virtually completed in 1000 years and subsequent generations of *Quercus* have little effect on these soil properties.

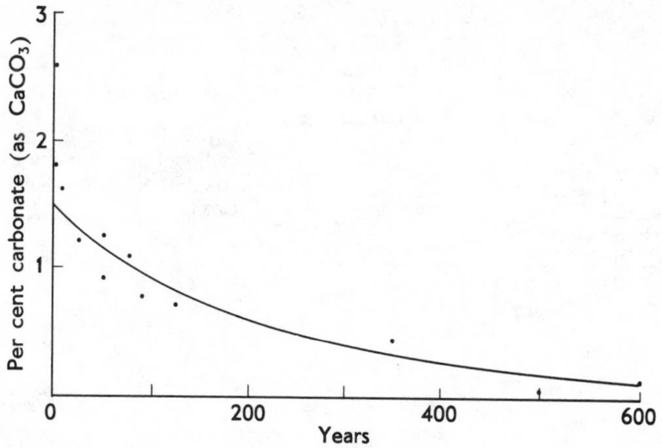

Fig. 3.9 Leaching of carbonates from the surface layers of dune soil. (From Olson, 1958; courtesy of *Bot. Gaz.*)

Again the relationships between the environment, the vegetational stages and the development of the succession through modification of the environment can only be described in a general way. The initial dune formation is clear and results in an ecological niche with low fertility which is in fact invaded by species with low nutritional requirements. The subsequent accumulation of soil nitrogen and organic carbon presumably control to some extent the time of entry of oak into the succession. What evidence there is suggests that *Quercus velutina* has relatively low nutrient requirements and accordingly is able to thrive in a habitat which remains at a very low level of soil fertility. The annual addition of leaf litter from the final stages of the succession appears to be equivalent to the amount of nutrient required to maintain the *Quercus* communities. Thus the soil conditions remain fairly constant over a long period of time (probably an indefinite period) and limit the further development of the succession. It appears very improbable that species with higher nutrient requirements will even enter the succession and the occasional and isolated hickory, beech and maple trees are restricted to sites with special topographic or drainage features. Accordingly the *Quercus velutina* represents the final stable vegetation of the area largely restricted by the edaphic features outlined above.

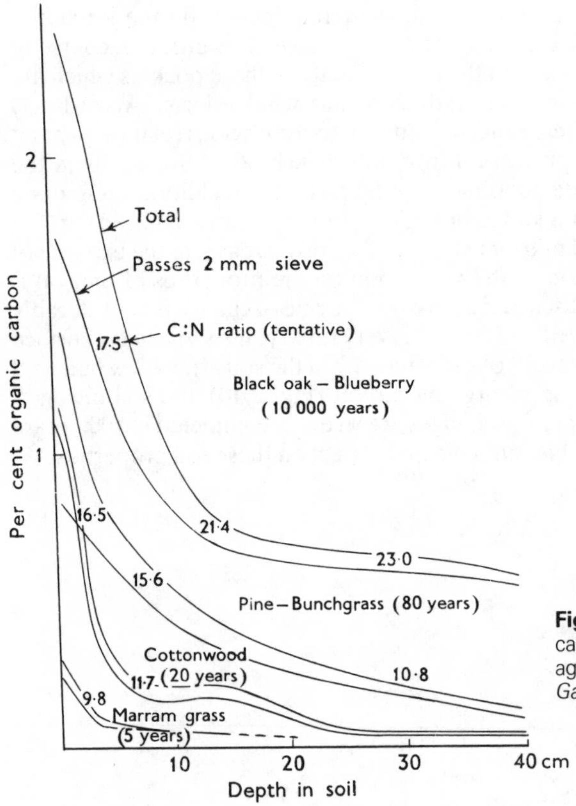

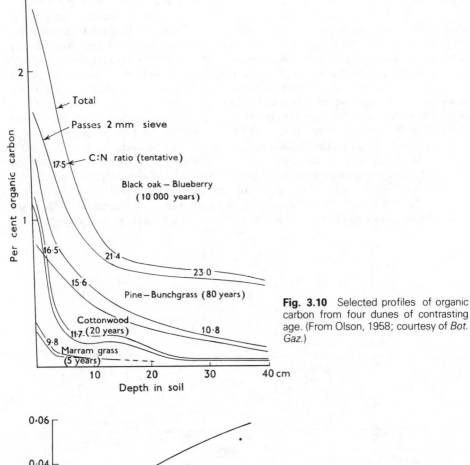

Fig. 3.10 Selected profiles of organic carbon from four dunes of contrasting age. (From Olson, 1958; courtesy of *Bot. Gaz.*)

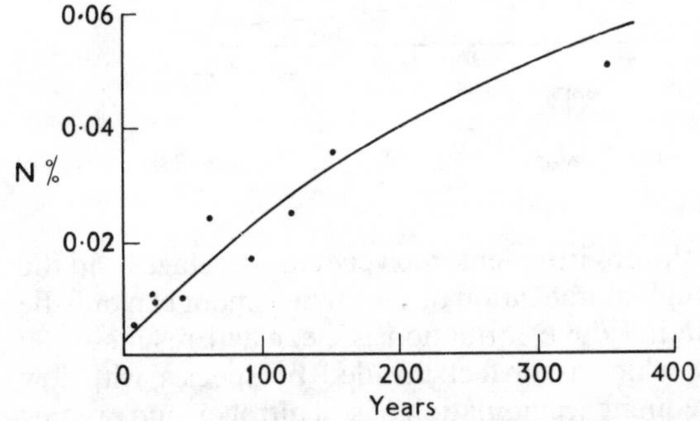

Fig. 3.11 Increase of soil nitrogen in surface layers of soil from dunes of different age. (From Olson, 1958; courtesy of *Bot. Gaz.*)

The classical view of the climax state

In both the examples of succession described above the vegetation has developed to a certain stage at which point it has reached a state of equilibrium with the environment. This final stage of the succession is related directly to the environment and is referred to as the climax. Numerous definitions of the climax have been made: Braun-Blanquet (1932) defined it as the development of vegetation and the formation of soil towards a definite end point determined and limited by climate. Similarly Odum (1953) suggests that a climax is the final or stable community in a successional series; it is self-perpetuating and in equilibrium with the physical habitat. Both these definitions seem very similar and the differences do not appear to be completely incompatible. However, they reflect the outlook on climax vegetation of two schools of thought which have developed since the initial usage of the word in an ecological context. The two theories concerned are referred to as the ***monoclimax*** and the ***polyclimax*** theories which respectively relate the control of climax vegetation to one factor, that of climate, or to a number of factors ranging from climate to less obvious edaphic or even biotic (i.e. animal) factors.

The existence of stable communities largely controlled by edaphic factors is obviously not denied by the protagonists of the monoclimax theory but they are regarded rather as 'exceptions' and categorized by the introduction of special terms – 'subclimax' and 'post-climax' for example (see below). [Whitaker (1953) lists and describes in detail the numerous terms used in relating edaphically controlled climax vegetation to the ultimate climatic climax.] It is claimed that over a long period of time these apparent exceptions will be completely modified by the action of the climate into true climax vegetation.

The polyclimax theory is not hampered by this necessity of introducing a large number of terms to describe apparent exceptions to the theory – an edaphic geological or physiographic climax is accepted for what it is.

The monoclimax theory was developed largely by Clements (1916) and some of the terms relating to climax vegetation clearly show the inherent difficulty of defining a time scale over which stability can be assessed.

Subclimax

This is the penultimate stage of a succession which persists for a long time but is eventually replaced by the true climax vegetation. For example, Gleason (1922) and Cooper (1926) give an account of the relic coniferous forest uncovered in the city of Minneapolis and which was shown to be immediately post-glacial. The present climax is deciduous forest which presumably has very slowly replaced the coniferous 'subclimax'.

Disclimax

This is the vegetation replacing or modifying the true climax after a disturbance of the environment. The majority of 'natural' grasslands in the British Isles below 500 m are maintained by the grazing of cattle and sheep, and if this factor was removed it is believed that a deciduous forest climax would result. Similarly the introduction of prickly pear cactus in Australia has formed a disclimax over wide areas.

Post-climax and Pre-climax

Climatic areas, latitudinally zoned from the equator to the arctic, are not discontinuous. There is a gradual change from one zone to the next reflected in equally gradual vegetational change. Any slight fluctuation of climate will modify the vegetation and such vegetation differences reflecting cooler and/or moister conditions than the average (usually topographically induced) are termed post-climax, drier and/or hotter, preclimax.

The monoclimax theory and the excessive use of this type of terminology has been severely criticized by many ecologists since rather than clarifying the position, the dozens of special terms complicate the issue out of all proportion. Conversely an explanation by polyclimax theory is readily available.

The real difference between these two schools of thought lies not in the possible existence of edaphically-stable communities but in the time factor involved in any consideration of relative stability of vegetation. Selleck (1960) expresses this very clearly: 'The rift (between the two schools of thought) occurs either in the assumption that, given sufficient time, climate is the over-all controlling factor of vegetation, or in the length of time considered adequate for stabilization to occur.' In other words given a sufficiently long period of time edaphic factors would be reduced to a common level by the action of climate and the vegetation as a whole would develop uniformity accordingly.

Some initial criticisms of the climax concept

Although the concept of the final end point of succession resulting in a vegetational climax, where complete equilibrium with the environment is attained and where there is thus no further directional change, was initially widely accepted, some criticisms of the theory were subsequently also widely established. Thus the concept implies environmental stability with the vegetation equilibrium accordingly also stable. It is impossible however, to define stability without reference to a time scale and since these are usually related to the life span of man himself, vegetational changes which occur over a 'geological time scale' are not readily appreciated. If the expectation of life was one minute then a population of groundsel (*Senecio vulgaris*) would be a climax vegetation. Conversely if our life expectation was measured in thousands of years the coniferous forest subclimax of North America would be merely a successional stage which precedes the establishment of deciduous forest. Such considerations are by no means as absurd as would be imagined. There is a continuous fluctuation of climate over large areas – the Pleistocene ice-age is an outstanding example – but other less marked changes have occurred in more recent times. Beschel (1961) gives data which show that remarkable glacier fluctuations have occurred over the last 2000 years (Fig. 3.12). The evidence is taken from a variety of sources, geomorphology, historical records and dendrochronology, and shows a series of fluctuations consistent in areas as far apart as Greenland and Africa. The control of glacial activity is dependent on the over-all climate of the area and it is clear that there is a continuous variation of climate which can be very marked over a long period of time. Such climatic fluctuations will produce a successional response in apparently stable 'climax' vegetation. Cowles (1901) was well aware of this problem: 'The condition of equilibrium is never reached, and when we say that there is an approach to the mesophytic forest, we speak only roughly and approximately. As a matter of fact we have a variable approaching a variable rather than a constant.'

Whitaker (1953) attempted to circumvent the problem of time scale by proposing an approach based on the 'prevailing climax' which was related to the maximum area covered. As Selleck (1960) points out this can be criticized since a minor constitutent of a forest understory for example could become the dominant species of the next generation.

One marked aspect of the theoretical climax relative to the stages of succession leading to it is the rate of change. The initial stages of succession are usually relatively rapid, the rate of change gradually decreasing until vegetational change is at a minimum and if we assume climatic and environmental stability all directional change ceases in the 'climax'. Again Cooper (1926), one of the early protagonists in this field, expressed this aspect very well: 'The climax period comes into being insensibly. Much, or even all, of a unit succession may be characterized by a gradual diminution of the rate of change. The climax itself being merely a continuation of this process, it is not possible to mark it off absolutely from the period of more active succession.' In general terms the rate of vegetation change would seem to offer a possible means by which climax vegetation could be delimited from the terminal stages of succession, but again the definition of the *relative* rates of change in the two appears to be insurmountable.

We are forced to conclude that climax vegetation is an abstract concept which is never reached in reality due to the continuous fluctuations of the climate (*directional*

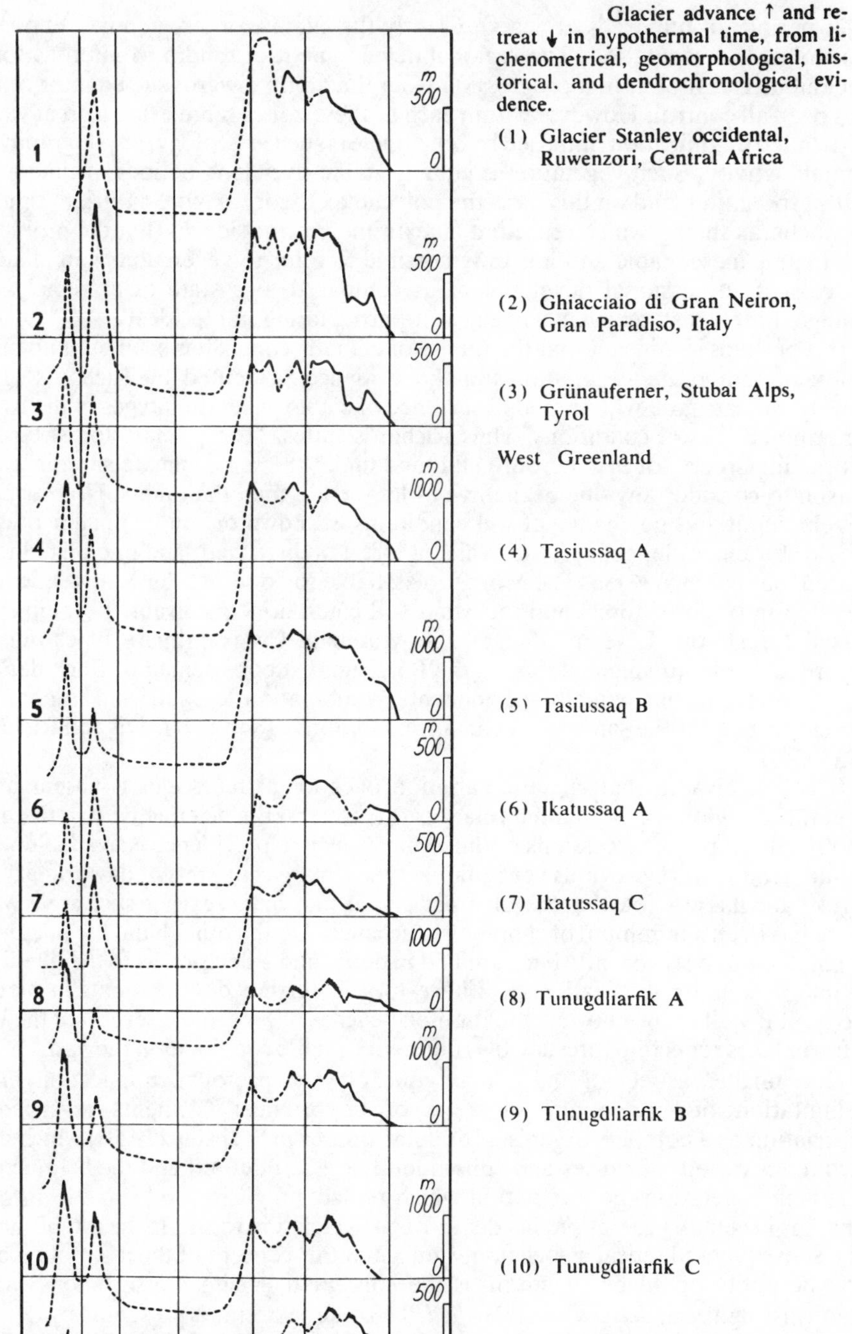

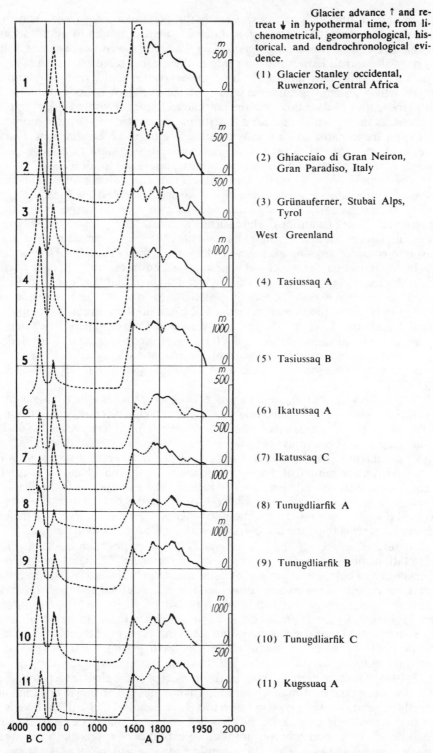

Fig. 3.12 Glacier advance and retreat in metres as summarized from lichenometrical, geomorphological, historical and dendrochronological evidence. (From Beschel, 1961; courtesy of Univ. Toronto Press.)

over periods of hundreds of years). Clearly the climate of a region has an over-all control of the vegetation. The regional development of tundra, coniferous forest, deciduous forest etc., in a progression from the arctic towards the equator reflects this over-all control. However within each of these zones there exists a great variety of detailed modification, imposed by other factors such as geology, physiography and human activity. Such vegetation is equally stable over long periods of time and is self-perpetuating and on this basis the polyclimax theory is more sufficient than the monoclimax theory which represent an extreme abstract ideal. That the problem is relative, is inescapable and it is only possible to refer to vegetation as in an active process of successional development as compared to a state of extremely slow change, barely distinguishable from a self-perpetuating stable 'ideal'.

It is perhaps significant that the theory of climatic climax has been built up on the study of temperate vegetation and the evidence presented by Richards (1952) clearly indicates the widespread occurrence of stable vegetation types in the tropics determined by soil conditions. Thus Richards states, 'The primary forest types of Moraballi Creek, for instance, are all found under the same climate and there is no reason to consider any one of the five as less stable than the others. No process of development and no change of soil conditions is known or can be imagined which would, for example, convert the Wallaba forest on bleached sand into mixed forest on red loam or vice versa. The Mora forest, liable to flooding, shows no tendency to develop into mixed forest and the Mora soil could not conceivably develop into a mixed forest soil. It seems that all the Moraballi Creek primary types must be regarded as of equivalent status, as the Polyclimax theory demands. Each depends on a different but equally permanent, combination of soil and topography developing under the same set of climatic conditions.' (See also Davis and Richards, 1933–34.)

It is conceivable that climatic variation over long time scales is absent in the equatorial region, or is minimal relative to the marked glacier activity in the arctic and resultant plant successions outlined previously. No evidence is available which would demonstrate previous vegetational types in areas of present day primary rain forests and it is possible that the 'abstract' climax may in fact exist in such areas where there has been a minimum of climatic disturbance. On the other hand the fluctuation of climate has only recently been studied in detail and a decision as to the likelihood of major equatorial variations will have to wait further developments in meteorology. It is well known however that raised beaches occur around several of the large African lakes reflecting (presumably) different climatic conditions in the past.

One further aspect of the climax concept with profound implication in the delimitation of 'community units' is of importance: Clements regarded an association as a complex organism [or quasi-organism, Tansley (1920)] which arose through a variety of causes and subsequently grew, matured and died. In actively changing vegetation the association which replaces a given seral stage is different whereas in climax vegetation the 'death' of an association results in the establishment of a subsequent identical association. Implicit in this concept of the association as an organic entity or 'living organism' is the concept that climax associations can be readily recognized, *defined* and *classified*. The other extreme is presented by Gleason (1917, 1926) who proposed the individualistic concept of the plant association, which basically relates the continuously variable environment to an equally continuously variable vegetation. Thus no two associations are identical and implicit here is the fact that accordingly they cannot be classified and often cannot be clearly defined. Embodied in these two concepts are the two extremes of approach to community classification in use today. This will be considered in some detail at a later stage (see Chapter 9). In addition to this aspect of the climax association as a distinct and recognizable entity Greig-Smith (1952*b*) points out: 'This (the complex organism of Clements) implies direct interaction between individuals both of the same and of different species.' Interaction between species or individuals of the same species can be conveniently measured by means of a χ^2-test or by an analysis of the 'pattern' of species distribution respectively (see Chapters 2 and 6). Greig-Smith investigated several sites in secondary forest ranging from areas recently disturbed (Site D) to areas disturbed many years before the investigation and only imperceptibly different from (undisturbed) primary forest (Site B). In addition an area of primary forest was sampled for comparison (Site E). The association between species is given in Table 3.1, 3.2 and 3.3 as the numbers of observed and expected joint occurrences of pairs of species. Clearly the maximum amount of association is present in Site D (most recently disturbed) and least evidence of association occurs in Site E (primary forest). This is clear evidence that the maximum interaction between species inherent in the 'complex organistic' concept of climax vegetation is not valid at least in tropical rain forest. Greig-Smith thus concludes that the individualistic concept of Gleason is a more satisfatory alternative in part at least. The evidence relating the pattern of distribution of the individuals of a species in an area to the 'stability' of the vegetation, suggests equally a minimum of interaction in stable vegetation. Greig-Smith (1961*b*) examined the different pattern scales in *Ammophila arenaria*, sampling young developing dunes, stable dunes and dune 'slacks'. His data (Fig. 3.13)

Table 3.1 Expected and observed numbers of joint occurrences of pairs of species in plots at site D (secondary forest). (From Greig-Smith, 1952*b*; courtesy of *J. Ecol.*)

Species	Total	*Pisonia cuspidata* 73	*Rudgea freemani* 72	*Amaioua corymbosa* 58	*Miconia prasina* 58	*Lacistema aggregatum* 54	*Alibertia acuminata* 42	*Protium guianense* 42	*Coccoloba latifolia* 35
Rudgea freemani		52.56							
	72	53							
Amaioua corymbosa		42.34	41.76						
	58	**38** *	44						
Miconia prasina		42.34	41.76	33.64					
	58	**34** ***	39	**24** ***					
Lacistema aggregatum		39.42	38.88	31.32	31.32				
	54	48	43	**37** *	**26** *				
Alibertia acuminata		30.66	30.24	24.36	24.36	22.68			
	42	**26** *	33	29	**19** *	**29** *			
Protium guianense		30.66	30.24	24.36	24.36	22.68	17.64		
	42	32	34	26	22	**30** **	17		
Coccoloba latifolia		25.55	25.20	20.30	20.30	18.90	14.70	14.70	
	35	27	23	**25** *	**13** *	21	11	13	
Vismia guianensis		16.79	16.56	13.34	13.34	12.42	9.66	9.66	8.05
	23	**13** *	16	15	10	10	12	8	7
Bactris sp.		14.60	14.40	11.60	11.60	10.80	8.40	8.40	7.00
	20	17	17	11	11	12	7	12	7
Isertia parviflora		13.87	13.68	11.02	11.02	10.26	7.98	7.98	6.65
	19	14	13	10	11	**6** *	6	8	4
Clusia minor		–	–	8.12	8.12	7.56	5.88	5.88	
	14	–	–	9	8	7	5	5	
Cassia fruticosa		–	–	7.54	7.54	7.02	5.46	5.46	
	13	–	–	8	5	**11** *	5	7	
Casearia spinescens		–	–	7.54	7.54	7.02	5.46	5.46	
	13	–	–	8	6	10	8	7	

The number against each species is the total number of plots in which that species occurred. The numbers against each pair of species are the expected (above) and observed (below) numbers of joint occurrences.

* $P = 0.01$–0.05. ** $P = 0.001$–0.01. *** $P = <0.001$.

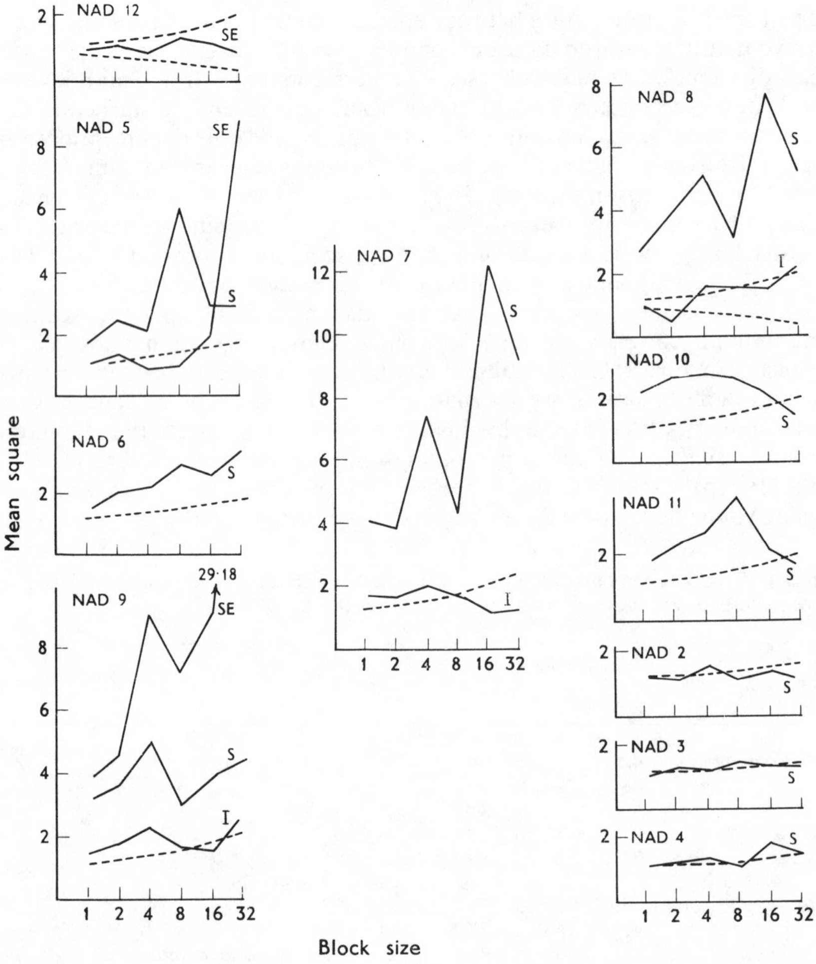

Fig. 3.13 The changing intensities of pattern of *Ammophila arenaria* in relation to dune age and stability. The dotted lines represent upper and lower 95 per cent limits for random distribution. (From Greig-Smith, 1961*b*; courtesy of *J. Ecol.*)

show almost complete absence of pattern in the early formation of dunes (NAD 12), a maximum of pattern in the stabilizing dunes (NAD 7, 8 and 9) and only slight indications of pattern in the most stable sites examined (NAD 2, 3 and 4). The amplitude of the peaks is proportional to the intensity of the patterns. In this example the pattern varies from one consisting of patches of slightly higher density which alternate with similar patches of lower density, to sites where the patches are more aptly described as 'clumps'. Again the evidence suggests random distribution in the final stable vegetation, that is to say no interaction between individuals of a species.

A similar set of data is given by Kershaw (1963) collected from associations of *Calamagrostis neglecta* in central iceland. It behaves here as a primary coloniser of open water and remains at a reduced level of performance and density while the silting up of the area continues. The recent volcanic nature of the rock in the area results in the annual deposition of vast quantities of cindery silt on the valley floors and by an inspection of the depth of water table below a stand of *Calamagrostis* it is possible to arrange a relative time scale for each stand. The deeper the water table and the more the silting, the longer the *Calamagrostis* has maintained itself in that

Table 3.2 Expected and observed numbers of joint occurrences of pairs of species in plots at site B (secondary forest). (From Greig-Smith, 1952*b*; courtesy of *J. Ecol.*)

	Cordia curassavica 60	*Rudgea freemani* 50	*Casearia spinescens* 45	*Bactris sp.* (stems) 35	*Fagara martinicensis* 33	*Bactris sp.* (seedlings) 32
Rudgea freemani	30.00					
50	34					
Casearia spinescens	27.00	22.50				
45	31	**30** * *				
Bactris sp. (stems)	21.00	17.50	15.75			
35	21	18	**11** *			
Fagara martinicensis	19.80	16.50	14.85	11.55		
33	21	15	18	10		
Bactris sp. (seedlings)	19.20	16.00	14.40	11.20	10.56	
32	17	14	16	12	10	
Acnistus arborescens	18.60	15.50	13.95	10.85	10.23	9.92
31	21	15	14	13	14	10
Casearia decandra	18.40	7.00	6.30			
14	8	9	**13** * * *			
Casearia guianensis	7.80	6.50	5.85			
13	10	9	**10** * *			

The number against each species is the total number of plots in which that species occurred. The numbers against each pair of species are the expected (above) and observed (below) numbers of joint occurrences.

* $P = 0.01 - 0.05$. * * $P = 0.001 - 0.01$. * * * $P = <0.001$.

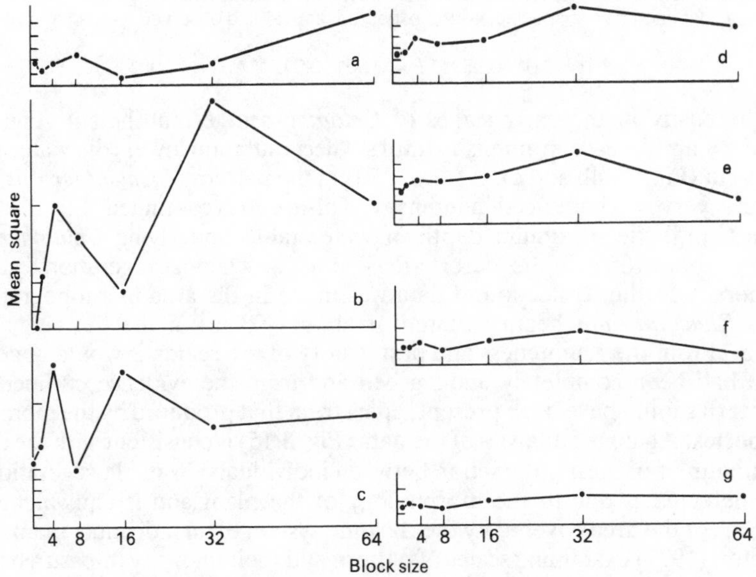

Fig. 3.14 The changing intensities of pattern of *Calamagrostis neglecta* in relation to age and stability. (From Kershaw, 1963; courtesy of *Ecology*.)

particular site. In detail such a method of dating would be open to severe criticism but it does give an indication of the relative age of stands. Samples were taken from numerous sites. The data (Fig. 3.14) show an identical trend. There is a minimal

Table 3.3 Expected and observed numbers of joint occurrences of pairs of species in plots at site E (primary forest). (From Greig-Smith, 1952*b*; courtesy of *J. Ecol.*)

	Myrcia granulata 59	*Rudgea freemani* 45	*Hymenaea courbaril* 37	*Rinorea lindeniana* 33	*Eugenia perplexans* 26
Rudgea freemani	26.55				
45	30				
Hymenaea courbaril	21.83	16.65			
37	25	17			
Rinorea lindeniana	19.47	14.85	12.21		
33	17	**20** *	**7** *		
Eugenia perplexans	15.34	11.70	9.62	8.58	
26	17	15	12	10	
Peltogyne porphyrocardia	14.16	10.80	8.88	7.92	6.24
24	16	8	6	5	7
Swartzia simplex	11.80	9.00	7.40	6.60	5.20
20	11	8	10	3	5
Calptranthes fasciculata	9.44	7.20	5.92	5.28	
16	8	8	3	7	
Clathrotropis brachypetala	9.44	7.20	5.92	5.28	
16	8	6	4	3	
Brownea latifolia	9.44	7.20	5.92	5.28	
16	7	6	6	5	
Eschweilera subglandulosa	8.85	6.75	5.55		
15	9	9	7		
Bactris sp.	8.26	6.30	5.18		
14	10	7	7		

The number against each species is the total number of plots in which that species occurred. The numbers against each pair of species are the expected (above) and observed (below) numbers of joint occurrences.

* $P = 0.01–0.05$. ** $P = 0.001–0.01$. *** $P = <0.001$.

pattern intensity in the early stages of *Calamagrostis* establishment (Fig. 3.14a) followed by an increase in intensity until silt accumulation lowers the water table to about 25 cm (Fig. 3.14b and c). At about 50 cm the pattern of *Calamagrostis* though still clear is very much reduced in intensity. This trend is continued (Fig. 3.14d-g) in the series until the maximum depth of water table underlying *Calamagrostis* is reached at about 105 m in the 'desert areas', after which most vegetation disappears altogether. A further collection of data was made in the area in a topographically uniform *Rhacomitrium* heath situated at about 790 m on the flat summit of a mountain. From the remoteness and past history of the region is would appear that the area had been completely undisturbed and from the evidence outlined above little pattern should have been present, apart from that produced by the morphology of the species. Again the analysis of the data (Fig. 3.15) is consistent with the concept of a minimum of pattern (interaction between individuals) in stable vegetation. The pattern detected is due to the morphology of the plant and is equivalent to the dimensions of the area covered by the rhizome system of an individual plant.

Brereton (1971) examining scales of pattern and their intensity (measured as peak height) showed for two salt marsh species *Puccinellia* and *Salicornia* a marked change in both pattern scale as well as pattern intensity. The sites sampled ranged from high water mark spring tides (stand 1) successively down through the marsh, to the lower limit of *Salicornia* at approximately the level of high water at ordinary neap tides (stand 6). The results (Fig. 3.16) are extremely clear cut and call for little comment.

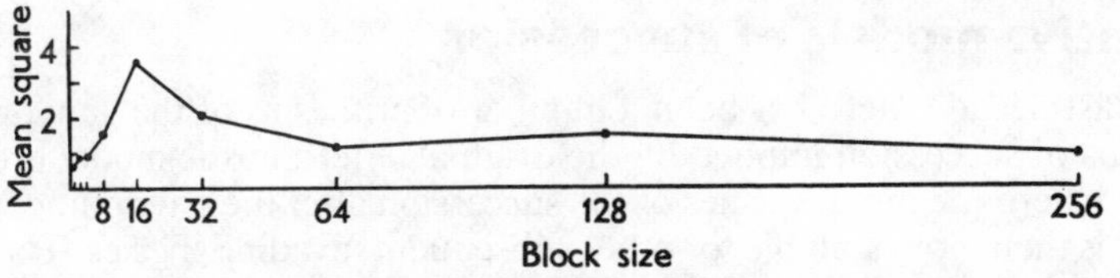

Fig. 3.15 General absence of pattern on *Carex bigelowii* growing in stable *Rhacomitrium* heath other than small scale morphological pattern (see text). (From Kershaw, 1963; courtesy of *Ecology*.)

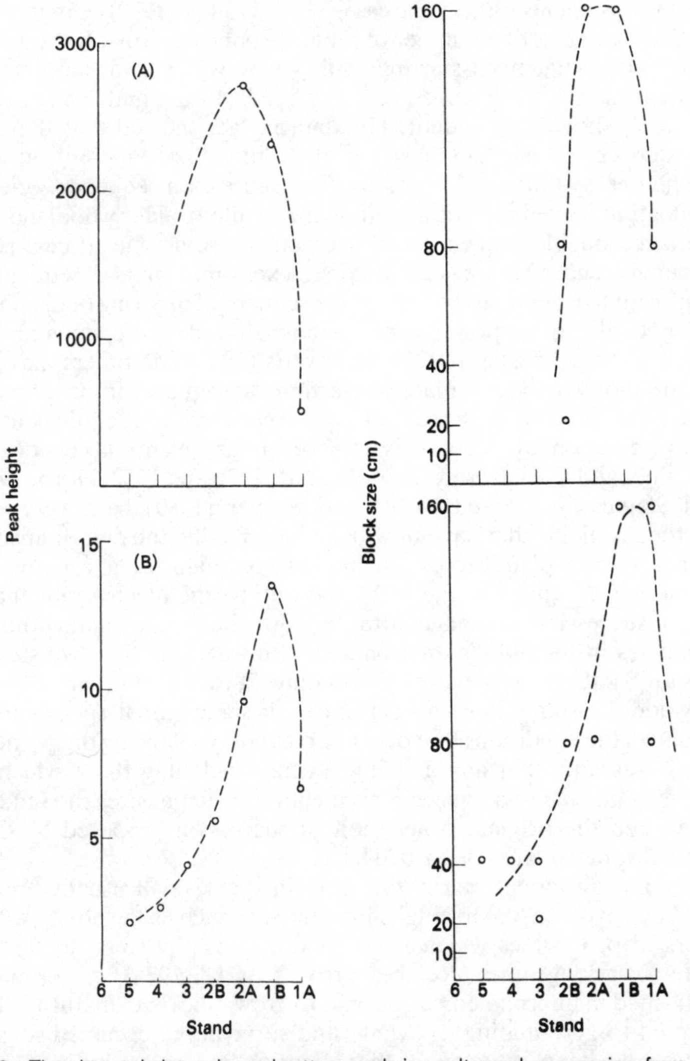

Fig. 3.16 The change in intensity and pattern scale in a salt marsh succession, from high water (1A) a low water mark (6). (A) *Puccinellia*. (B) *Salicornia*. (From Brereton, 1971; courtesy of *J. Ecol.*)

The evidence is consistent; it is drawn from a variety of habitats ranging from sub-arctic moss-heath to tropical forest and indicates that the original thesis of Greig-Smith's is correct enforcing a re-evaluation of Clements' analogy of the climax vegetational state with a complex organism.

Alternative models of succession

Over the last decade there has been a number of criticisms of the generality of the mechanisms of succession implied by the original deterministic model proposed by Clements: where the initial species of the succession alter the environment in such a way that it is then more suitable for other subsequent invading species. The corollary of this is that the actual sequence is predictable and accordingly the form and composition of the climax vegetation virtually pre-defined; the Clementsian model is thus completely deterministic. However, a body of evidence has accumulated indicating that this is not necessarily the case. Thus, Walker (1970) examined the peat stratigraphy at the margins of a number of replicate ponds in Great Britain where the classical succession sequence, starting with open water and running through submerged aquatics to floating-leaved aquatics, reed-swamp, and finally to an alder-willow woodland, should be evident. His data in fact showed that there was no consistent sequence at all, emphasizing the difficulties inherent in ordering successional phases by 'intuitive' methods. The succession sequence, open-water; submerged, floating-leaved aquatics; reed swamp; willow, alder woodland certainly is intuitively a reasonable sequence and can actually sometimes occur. However, since this primary succession takes so long, experimental and sequential field evidence is difficult to obtain, and the intuitive sequential ordering of disjunct stages of both this, or equally, other primary successions can lead to erroneous conclusions. Egler (1954), Drury and Nisbet (1973), Colinvaux (1973) and others, have all now questioned the validity of the completely deterministic sequence in any primary plant succession and the validity of these criticisms have forced an equivalent level of rigorous re-examination of determinism in secondary plant successions as well (Hulst, 1978; Peet and Christensen, 1980; Pickett, 1976 and 1982; Horn, 1974; etc). Connell and Slatyer (1977) and Noble and Slatyer (1980) have very elegantly summarized the problems that have now been identified in the deterministic model of primary or secondary plant succession and have outlined two alternative models. Their three models are given in Fig. 3.17 and outline the mechanisms that would bring about a successional change after a perturbation, assuming no further significant changes in the abiotic environment. Between the first two steps in the diagram (A to B) there is a major dichotomy between alternative models of succession. Model 1 assumes that only certain 'early successional' species are able to colonize the site in the conditions that occur immediately following the perturbation. Models 2 and 3 assume that any arriving species, including those which usually appear later, may be able to colonize. This dichotomy emphasizes the fundamental difference between the original conception of succession proposed by Clements (1916) and the alternative ones described here.

Up to this point all models agree that certain species will usually appear first because they have evolved 'colonizing' characteristics such as the ability to produce large numbers of propagules which have good dispersal powers, to survive in a dormant state for a long time once they arrive (Marks, 1974), to germinate and become established in unoccupied places, and to grow quickly to maturity. They are not well adapted to germinating, growing, and surviving in occupied sites, where there is heavy shade, deep litter etc., so that offspring seldom survive in the presence of their parents or other adults. Thus in all models, early occupants modify the environment so that it is unsuitable for further recruitment of these early successional species. In the ***deterministic*** model (Model 1) this process of reaction continues as the classical Clements sequence.

In Model 2, the modifications wrought on the environment by the earlier colonists neither increase nor reduce the rates of recruitment and growth to maturity of later colonists (steps C and D). Species that appear later are simply those that arrived

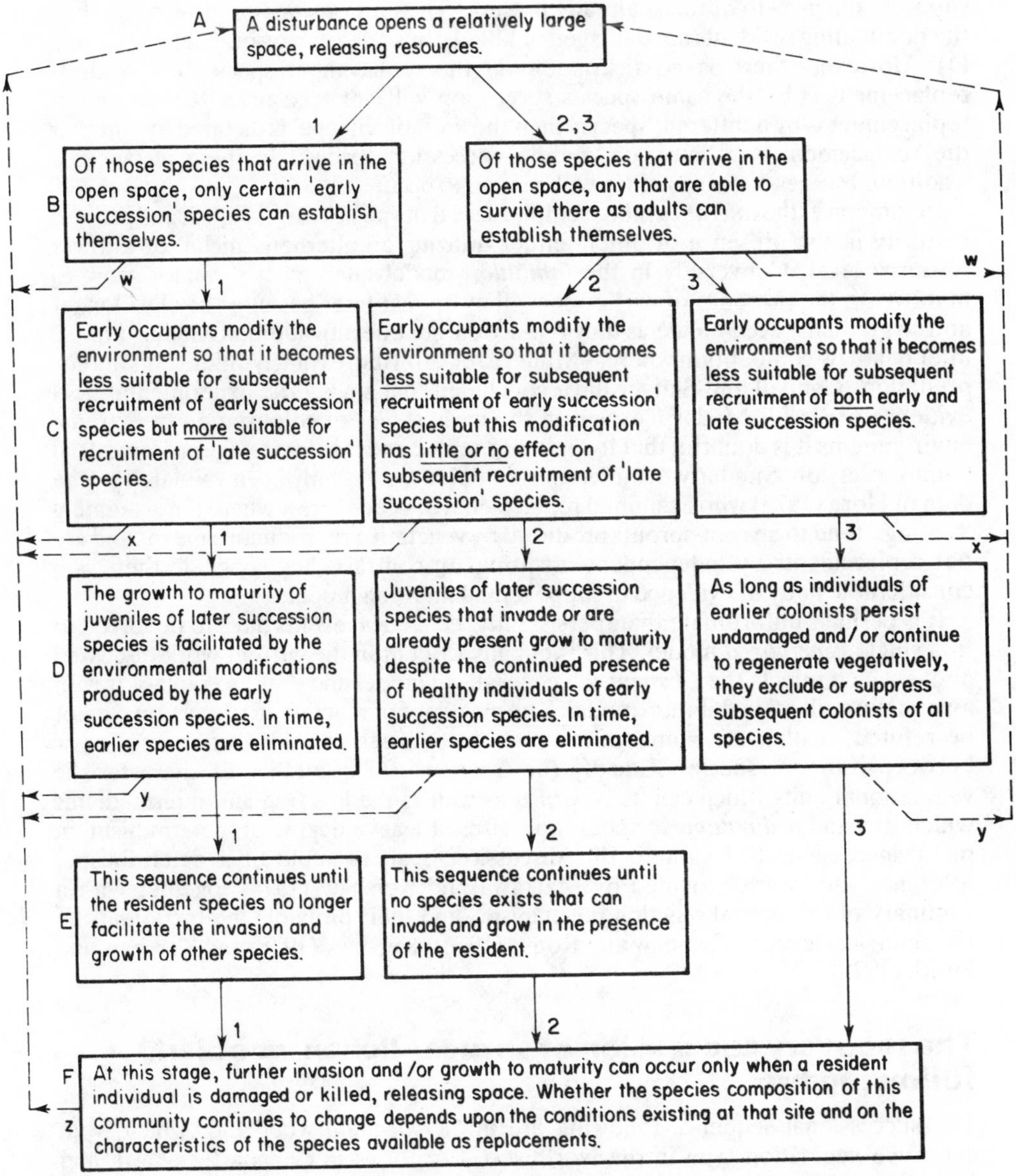

Fig. 3.17 Schematic representation of three possible mechanistic sequence describing plant succession. The dashed lines represent interruptions of the process, in decreasing frequency in the order of w, x, y and z. (From Connell and Slatyer, 1977; courtesy of *American Midland Naturalist*.)

either at the very beginning or later and then grew slowly. The sequence of species is determined solely by their life-history characteristics. In contrast to the early species, the propagules of the later ones are dispersed more slowly and their juveniles grow more slowly to maturity. They are able to survive and grow despite the presence of early-succession species that are healthy and undamaged. Connell and Slatyer refer to this model as the ***tolerance*** model and it stands as an intermediate case between the first and third, the ***inhibition*** model.

In contrast to the first model, the third holds that once earlier colonists secure the space and/or other resources, they inhibit the invasion of subsequent colonists or suppress the growth of those already present. The latter invade or grow only when the dominating residents are damaged or killed, thus releasing resources (steps C and D). However, there is no restriction on the replacement species involved. If replacement is by the same species succession will not take place. Conversely if replacement is by a different species then the level of 'change' is dictated by whether the replacement is a 'late', or 'middle'-succession species. In these instances a traditional successional sequence will be seen to occur.

In summary, the *tolerance* model emphasizes that species can establish and grow to maturity in the presence of others either utilizing an alternate 'niche' or a lower resource level. Conversely in the *inhibition* model later species cannot grow to maturity in the presence of earlier ones; they appear later because they live longer and so gradually accumulate as they replace earlier ones. In addition Models 1 and 2 infer death by competition whereas Model 3 relies on more cataclysmic events and/or predation. Connell and Slatyer have concluded that since little experimental field evidence exists for Model 1 except in the early stages of colonization of extreme environments it is doubtful that it applies to replacement in later stages of terrestrial plant succession. Similarly the tolerance model so far has only been validated by the data of Horn (1975) who examined replacement in beech forest where unfortunately 'saplings' tend to be root-sprouts produced by vegetative reproductive means and are not replacements by independent offspring. In contrast they conclude there is a considerable body of evidence to support the inhibition model.

It is perhaps unfortunate that, persistently, ecological events have to be forced to fit a single *generalized* model. This is equally evident in the current work discussed above which rejects the *Clementsian* generalized model and seeks to replace it with an equally pedantic inhibition model. The existence of reaction in succession cannot be refuted, neither can competitive reactions between individuals of a species or between different species. Equally the frequency of occurence of characteristic vegetational units which can be *classified* as well as the less frequent intermediates which demand *ordination*, further emphasises at least a degree of 'determinism' in plant succession. Relevant to this discussion is an example of a deterministic-tolerance succession provided by Maikawa and Kershaw (1976) together with a summary of the partial reaction mechanisms and their probable interaction at the physiological level by Kershaw and Rouse (1976), Kershaw (1978) and Kershaw and Smith (1978).

The recovery succession of spruce–lichen woodland following fire

The successional sequence following fire in *Stereocaulon* woodland, which is an extensive vegetation type in the Northwest Territories in Canada (Maikawa and Kershaw, 1976), offers a useful example here, since the data are derived from an area of topographic and soil uniformity but from sites burnt at specific times estimated quite accurately from fire scars readily available at the periphery of each burn. Johnson and Rowe (1975) have examined fire reports prepared by the Northwest Forestry Service, for a region of subarctic forest which lies to the East of Great Slave Lake, and show that fire is a frequently occurring but natural event with almost all of the fires caused by lightning. Four phases of recovery are recognized: Phase 1, the *Polytrichum* phase with *P. piliferum*, *Biatora granulosa* and *Lecidea uliginosa* forming the major cover over the recently burnt surfaces. Phase 2, the *Cladonia* phase, developing after 15–20 years with the establishment of a number of *Cladonia* species, in particular *Cladonia gonecha*, *C. uncialis* and *C. stellaris*. Phase 3, the *Picea mariana–Stereocaulon paschale* woodland phase, developing after 60 to 70 years, forming the mature woodland phase and dominating large areas of the region.

Finally, Phase 4 which is apparent after 130–150 years, by which time the canopy closes and the luxurious carpet of *Stereocaulon paschale* is replaced by a carpet of mosses dominated by *Hylocomium splendens, Pleurozium schreberi* and *Dicranum* species. Only rarely does this final closed canopy spruce – moss woodland develop, since the *Stereocaulon* woodland is extremely fire-susceptible and is usually reburnt before the final phase of the succession is initiated. It is important to recognize that *Picea mariana* as well as *Vaccinium vitis-idaea* enter the succession immediately, the former arriving as seed and the latter surviving the fire as an underground root-stock (Fig. 3.18).

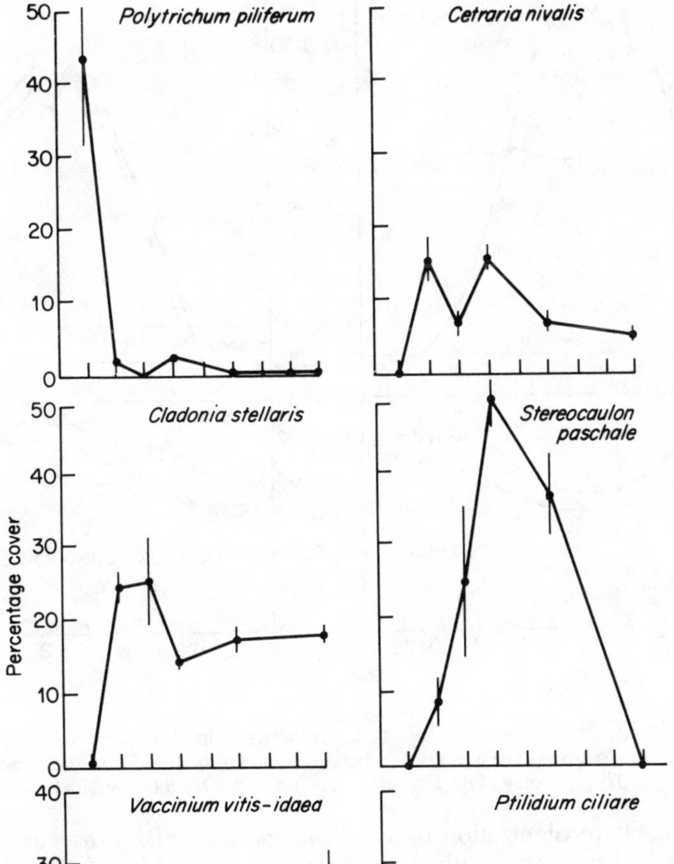

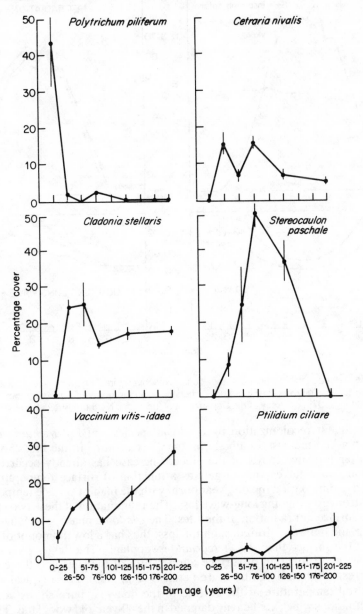

Fig. 3.18 The changing species abundance (% cover) the successional sequence following forest fire in spruce – lichen woodland in northern Canada. (From Kershaw and Rouse, 1976; courtesy of the Department of Indian and Northern Affairs.)

During the vegetational succession there is a concurrent and equally marked development of surface microclimate. Kershaw *et al*., (1975), Kershaw and Rouse (1976), Rouse (1976) and Kershaw (1978), have described the surface microclimate of freshly burned plots and contrasted this with 23- and 80-year old recovery phases, representative of the *Cladonia* and *Stereocaulon* phases of the recovery sequence, respectively. Considerable differences develop in surface albedo following fire with values dropping from 20% to 5% on freshly burned surfaces (Fig. 3.19).

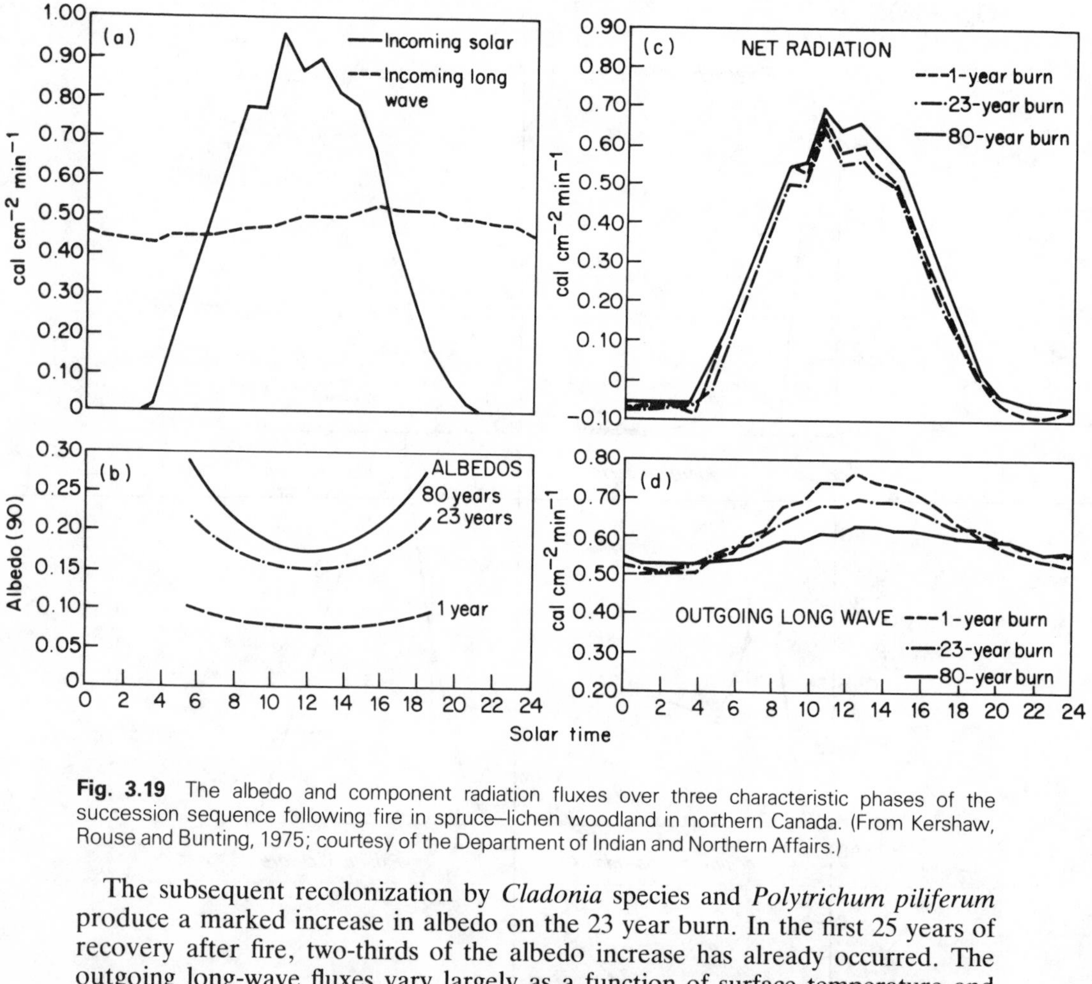

Fig. 3.19 The albedo and component radiation fluxes over three characteristic phases of the succession sequence following fire in spruce–lichen woodland in northern Canada. (From Kershaw, Rouse and Bunting, 1975; courtesy of the Department of Indian and Northern Affairs.)

The subsequent recolonization by *Cladonia* species and *Polytrichum piliferum* produce a marked increase in albedo on the 23 year burn. In the first 25 years of recovery after fire, two-thirds of the albedo increase has already occurred. The outgoing long-wave fluxes vary largely as a function of surface temperature and during the day time period the one year burn, with the highest surface temperature, shows the strongest outgoing long-wave loss. Then summation of these two effects is seen in the surface net radiation of the sites. The *Cladonia* phase, which has both a high albedo and marked infrared radiation loss, thus has a lower amount of energy available at the surface than the *Stereocaulon* woodland. The long-wave radiation losses are simply a function of surface temperature of the sites and there are, of course, corresponding soil temperature patterns (Fig. 3.20). For a typical summer period the soil temperature at the surface of the one-year burn shows very high diurnal fluctuations which are heavily damped in the 80-year old woodland. The soils in the lichen woodland are colder at all depths, but of more significance to the developing lichen cover, surface maxima and minima calculated from the long wave

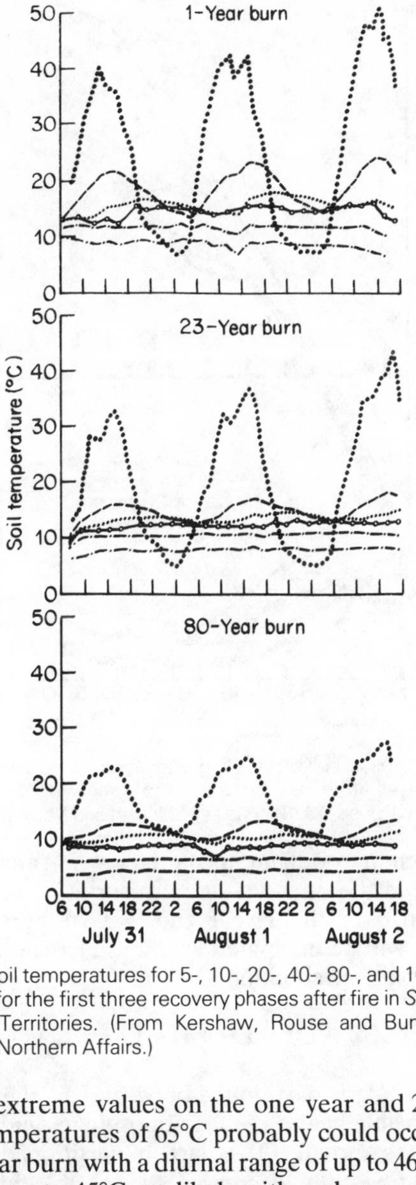

Fig. 3.20 Mean hourly soil temperatures for 5-, 10-, 20-, 40-, 80-, and 160-cm depths for the period in July 31 to August 2, 1975 for the first three recovery phases after fire in *Stereocaulon* woodland, Abitau Lake region, Northwest Territories. (From Kershaw, Rouse and Bunting, 1975; courtesy of the Department of Indian and Northern Affairs.)

radiation data show extreme values on the one year and 23-year burned surfaces. Maximum surface temperatures of 65°C probably could occur during high radiation periods on the one year burn with a diurnal range of up to 46°C. On the 23-year burn, surface temperatures up to 45°C are likely with early morning ground frost even in July, as a result of extreme long wave radiation cooling.

These high surface temperatures occurring during the early successional stages following fire reflect the very xeric conditions typical of these intial phases. Under these dry conditions of the *Cladonia* recovery phase, the majority of the net radiation is dissipated as a sensible heat flux. Thus the 1976 soil moisture and evapo-transpiration characteristics of the *Stereocaulon* lichen woodlands on the Northwest Territories (Kershaw and Rouse, 1976) show the same constant evaporation and evapo-transpiration to the end of June when there is a rapid and substantial divergence, the 81-year burn maintaining much the same rate through to the end of the measurement period, whereas the 1- and 24-year burns show a large decrease in evaporation during July and August (Fig. 3.21). This decrease reflects the initially drier soils of the 24-year old burn as well as the importance of evapo-transpiration

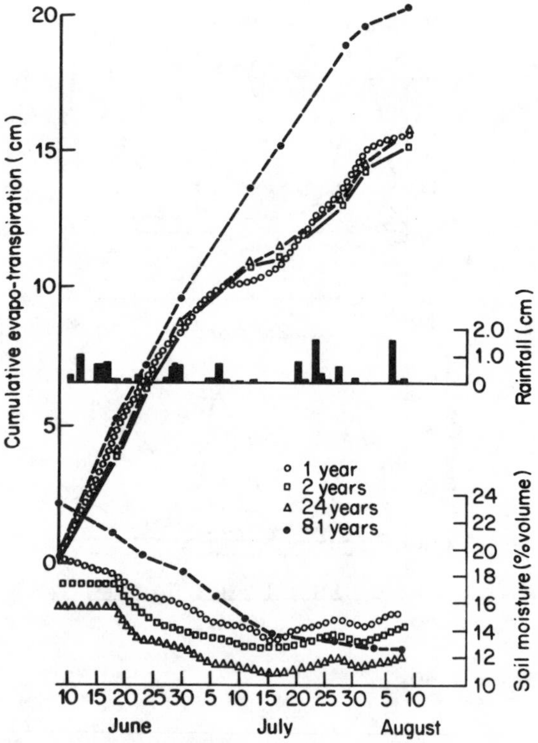

Fig. 3.21 The seasonal patterns of soil moisture, precipitation, and cumulative evapotranspiration in 1975 for th first recovery phases after fire in *Stereocaulon* woodland, Abitau Lake region, Northwest Territories. (From Kershaw and Rouse, 1976; courtesy of Indian and Northern Affairs.)

from the spruce of the lichen woodland; the initial low soil moisture values reflecting the earlier snowmelt from the open young burn surface. As a direct result of this decrease in evaporation over the 1-year and 24-year burns the energy usually involved in evaporation will be dissipated as a sensible heat flux. This can be seen readily from the energy balance equation

$$Q^* = LE + H + G$$

where the net radiation, Q*, is partitioned between LE, the latent heat of evapotranspiration, H, the sensible heating of the atmosphere, and G, the heating of the subsurface. Since G is very small, Q* is largely partitioned between LE and H. Accordingly, if the latent heat flux decreases, the sensible flux must increase. As a result of this, in July and early August after the decline of the evaporation from the 1- and 24-year burns, the sensible heat flux will be considerably greater than over the lichen woodland. This is visually evident as dust devils which form frequently over the burned surfaces on hot summer days. These severe conditions may form one of the essential limitations to higher plant growth and allow the development of the lichen surface without competition from such potentially faster growing species. Surface temperatures are finally ameliorated in the developing *Stereocaulon* woodland as a result of increased shading by the steadily growing spruce trees and by the effective mulching properties of the lichen mat with the resultant maintenance of soil moisture at field capacity evident in Fig. 3.21. This allows evaporative cooling with a diminished sensible heat flux and results in lower surface temperatures.

The results are confirmed by direct measurement of surface, thallus and air temperatures. Thus air temperature 2 mm above *Biatora granulosa* crusts reach about 48°C, contrasting markedly with equivalent air temperatures at the surface of the *Stereocaulon paschale* mat even in very open woodland. As the tree density slowly rises and the lichen surface becomes increasingly shaded, this temperature contrast is even more marked, with the maximum temperature occurring at any one site being of very restricted duration.

Kershaw and Smith (1978) have shown that *Stereocaulon paschale* is very sensitive to surface temperatures greater than 30°C. Experimental replicates of *Stereocaulon paschale* stored air-dry at 35°C day and 15°C night temperature show a 50% decline in their rate of net photosynthesis after two weeks (Fig. 3.22). At higher stress

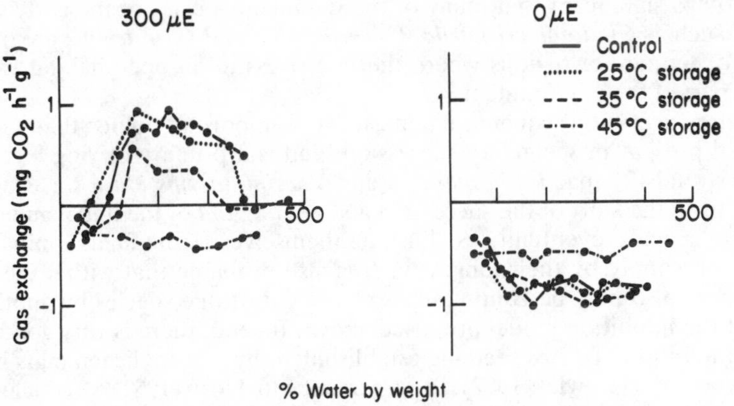

Fig. 3.22 The respiration and net photosynthetic response of *Stereocaulon paschale* at 25°C with 0 and 300μE m^{-2} s^{-1} illumination after the experimental replicates had been stored air dry for 14 days at a range of temperature stress levels. Mean gas exchange rates are given as milligrams CO_2 per hour per gram with a maximum SE of 0.6 mg CO_2 h^{-1} g^{-1}. Thallus saturation is expressed as a percentage of the final ovendry weight of each replicate. (From Kershaw and Smith, 1978; courtesy of the *Canadian Journal of Botany*.)

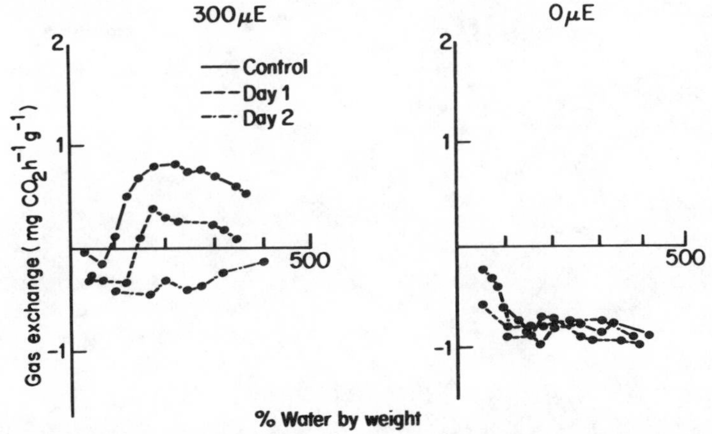

Fig. 3.23 The respiration and net photosynthetic response at 25°C with 0 and 300 μE m^{-2} s^{-1} illumination after the experimental replicates had been stored air dry for 1 and 2 d at 45°C day, 15°C night temperatures. Mean gas exchange rates are given as milligrams CO_2 per hour per gram with a maximum SE of 0.6 mg CO_2 h^{-1}g^{-1}. Thallus saturation is expressed as a percentage of the final ovendry weight of each replicate. (From Kershaw and Smith, 1978; courtesy of the *Canadian Journal of Botany*.)

temperatures (45°C day and 15°C night temperature) a single 12-hour period effects a 50% net photosynthetic decline and a further 12-hour stress period virtually eliminates all photosynthetic activity (Fig. 3.23). It is evident then that the surface temperature of the earlier stages of the succession is too excessive for the survival of *Stereocaulon paschale* and its entry into the succession is delayed for 50–80 years, during which time the harsh thermal regime is modified. This occurs in two ways. There is a gradual accumulation of an organic layer and thus an enhanced retention of soil moisture, resulting in a significant reduction of the sensible heat flux (Rouse, 1976; Kershaw, 1977). At the same time tree growth leads to a sharp decrease in direct radiation and the combination of these two processes results in a much cooler average surface temperature which can be successfully colonized by *Stereocaulon paschale*. Thus the long successional sequence leading to the final establishment of *Stereocaulon* woodland is primarily a function of the heat sensitivity of *Stereocaulon*. It is perhaps significant that many of the dominant species of the early succession phases, such as *Cladonia cristatella, C. verticillata*, and *C. unicialis*, are often more typical of temperate regions where thermal stress in an open habitat would be a normal part of the environment.

This fire-succession sequence emphasizes two important points: that, (1) reaction is indeed present in secondary succession, and is a primary driving force *in some instances*, and (2) that *Picea mariana* and *Vaccinium vitis-idaea* for example, are present from the start of the succession and are *tolerant* of the initial and secondary dominant species, eventually establishing themselves as the final dominants of the succession, simply by their longevity. It is also probable that within the *Cladonia* phase there also may be continuous replacement of one species by another in the sense of the inhibition model proposed above. Indeed, there is very good evidence for the inhibition of *Picea* seedling establishment by mature lichen mats in spruce–lichen woodland (Cowles, 1982). In agreement with Horn (1981) we conclude that all three models originally formally proposed by Connell and Slatyer (1977) will legitimately apply to vegetational succession, either individually to specific stages of a succession, or equally in *combination* to any stage or stages. We also emphasize that reaction may be considerably more subtle in its effects than has been traditionally recognized.

4

Cyclic Vegetational Change

It is commonplace to regard vegetation which is stable and in equilibrium with its environment as a static entity, which having gone through various stages of succession has finally reached an ultimate level of stability and ceases to change any further. This superficial impression is misleading since in addition to the external environmental factors which regulate the over-all composition of the vegetation, there exists a series of intrinsic factors which continues to induce vegetational change. This intrinsic vegetational change is not manifested as a continuation of succession but as a cycle of events at any particular point in a community. These events are replicated many times over the whole of the community, and are apparent as a series of phases present at a number of points at any instant in time. The over-all summation of these phases constitutes the composition of the community. Thus while plant succession is a *directional change*, cyclic changes represent *fluctuations about a mean value*.

Hummock and hollow cycle

Watt (1947*a*) outlined a number of examples of cyclic fluctuations containing several marked phases which are now regarded as classic examples of cyclic change. The regeneration complex which occurs on actively growing raised bog (the 'hummock and hollow cycle' as it is often called) is an outstanding example of a series of dynamic phases which are related to each other in time. The phases form a mosaic of patches, each at a different stage from its neighbours. Oswald (1923) and Godwin and Conway (1939) have shown the relation of the phases to each other by examination of the stratigraphy of the underlying peat. The initial phase is represented by *Sphagnum cuspidatum* which invades small pools of water. This species is invaded by *S. pulchrum* which in turn is replaced by *S. papillosum*. The latter species build up low hummocks which are eventually colonized by *Calluna vulgaris, Erica tetralix, Eriophorum vaginatum* and *Trichophorum caespitosum*. At a later stage the lichen species *Cladonia arbuscula* (*C. sylvatica*) enters the sequence, each sequence being abundantly preserved as stratified deposits of peat. At this stage the sequence is identical to a 'micro-succession' and each of the subsequent stages is dependent on the reactions on the environment by its predecessors. However, as the *Calluna* ages and dies the hummock is eroded and a pool of open water re-established largely by the activity of the adjacent phases building up hummocks. Thus, under these conditions, an area of raised bog is a mosaic of patches each of which represents a phase of a cycle; each patch will progress eventually and repeatedly through the same cycle of events and although a given (small) area will change over a period of time the whole community remains essentially the same.

A similar example is drawn by Watt from grassland, where the vegetational variation is related to microtopography. Four phases are present, descriptively termed 'hollow', 'building', 'mature' and 'degenerate' (Watt, 1940). The hollows are invaded by seedlings of *Festuca ovina* around which accumulates a variety of wind-borne particles in addition to humus remains of the parent plant. Over a period of time a small hummock is formed, representing the 'building' phase (Fig. 4.1). In the

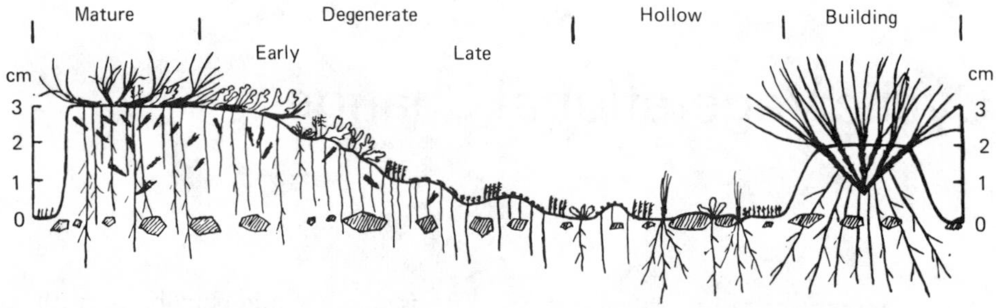

Fig. 4.1 Diagrammatic representation of the *Festuca ovina* erosion cycle indicating 'fossil' shoot bases in the soil. (From Watt, 1947*a*; courtesy of *J. Ecol.*)

mature phase the hummocks reach a height of about 4 cm and are colonized by several lichen species notably *Cladonia alcicornis* and *C. rangiformis*. The lichen species slowly assume a more dominant role in the early stages of the 'degenerate' phase and are gradually replaced by crustaceous lichen species. Erosion of the hummock follows completing the full cycle of events back to the 'hollow' phase. Again the over-all structure and composition of the community remains unchanged since each phase is represented by a mosaic of cyclic units related spatially to each other as well as representing a time sequence.

The occurrence of cyclical phases is not restricted to perennial herbaceous species and Watt (1947*a*) also describes a cyclical development which occurs over a long period in beech woodland on chalk plateau. Following the death of a full grown tree the gap that is left is initially devoid of vegetation and represents the first phase of the cycle. This is invaded initially by *Oxalis* and subsequently by *Rubus*, representing the next two phases of the cycle. Concurrently beech and ash seedlings become established forming a 'reproduction circle' (Watt, 1925) with the ash in the centre surrounded by a circle of young beech originating from seed shed by the surrounding mature beech. Eventually the ash is over-topped by the beech saplings and the gap finally closed. The cycle is finally completed by the ultimate degeneration and death of the beech colonizers initiating the same cycle again. The time scale in this example is very long but the nature of the process is identical.

In general the sequence of events is similar in all the examples given by Watt (1947*a*) and these he descriptively terms pioneer, building, mature and degenerate. The productivity of each phase appears to follow a generalized curve of increasing productivity in the early phases of a cycle, leading to an optimum in the mature phase and followed by reduced productivity in the degenerate phase (Fig. 4.2). Billings and

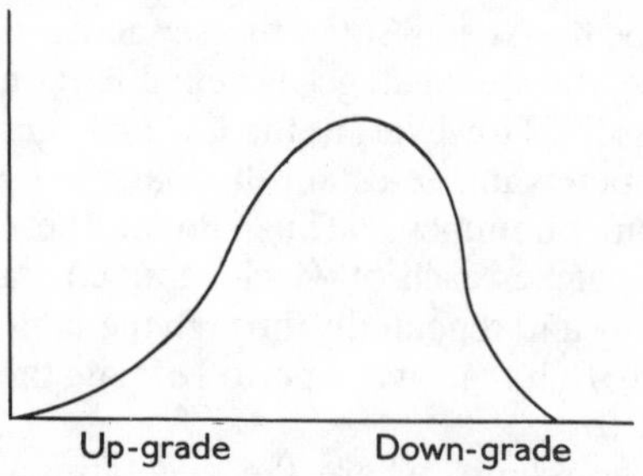

Fig. 4.2 Generalized productivity curve. (Redrawn from Watt, 1947*a*.)

Mooney (1959) give an account of an extremely interesting cycle in tundra vegetation at 3600 m in the Medicine Bow Mountains of southern Wyoming. The area as a whole is subjected to intense soil frost action and the valley bottom is covered with actively-forming frost hummocks of peat and stone polygons. The occurrence of frost hummocks and polygons is extremely characteristic of arctic and sub-arctic zones, where intense frost produces a number of marked topographical features. Billings and Mooney relate the formation of polygons in this area to the degradation of the frost hummocks of peat. When a hummock rises to a certain level so that it protrudes above the general level of the snow in winter, the species on the crest of the hummock are killed and eroded away by the repeated action of wind and wind-borne ice/snow particles. With continued exposure the hummock itself is eroded exposing rocks and silt which are pushed up concurrently from below by further frost action. Eventually the hummock is completely eroded to its base and this stage appears as a stone polygon surrounded with a 'rim' of peat.

The water table in the whole area fluctuates from year to year; the hollows usually are fairly wet and often whole areas of polygons are completely submerged. During these submerged periods the hummock formation appears to be initiated by *Carex aquatilis* which invades the rocky intervals of the polygons. The general sequence of events is shown in Fig. 4.3. Following the establishment of *C. aquatilis, Sedum rhodanthum* and several bryophyte species become established. Eventually these species are replaced by *Geum turbinatum* and *Polygonum viviparum* which form a closed turf with other additional species on the mature hummocks. The erosion commences at this stage of the cycle and subsequent drowning of polygons initiates the next cycle. Thus there is a continuous cycle of change, hummocks alternating with the centre of polygons and the margin of the polygons alternating with subsequent hummocks. The general phasic development of this cycle is strikingly similar to the hummock and hollow cycles discussed by Watt (1947*a*). The mechanism is partly different and the species involved obviously different, but the same general principle is common to all three cases. The pioneer phase in this instance is dominated by *Carex aquatilis*, the building phase by *Carex* and *Sedum* and the mature phase by *Geum* and *Polygonum*.

Marginal effects

Cyclic phases as described above are partly dependent on intrisic properties of the plants themselves and partly dependent on intrinsic properiesof thecommunity,both of which are modified and controlled to a greater or lesser extent by the environment. Thus the ability of *Sphagnum papillosum* to form hummocks is an intrinsic property of that species, the availability of the other species involved in the cycle is an intrinsic property of the community, and the final partial erosion of the hummock is a result of environmental control. The downgrade phases of the cycle are terminated by erosion of the hummock but it is interesting to speculate as to the actual initiation of the downgrade phase. A possible explanation lies in the relationship between the general vigour of a perennial plant and its age. This important concept was originally outlined by Watt (1947*b*) in relation to the vigour of bracken invading grassland where the marginal zone was considerably taller and apparently more vigorous than in the 'hinterland'. The difference in height of the fronds in the margin and hinterland is considerable (Fig. 4.4). The immediate possibility which would account for this 'marginal effect' would be the presence of some factor necessary for the full development of the fronds which is markedly reduced in quantity by the advancing front and is in short supply in the hinterland. However, analysis of the soil (Table 4.1) shows no consistent trend to support such a hypothesis and furthermore the addition of different combinations of fertilizers containing nitrogen, potash and phosphorus to sample plots in the hinterland had no significant effect.

The significant feature is the progressive increase of rhizome age back from the margin to the hinterland where the oldest part of the rhizome decays and lateral

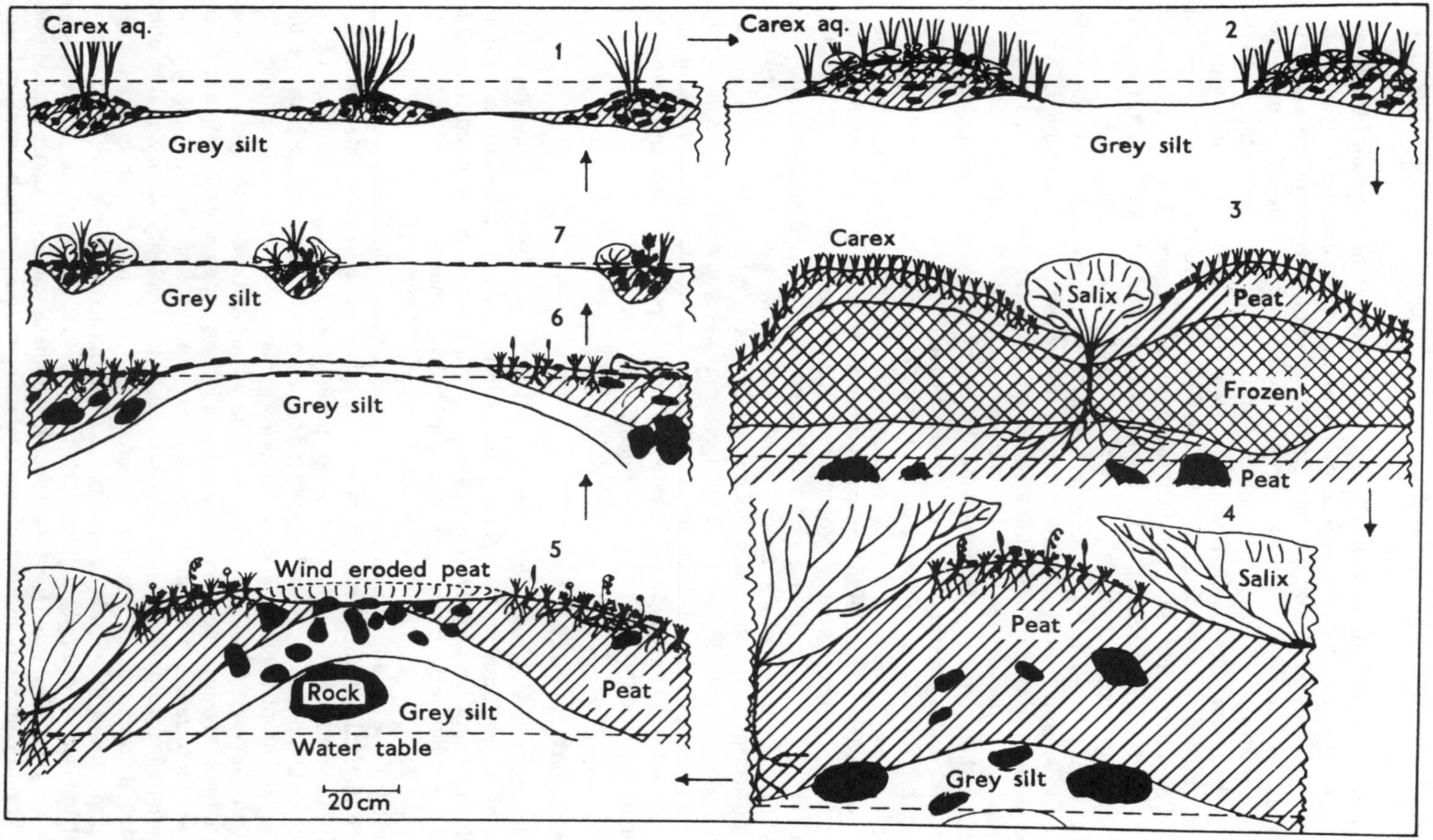

Fig. 4.3 Diagrammatic representation of the hummock, frost scar and polygon cycle. (From Billings and Mooney, 1959; courtesy of *Ecology*.)

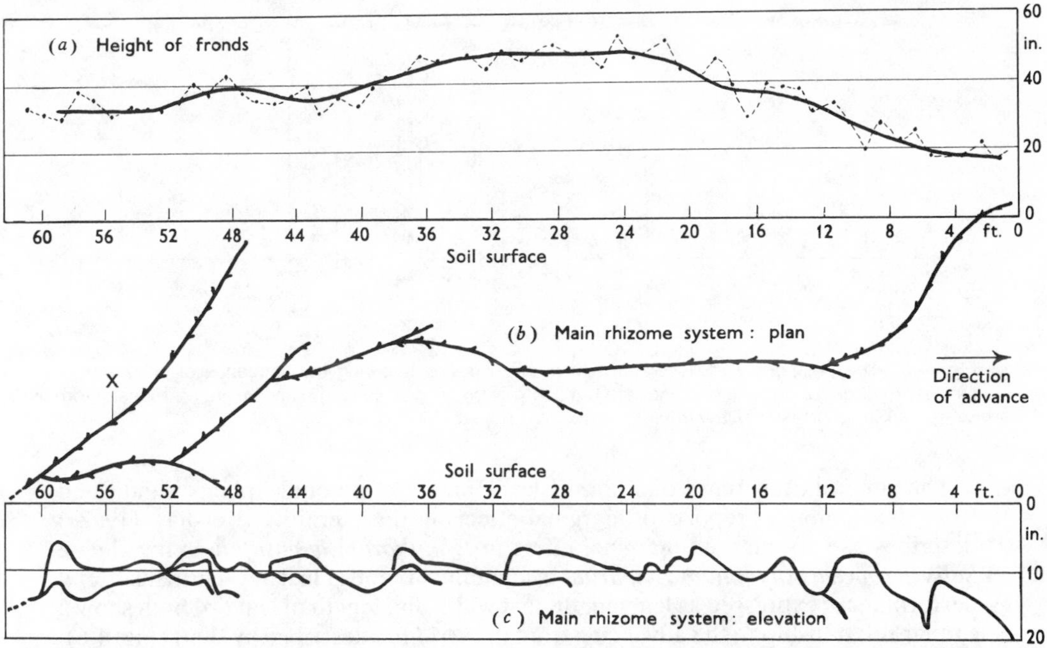

Fig. 4.4 (**a**) the average height of fronds across the advancing margin of bracken; (**b**) plan and (**c**) elevation of main rhizome. (From Watt, 1940; courtesy of *New Phytol.*)

Table 4.1 Soil data from surface layers (0 –7.6 cm) in different zones of bracken invading grassland. (From Watt, 1940; courtesy of *New Phytol.*)

	In grassland	In zone of maximum height	In dense uniform bracken	In hinterland
pH	4.2	4.0	3.8	4.0
Carbon per cent	1.785	1.425	1.740	1.610
Carbon/Nitrogen	19.40	19.79	19.12	19.63
Exchangeable Calcium in marginal effect	1.44	1.28	1.28	1.36
Total exchangeable bases	2.40	2.00	1.61	1.61

shoots become separated as individuals. Thus the margin has a large number of young even-aged rhizomes orientated parallel to each other. These younger parts of the rhizome produce larger and more vigorous shoots resulting in the marked marginal effect. Conversely in the hinterland the rhizomes are initially much older and the shoots smaller. As the main rhizome decays lateral shoots of different ages are separated as individuals, each of which will subsequently follow the same progressive change of vigour as each individual ages. Thus the hinterland consists of very unevenly aged groups of individuals at all stages of the cycle, forming a mosaic of phases. In addition to the height of frond being at its maximum in the mature phase the length of petiole above ground follows the same trend and the length of petiole below ground level follows an inverse trend (Fig. 4.5). There is a marked relationship between the age of the rhizome and its morphology, and also between the vigour or performance of the aerial shoots from young and older parts of the same rhizome. This feature may explain the onset of the downgrade stage of the cyclic phases outlined above. As the principal species of the mature phase age, their performance falls markedly and the degenerate phase commences.

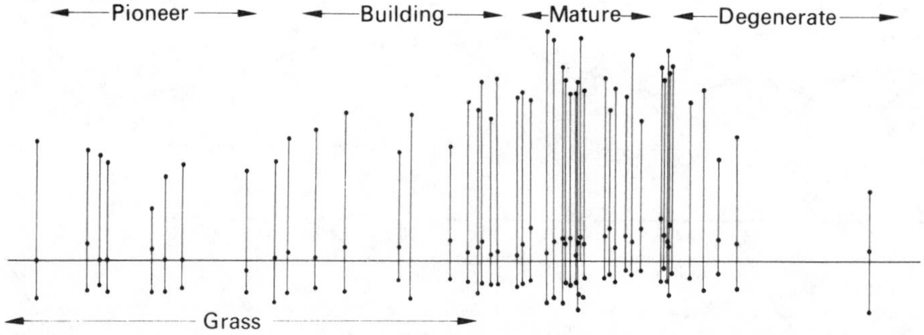

Fig. 4.5 The positions and dimensions of bracken fronds showing the sequence of phases from pioneer to degenerate. The dots indicate the depth of origin, petiole length and height of fronds. (From Watt, 1947*b*; courtesy of *New Phytol.*)

The general occurrence of a phasic development of perennial plants is indicated from the numerous reports of marginal effects in the literature. Kershaw (1962*a*) describes as very marked marginal effect in *Eriophorum angustifolium* invading a *Salix herbacea*/*Polygonum viviparum* community in central Iceland. The density and performance (expressed as leaf length, leaf width and length of leaf tip) both show a similar relationship to the advancing front in *Pteridium* described by Watt (Fig. 4.6).

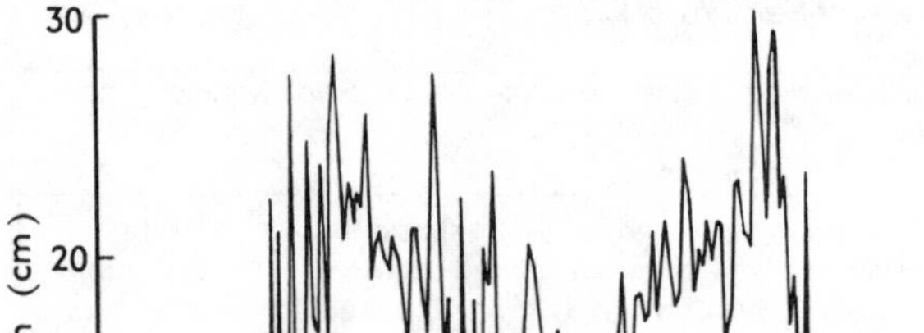

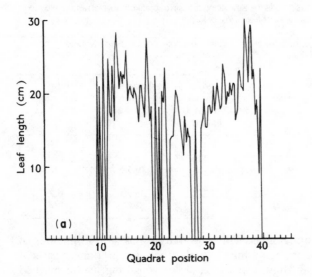

Fig. 4.6(a) The variation of performance of *Eriophorum* (leaf length) along a transect running through the circular advancing margin.

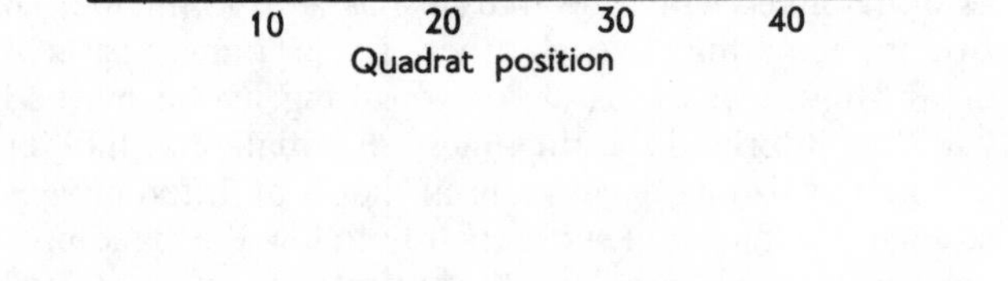

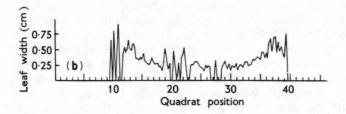

(b) The variation of performance of *Eriophorum* (leaf width) along a transect running through the circular advancing margin.

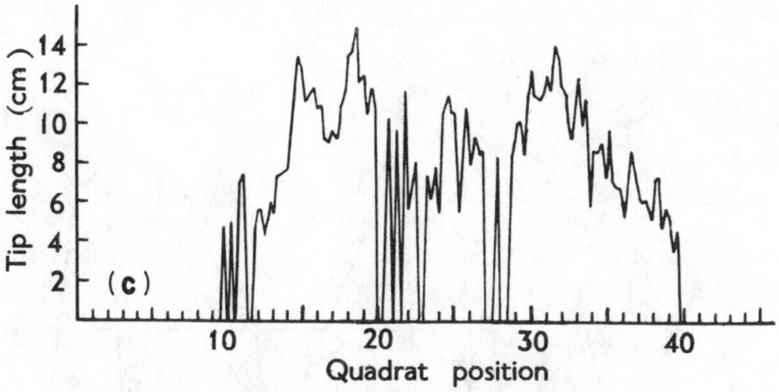

(c) The variation of performance of *Eriophorum* (tip length) along a transect running through the circular advancing margin. The length of tip is inversely proportional to performance. (From Kershaw, 1962*a*; courtesy of *J. Ecol.*)

Similarly Ovington (1953), Caldwell (1957) and Pénzes (1958, 1960) have described numerous examples of advancing fronts of even-aged rhizomes at an optimum level of performance relative to the hinterland with its very much lower level. Watt (1955) shows that *Calluna* follows the same phase development and furthermore where *Pteridium* is present there is an inverse relationship between the two species. Where *Calluna* is at its building/mature phase, competing bracken is at its minimum development. Similarly where *Calluna* is in its degenerate phase, bracken is at the building or mature phase. Thus concurrently the competitive ability changes with the phasic series pioneer, building, mature and degenerate. The phasic development of *Pteridium* and *Calluna* have been confirmed by Anderson (1961*a*, 1961*b*) and Nicholson and Robertson (1958).

The interaction between phasic development of perennial species and their competitive ability has been well documented by Watt (1981) who monitored the fluctuating abundance of a number of grassland species between 1936 and 1973 following the cessation of rabbit grazing. The changes are summarized in Fig. 4.7 for

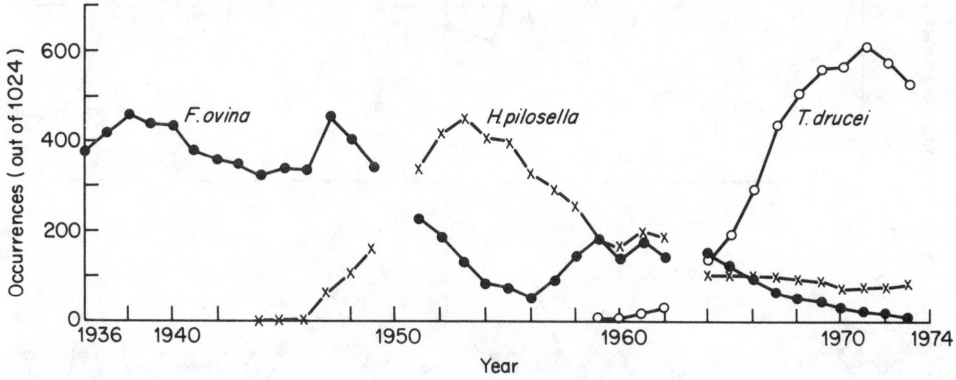

Fig. 4.7 Changes in cover of the three dominant species of grassland and after the cessation of rabbit grazing. (From Watt, 1981; courtesy of the *J. Ecol.*)

the three dominant species, *Festuca ovina, Hieracium pilosella* and *Thymus drucei*. The change of abundance of each species is primarily related to the year it enters the plot and secondarily to differences in their competitive abilities. Each species differs specifically in its life-form of course, but also the level of competitive ability changes continuously as a function of the stage of phasic development of each component.

Thus the original population of *Festuca ovina* at the start of the experimental period was in the mature phase and since its early and degenerate phases are both susceptible to frost-heaving the initial population increase, subsequent decline and the final increase are a response to the initial release of grazing pressure, a

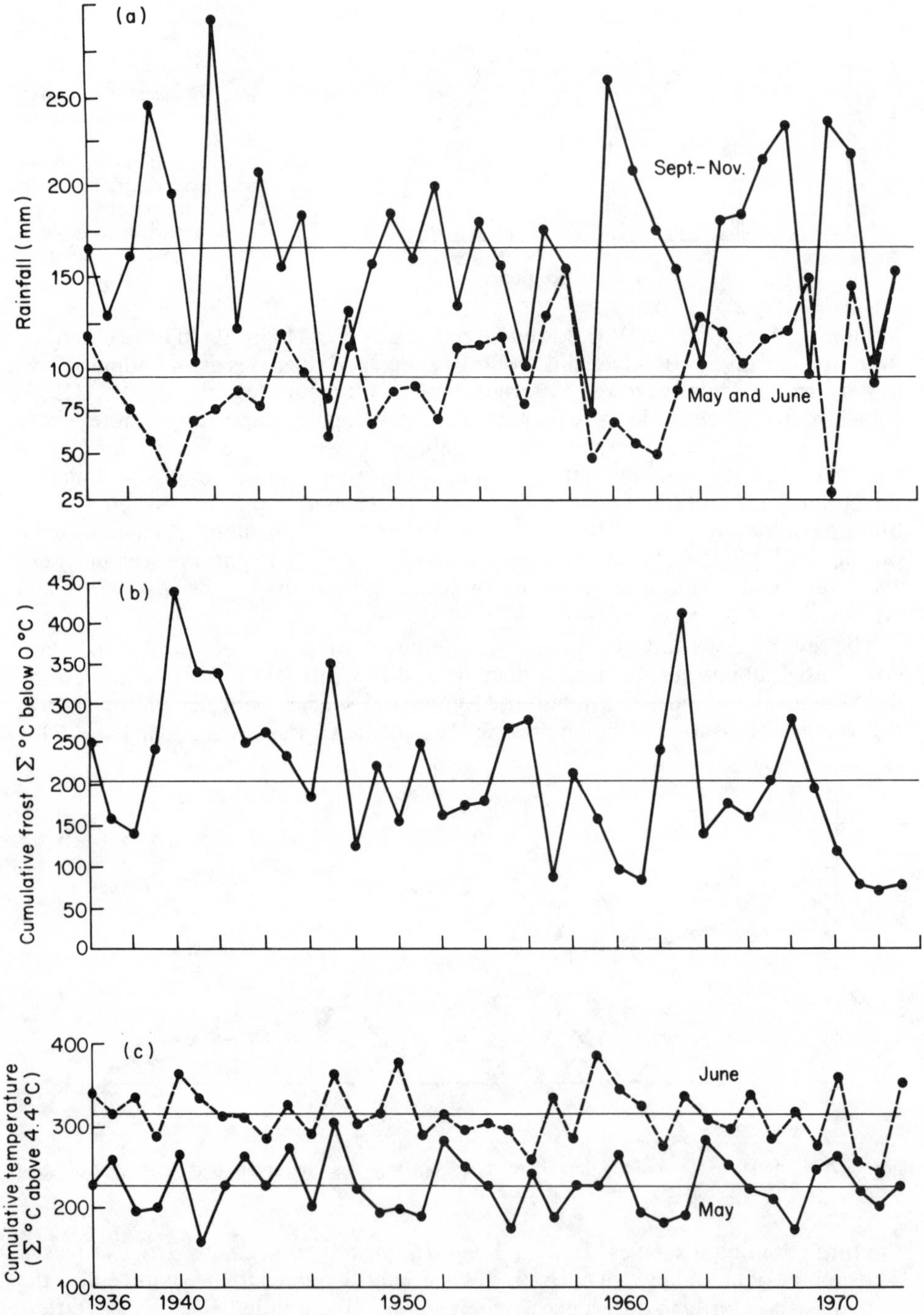

Fig. 4.8 Meteorological data from Elvedon. Horizontal lines are mean values. (**a**) Rainfall for May and June (mean 94 mm) and for September, October and November (mean 166 mm). (**b**) Cumulative degrees (°C) of ground frost for December of the named year, and January, February and March of the following year (mean 206 °C). There was one observation, at 09.00 hours G.M.T. on each day, of a thermometer on the ground. (**c**) Cumulative degrees (°C) of temperature over 4.4°C (40°C) for May (mean 226 °C) and June (mean 317 °C). There was on observation, at 09.00 hours G.M.T. on each day, of a thermometer at 150 cm height in a Stevenson screen. (From Watt, 1981; courtesy of the *J. Ecol.*)

cumulative sequence of severe winter frosts, and a final sequence of more moderate winter periods (1947–48) respectively (Fig. 4.8). The final decline of *Festuca* coincides with its degenerate phase, and is presumably accelerated by the competitive vigour of *Hieracium*. Subsequently the variation of macroclimate from year to year has no appreciable effect on the curve of abundance change in *Hieracium* which simply progresses through the classically upgrade and downgrade cycle, or pioneer, building, mature and degenerate phase of *Hieracium* which exposes bare soil and allows the entry of seedling individuals of *Thymus drucei*. Watt (1981) suggests however the supression of *Hieracium pilosella* also appears to involve active competition between the two species and is not solely a function of the degenerate phasic stage reached by the even aged *Hieracium*.

Thus we now have long-term data that clearly confirms the importance of phasic vegetation cycles probably mediated through corresponding cycles of competitive ability. It is also very probable that some of the demographic changes discussed in Chapter 6 similarly involve phasic changes of perennial species.

Interaction of age and performance

Confirmation of the relationship between age of an individual and its competitive ability is given by Kershaw (1962*a* and *c*). The effect of the advancing front of *Eriophorum* (see p. 66) was related to the density of *Salix herbacea*, the density of *Salix* being inversely related to the density and performance of *Eriophorum* (Fig. 4.9). In effect where the density of *Salix* was at a minimum *Eriophorum* had a maximum level of performance and abundance, indicating the greater competitive ability of *Eriophorum* in its building and mature phases. Thus in general terms there exists a

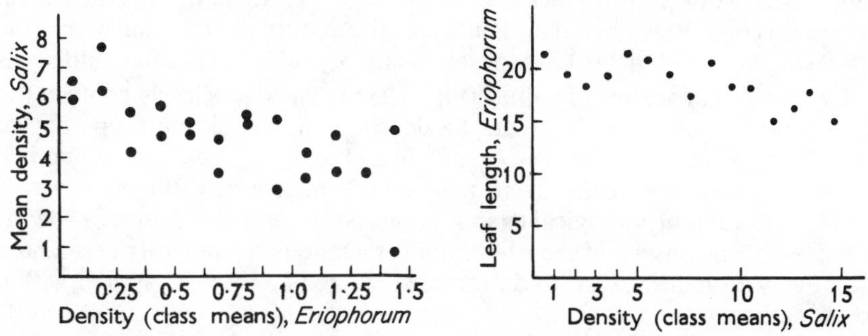

Fig. 4.9 (**a**) Relationship between the densities of *Eriophorum* and *Salix herbacea*. (**b**) Relationship between the density of *Salix herbacea* and performance (leaf length) of *Eriophorum* (see text). (From Kershaw 1962*a*; courtesy of *J. Ecol.*)

direct relationship between age, performance and competitive ability, potentially, for most if not all perennial plants.

The quantitative assessment of the age/performance relation presents some difficulties since it is often impossible to determine the age of most herbaceous perennials. Data for two species only are available (Kershaw 1960*b*, 1962*c*), for *Alchemilla alpina* and *Carex bigelowii* the assessment of age being based on ring counts and the rhizome morphology respectively. The mean leaf diameter and leaf number per individual of *Alchemilla* were used as a measure of performance, each individual being aged by a count of growth rings. A considerable variation of performance is present (Fig. 4.10) between individuals of the same age reflecting local environmental effects, but the total data show a clear rise and fall in the

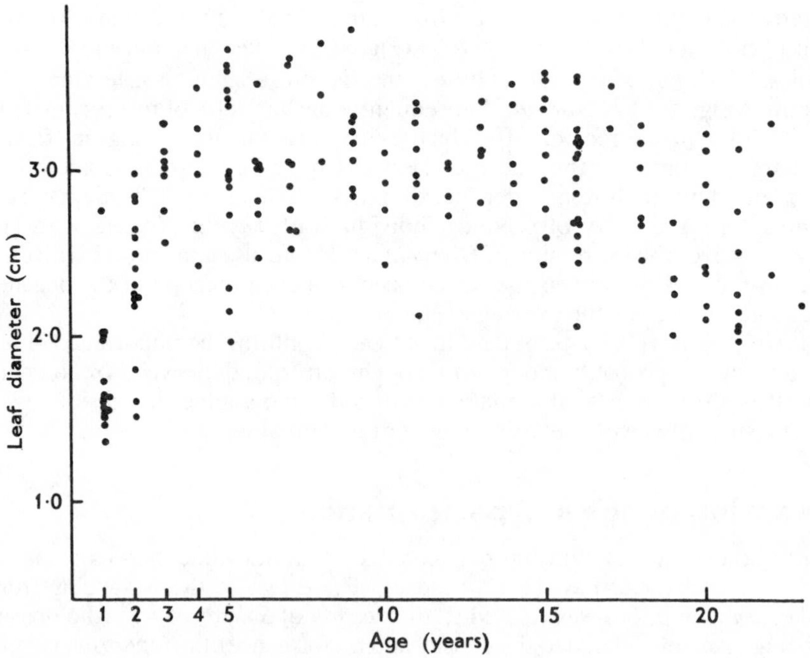

Fig. 4.10 The scatter of mean values of leaf diameter (performance) for individual plants of *Alchemilla alpina* of all ages. (From Kershaw, 1960*b*; courtesy of *J. Ecol.*)

relationship between performance and age. A regression of the form $Y = a + bx + cx^2$ is highly significant and indicates the mature phase to fall within the age groups 9–15 years (Fig. 4.11*a*). The relationship between leaf number and age is similar but with a larger scatter of values (Fig. 4.11*b*). The data clearly confirm the relationship initially proposed by Watt to describe the phasic development of *Pteridium*. The data for *Carex bigelowii* are similar (Fig. 4.12). Leaf length and leaf width are used as measure of the performance and the age of a rhizome system determined by careful morphological examination. At the end of a season's growth, the aerial parts die, the basal leaf sheaths remain as a humified 'knot' on the rhizome and it is relatively straightforward to determine the age of a rhizome system with a fair degree of accuracy.

It seems highly probable that similar growth curves will exist for many other perennial species. For this reason phasic development and relationships between other species are likely to be widespread and the lack of additional evidence reflects the difficulties of detecting such cyclic relationships. Phasic development involving several species in a definite sequence seems unlikely to be as widespread as a mosaic of density phases reflecting the age/performance interaction with the attendant variation of competitive ability and the potential effects on associated species. Thus a stable community will be in a constant state of phasic fluctuation, one species becoming locally more abundant as another species reaches its degenerate phase, the fluctuation in density being however slight and not visually distinct. Only by careful quantitative measurement of abundance and performance can phasic developments be detected in vegetation where there is a complete cover, and then only in those species which possess some morphological or anatomical feature which enables their age to be assessed. Where the vegetational cover is discontinuous, and the degenerate stage is eroded and followed by a 'bare' phase, cyclic relationships become visually more obvious

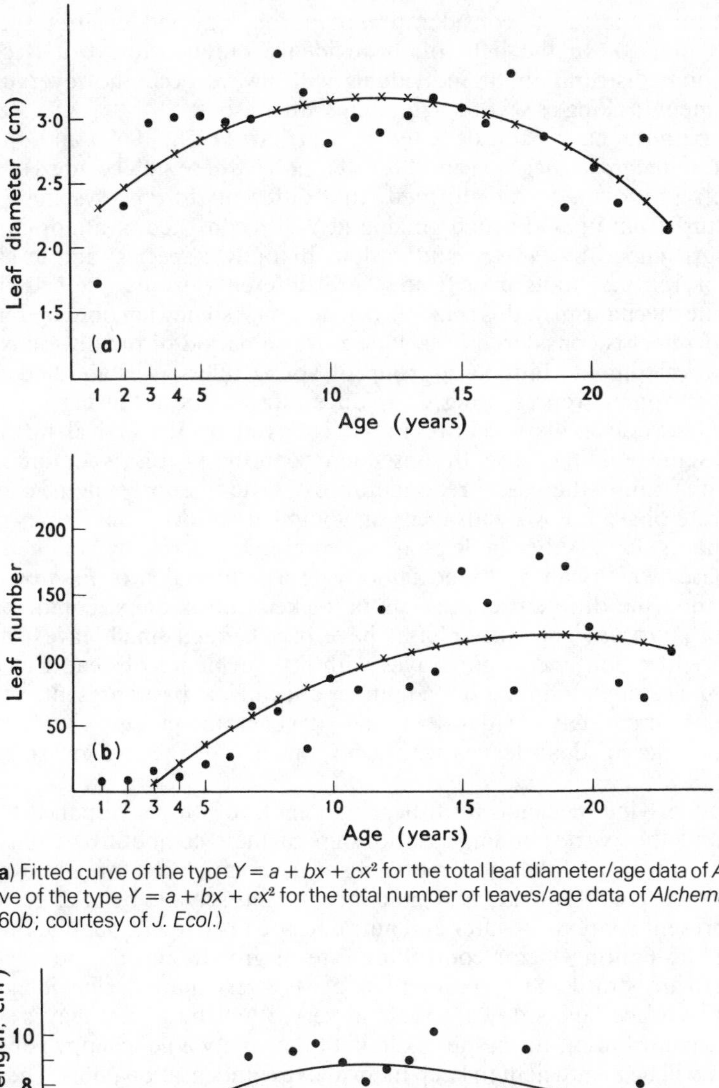

Fig. 4.11 (**a**) Fitted curve of the type $Y = a + bx + cx^2$ for the total leaf diameter/age data of *Alchemilla* (**b**) Fitted curve of the type $Y = a + bx + cx^2$ for the total number of leaves/age data of *Alchemilla*. (From Kershaw, 1960*b*; courtesy of *J. Ecol.*)

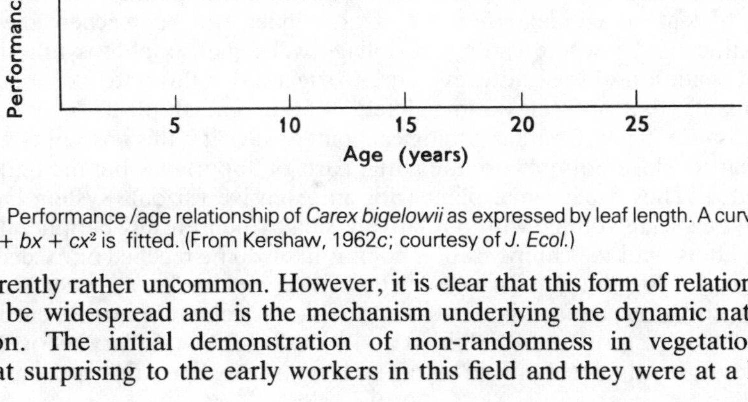

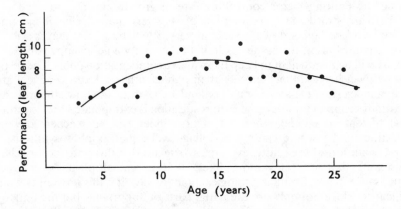

Fig. 4.12 Performance /age relationship of *Carex bigelowii* as expressed by leaf length. A curve of the type $Y = a + bx + cx^2$ is fitted. (From Kershaw, 1962*c*; courtesy of *J. Ecol.*)

but apparently rather uncommon. However, it is clear that this form of relationship is likely to be widespread and is the mechanism underlying the dynamic nature of vegetation. The initial demonstration of non-randomness in vegetation was somewhat surprising to the early workers in this field and they were at a loss to account for it apart from consideration of morphology and localized seed dispersal (see Chapter 8). In the light of the evidence outlined above it is clear that a non-random distribution of individuals will always occur however uniform the environment, as long as several age groups are present.

Such patterns have been detected by Kershaw (1958; 1962*b* and *c*) in *Agrostis tenuis, Calamagrostis neglecta* and *Carex bigelowii* where small clumps of tillers a few centimetres in diameter are formed from different rhizome systems (p. 153). A typical turf from upland rough grazing in Wales consisted of a mosaic of *Agrostis* tillers surrounded by *Festuca* and various bryophyte species, and in all cases the groups of *Agrostis* shoots arose from several different rhizome systems (Figs 8.15 and 8.16). The mechanism of this rhizome orientation is somewhat obscure but if any one group of tillers is considered in isolation over a period of time it follows a typical phasic development. Initially a group of young tillers is established from a few pioneer rhizomes, representing the pioneer stage. The following year the clump produces several axillary shoots and is enlarged by the arrival further pioneer rhizomes; the building phase. In subsequent years this process is continued through a mature phase until there is a preponderance of dead tiller bases in the clump and the degenerate phase follows with increasing numbers of dead tiller bases and weaker aerial shoots. Finally the whole phase is represented merely by humified remains of shoot bases which can be found underlying a small patch of *Festuca*. By careful examination the different phases can be picked out by the size and colour of the leaves of *Agrostis*. The early phases have bright green small leaves, the building phase bright green and larger leaves with little evidence of dead leaf bases. The mature phase though similar has a number of dead leaf bases present, a trend which becomes increasingly evident in the degenerate phase where there is a preponderance of dead leaves with a few small and often yellowish-green aerial leaves.

The underlying mechanism of these varying levels of performance of individual plants and the corresponding relationship of their competitive ability has been obscure. However a fuller understanding of the competition for light by crop plants has been achieved (Donald, 1961) and these results seem to be of direct significance in this present context. If water and nutrients are in adequate supply, then light will become the limiting factor controlling rate of growth and the production of dry matter. In any stratified vegetation there exists a very sharp decline in light intensity from the surface layers downwards and very often the lowest leaves will make a negative contribution to the net assimilation of the whole 'canopy' since the light intensity will be insufficient to keep them above compensation point. The respiration of these lower leaves in fact exceeds their photosynthesis. Expressing leaf area as a leaf area index (i.e. leaf area per unit of land area), it can be shown theoretically that if the relationship of photosynthesis to respiration is extrapolated for a continued increase of leaf area index, then a leaf area index will be reached at which the respiration of the whole canopy of foliage will equal its photosynthesis. Under natural conditions there is little evidence to suggest that this state is ever achieved in crop plants, but it seems more likely that perennial plants may reach this compensation point. From an ecological point of view it is this possibility which is of significance, since not only are the aerial parts of importance but the underground parts also. Thus a perennial plant with an extensive rhizome system has a large actively respiring volume with a relatively small (theoretically peripheral) zone of aerial shoots, and the compensation point is likely to be reached or exceeded in old individuals. *Carex bigelowii* is a case in point. A complete rhizome system 20 years of age will cover one square metre with a complex system of branches, each ultimate branch having a terminal aerial shoot and only one or two axillary shoots in the last two years' growth increment. In such a system it is very likely that the compensation point has been reached or exceeded and the plant is in its 'degenerate phase'. Conversely a young plant of 6–12 years of age has a much less extensive rhizome system and with a normally developed 'leaf canopy' is at its optimum level of growth.

Went (1957) has referred to the build-up of non-photosynthesizing tissues in old strawberry plants until the whole plant is barely at compensation point at a high light intensity and it seems probable that the ratio between the photosythesis and respiration of a perennial plant largely controls its general level of performance and its ability to compete with its neighbours. This possible explanation is equally satisfactory when applied to the parts of a fragmented degenerate individual. In *Alchemilla, Agrostis, Pteridium* and *Carex* the older parts of the rhizome slowly decay and as the zone of decay reaches a lateral branch this becomes separated off as an individual. The oldest plants of low peformance thus release 'propagules' with, initially at least, a low performance but with a morphologically optimal age. A propagule from a 25-year-old parent is likely to be 8–12 years of age. However the average performance of all individuals in this age group is optimal and accordingly it would appear that the level of performance of such a propagule immediately rises to its 'morphological age'. This seems quite logical when considered as a small rhizome system being severed from a very large system which was respiring actively. Immediately the photosynthesis/respiration ratio becomes unequal, the terminal leaves provide an excess of metabolites relative to the now small rhizome system, and active growth can take place at once.

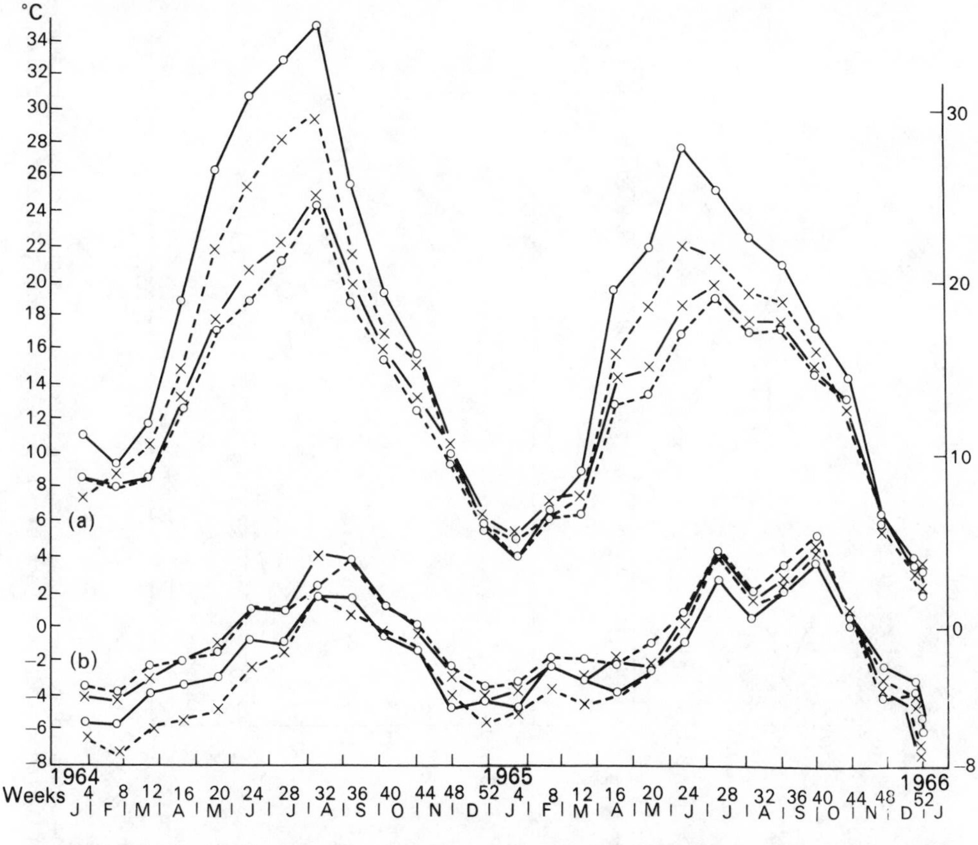

Fig. 4.13 Surface maximum and minimum temperatures for the period January 1964 –January 1966. (**a**) Mean maximum temperatures; (**b**) Mean minimum temperatures. —○—, Pioneer; —×—, building; -- ○ --, mature; -- × --, degenerate. (From Barclay-Estrup, 1971; courtesy of *J. Ecol.*)

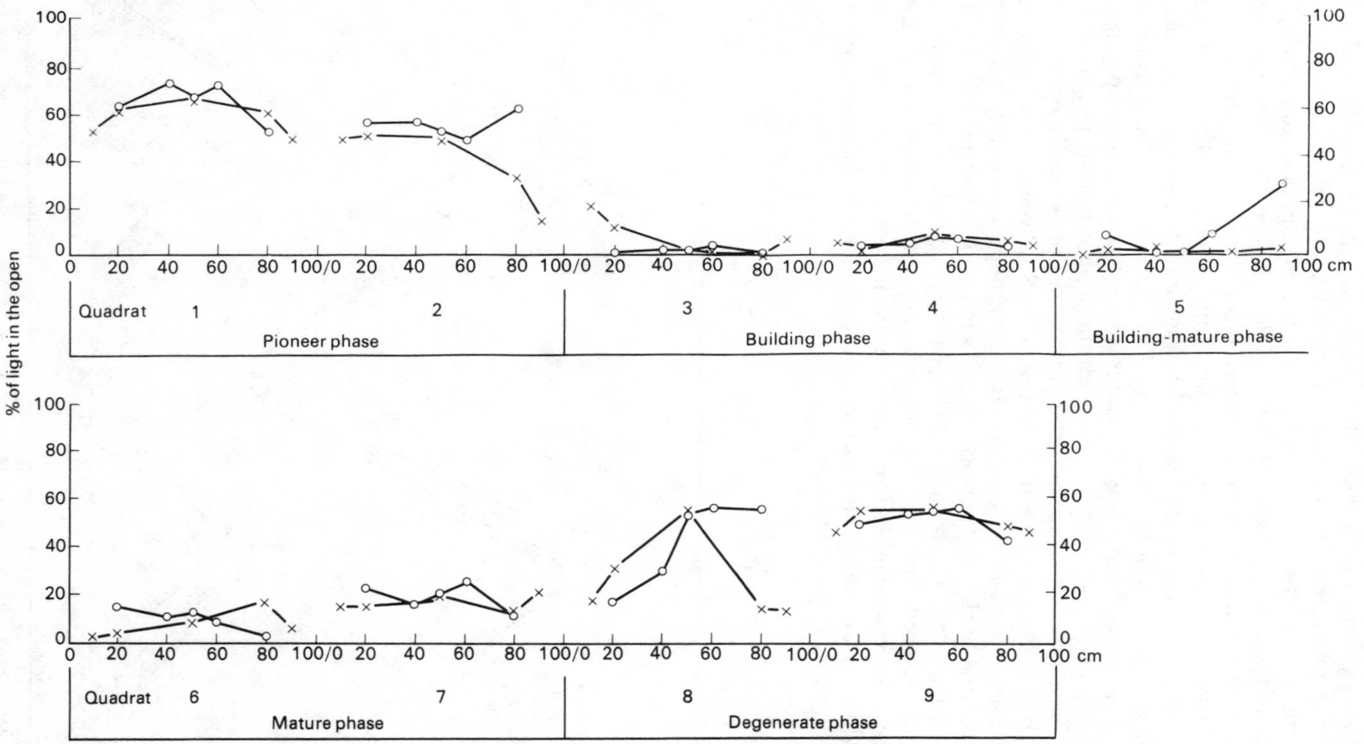

Fig. 4.14 Means of all ground-level photocell measurements. Each graph shows the light values along two lines crossing the quadrat in a south-north direction (○) and an east-west direction (×). (From Barclay-Estrup, 1971; courtesy of *J. Ecol.*)

In all the cycles described above, the species involved are in one sense, part of a truncated succession. There is no directional change, the cycle eventually returning to its original position. However, just as in plant succession where there is a marked reaction of the species on the environment, there is a similar reaction on the environment throughout the cycle of phases. Furthermore, because of the phasic sequence the level of reaction will also vary sequentially. Barclay-Estrup (1971) examined the microclimate and its variation in relation to the phasic development of *Calluna vulgaris* (Watt, 1955), and showed the existence of considerable differences between the different phases. Soil surface temperatures during the summer months were much higher under the building phase than elsewhere and followed a decreasing sequence through the building phase and mature phase to the degenerate phase (Fig. 4.13). Conversely in the winter months the lowest temperatures recorded are under the pioneer and degenerate phases where the low biomass of the *Calluna* allows a greater loss of radiant energy from the soil. Similar considerations apply to soil temperature although the differences are not as clear cut. Interception of light is also related to the phasic sequence and during the building/mature phases, the light at the soil surface falls to 10–20% of full illumination, as opposed to 60–80% in the pioneer phase, and 40–60% in the degenerate phase (Fig. 4.14). Rainfall interception follows the same sequence with maximum interception by the canopy of the mature and building phases, decreasing through the degenerate phase to the pioneer phase where cumulative precipitation is fairly close to full precipitation measured in the open (Fig. 4.15).

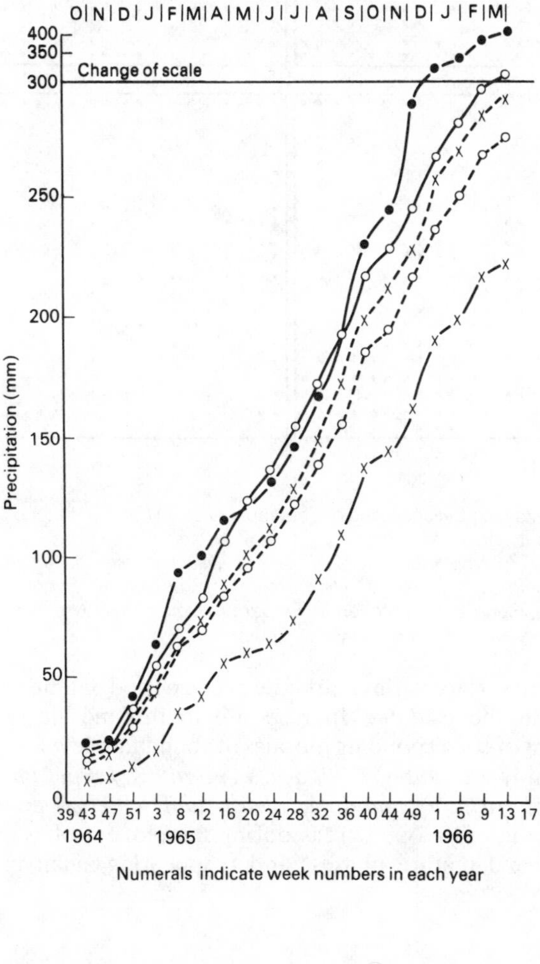

Fig. 4.15 Cumulative micro-precipitation under *Calluna* each phase. —O—, Pioneer; —×—, building; ---O---mature; ---×---, degenerate; —●—, open. (From Barclay-Estrup, 1971; courtesy of *J. Ecol.*)

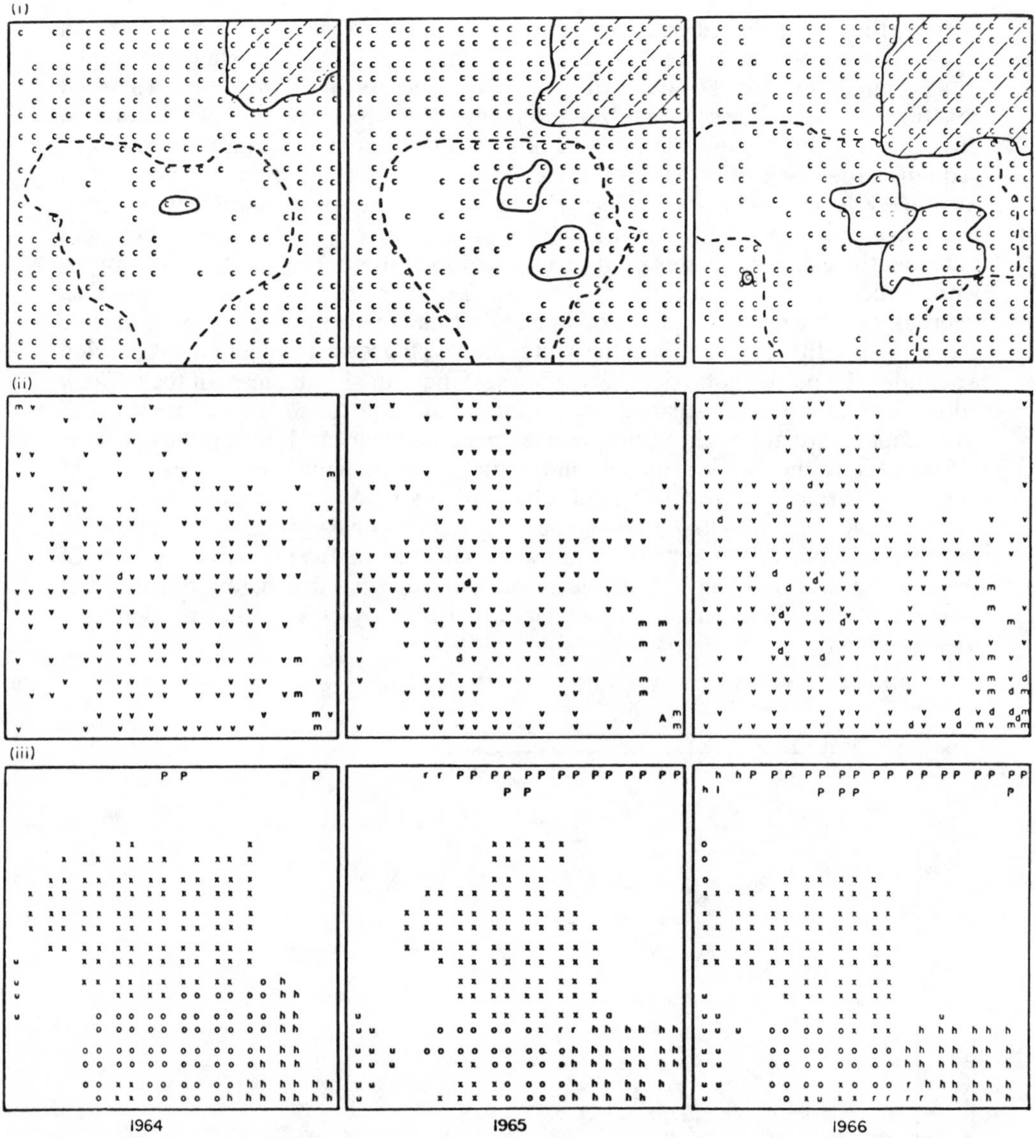

Fig. 4.16 Periodic maps of quadrat 8, containing *Calluna* in the degenerate phase. (i) *Calluna*: hatched areas, adjacent building plant; unshaded outlined areas, pioneer plant; outlined area (broken line), gap boundary. (ii) Other vascular plants. (iii) Bryophytes, C, *Calluna*; d, *Deschampsia flexuosa*; h, *Plagiothecium undulatum*; m, *Vaccinium vitis-idaea*; o, *Hylocomium splendens*; p, *Hypnum cupressiforme*; r, *Aulacomnium palustre*; u, *Pleurozium schreberi*; v, *Empetrum nigrum*; x, *Polytrichum commune*. (From Barclay-Estrup and Gimingham, 1969; courtesy of *J. Ecol.*)

Clearly the microclimate at the soil surface varies markedly over a period of time in response to the phasic changes in the *Calluna*. In response to this mosaic of microclimate there is a development of corresponding mosaics of abundance levels in associated species (Fig. 4.16). Notably *Vaccinum vitis-idaea, V. myrtillus, Empetrum nigrum*, and the moss species *Polytrichum commune, Pleurozium schreberi* and *Hylocomium splendens* dominate the more open areas during the pioneer phase, disappearing during the building and mature phases, and reappearing again in the degenerate phase. These results amplify the previous discussion on cyclic phases as a causal mechanism of pattern in vegetation. Thus the phasic development of *Calluna* affects other associated species through its changing levels of competitive ability. This is evident as a direct interaction with the environment inducing a change in the microclimate. The resultant spatial patterns of the bryophyte species especially are controlled by this changing microclimate.

5

The Causal Factors of Positive and Negative Association between Species

Most, if not all, data on the relative abundance of species in a given area will contain one or several correlations of species. The presence of such correlations reflects the pattern or, more usually, several scales of pattern in the vegetation. The existence of pattern in vegetation and the general factors underlying such patterns are dealt with in Chapter 7, but it is necessary to consider at this point in more detail some of the factors that are operative. These causal factors can be studied either by using the techniques available for the detection of pattern (Chapter 7), or by the use of association, correlation, or regression techniques. The two methods of analysis are very closely linked and are only treated separately here for the sake of clarity. The causal factors are split into four sections for convenience, each section represents the level of detail at which it is studied rather than any rigid classification.

Similar or dissimilar environmental requirements

This is probably one of the fundamental reasons for negative or positive relationships that exist between species. On a large scale and with an obvious difference of species composition in two areas such correlations between species lead to the the separation of groupings which typify a 'community' (see below and Chapter 9). On a smaller scale such relationships are rarely as obvious and usually need a careful quantitative approach to detect their existence.

An example of such an environmentally-controlled relationship is taken from an area of hummocky grassland associated with a calcareous mire. The data for *Carex lepidocarpa, C. flacca* and *Potentilla erecta* (Table 5.1) were obtained in the form of

Table 5.1 The relation between topography and the abundance of *Carex lepidocarpa, C. flacca* and *Potentilla erecta*.

Species	**Mean frequency** Hummock	Hollow	**SE**	**Difference between frequencies**	**t**	**p**
Carex lepidocarpa	3.15	5.34	0.4163	2.19	5.260	0.001
C. flacca	0.82	2.0	0.2963	1.18	3.986	0.001
Potentilla erecta	4.19	1.38	0.1967	2.81	14.264	0.001

frequency, from 150 quadrats positioned by using random co-ordinates (approximately half the sample was related to a hummock). The apparent preference of *Potentilla* for the hummocks and conversely the preference of the *Carex* species for the hollows is readily tested by means of a *t*-test which shows a highly significant difference between the species. This difference is not readily visible on even a close examination of the area though an explanation in general terms is not difficult.

The 'hollows' act as runnels for base-rich water after a period of heavy rain and the pH of the soil is thus neutral to slightly alkaline. Conversely the tops of the hummocks are not affected by calcareous run-off and with leaching over a period of years, together with the steady accumulation of humus, have a pH of about 5.6–5.8 (slightly acidic). It is most probable that the distribution of these species is not governed directly by the pH of the soil, but by other factors which are in turn correlated with pH. The relationship between the environment is thus most easily appreciated by relating their abundance in a sample quadrat to pH. This relationship of abundance and topography obviously leads to correlations (positive or negative) between pairs of species. In this example the distribution of the species is not sharply controlled by topography but merely influenced differentially by hummocks and hollows. The relationship accordingly cannot be appreciated merely by an inspection of the area.

Harper and Sagar (1953) have shown that the spatial distribution of *Ranunculus repens, R. bulbosus* and *R. acris* shows a marked correlation with drainage. They investigated the relative abundance of the species along a transect orientated at right angles to furrows and ridges in a permanent pasture. The results (Fig. 5.1) show quite clearly the differential environmental requirements of the three species, *Ranunculus repens* being abundant in the furrows, *R. bulbosus* more abundant on the ridges and *R. acris* showing an intermediate distribution on the sides of the ridges. Seedling establishment under experimental conditions at three levels of drainage suggest it is this part of the life cycle, i.e. germination and early establishment, which is susceptible to the drainage of the area as a whole (Fig. 5.2).

These results from a markedly variable habitat can be related to the non-random distribution of *Ranunculus* in more homogeneous grassland. Micro-variation of drainage will modify other nutritional factors to a greater or lesser extent, which will in turn influence the relative abundance of each species in a small area, and lead to negative correlation between them and to marked non-randomness in the population (see also Chapter 7).

Both these examples are related to micro-topography and to its effects on abundance of species either directly or, more probably, through other interrelated factors. They illustrate how slight environmental variations can lead to the production of patterns in vegetation and to the existence of correlation or association between species. The choice of examples relating micro-topography and the distribution of species by no means excludes the importance of other variable factors which may operate, but it merely underlines the scarcity of data on such interrelationships.

Modification of the environment by one species allowing establishment by other species

Considerable interest was aroused by the publication of a paper by Went in 1942 which demonstrated a marked relationship between the distribution of annual species and perennial shrub species in a desert area in California. His data showed a general positive relationship between shrubs and annual herbs, but a more striking relationship between annual species and dead *Encelia farinosa* bushes than with living bushes. This he suggested was due to the production of a toxic substance by *Encelia* which accordingly limited the herb population closely adjacent to a living individual bush. Gray and Bonner (1948) subsequently isolated such a substance from the leaves of *Encelia*. Later workers in this field have suggested other equally plausible explanations of the phenomenon in terms of the relative build-up of humus content underneath *Encelia* and other species of shrubs. Whatever the causal explanation (see below for a detailed account of the presence of toxic substances in plants) this work did focus attention on the relationship between an individual and its environment and the possible effects of this interaction on other species within the community.

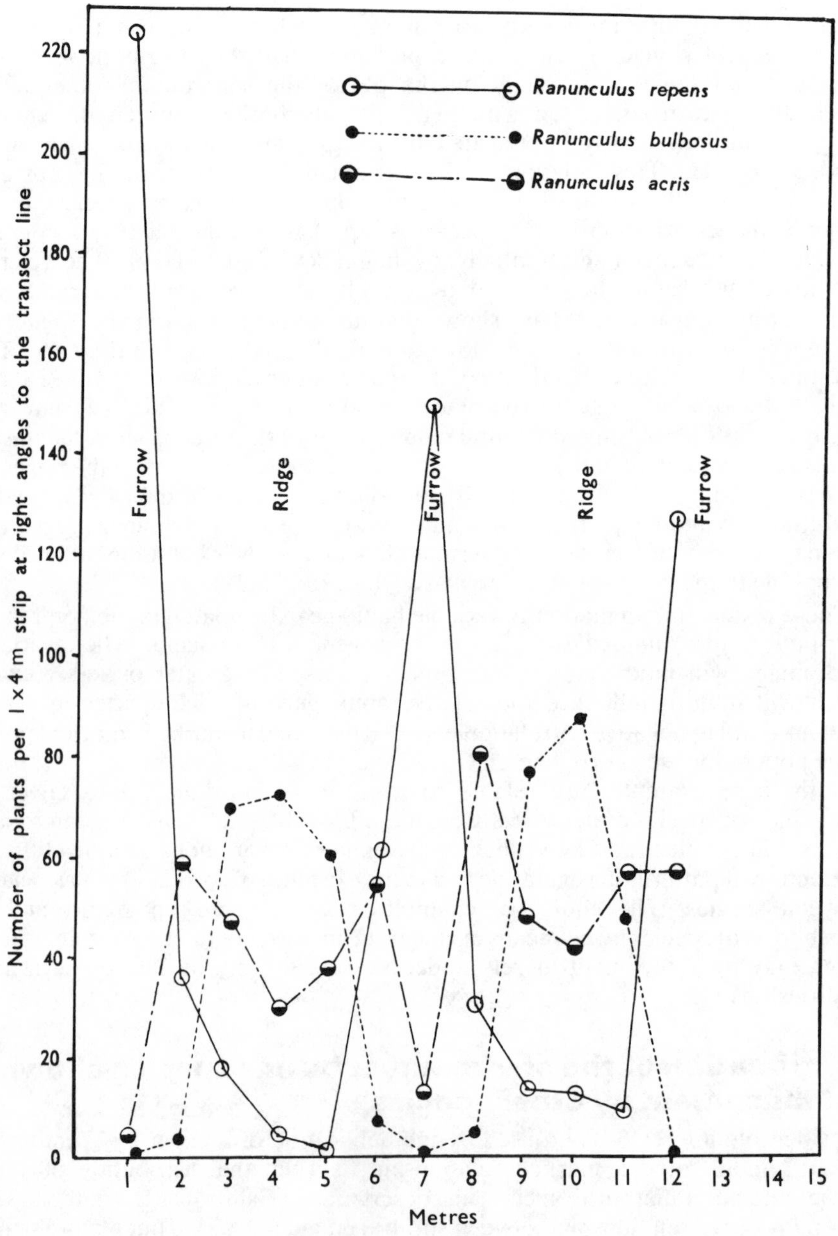

Fig. 5.1 The distribution of three species of *Ranunculus* on ridge and furrow grassland. (From Harper and Sagar 1953; courtesy of British Weed Control Council.)

A clear example of this type of relationship taken from stony desert in Iraq is given by Agnew and Haines (1960). There is a considerable deposition of wind-borne material round the base of perennial plants in this area, which leads to the formation of low mounds around the individual shrubs. *Astragalus spinosus* is an outstanding example of this and there is a marked positive correlation between this species and small annual or perennial herbs, both in the sense of abundance as well as luxuriance. A typical diagram of such mounds is given in Fig. 5.3. *Astragalus spinosus* is a straggling shrub with tough branches and an abundance of sharp spines which are

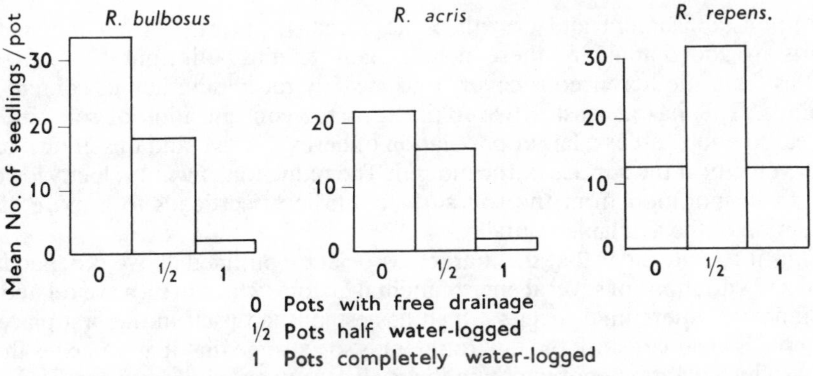

Fig. 5.2 The early establishment of three species of *Ranunculus* sown in pots with the watertable maintained at different levels. Sowings were made on June 26 1953 and counts made on July 28 1953. (From Harper and Sagar 1953; courtesy of British Weed Control Council.)

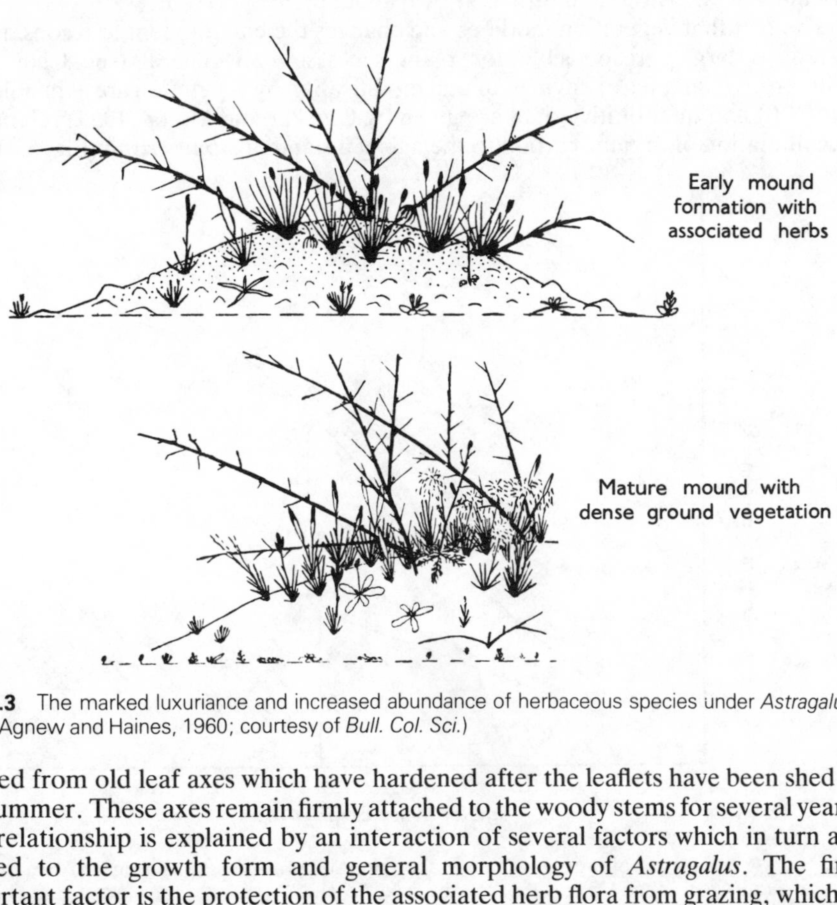

Fig. 5.3 The marked luxuriance and increased abundance of herbaceous species under *Astragalus*. (From Agnew and Haines, 1960; courtesy of *Bull. Col. Sci.*)

formed from old leaf axes which have hardened after the leaflets have been shed in the summer. These axes remain firmly attached to the woody stems for several years. The relationship is explained by an interaction of several factors which in turn are related to the growth form and general morphology of *Astragalus*. The first important factor is the protection of the associated herb flora from grazing, which is provided by the spiny branches. This factor is of considerable importance in the Middle East where there is very heavy grazing by goats, sheep and camels even in apparently remote desert areas. Grazing will markedly affect both the luxuriance and abundance of individuals of a species. The second factor is the actual presence of wind-borne material which provides a limited area of soil capable of supporting plant growth. The desert area is largely stony and the soil mounds thus provide an excellent site for establishment and growth of herbaceous species. Further increments of debris are added annually; these include plant remains both from the *Astragalus* as well as from the herbaceous cover, thus steadily raising the humus content of the mounds. This has at least a two-fold effect; the concentration of soil nitrogen is increased and enables a larger population of herbs to exist, and this in turn reduces wind velocity at the surface of the mound. The reduction of wind velocity lowers the rate of evaporation from the soil surface and in effect leads to a more efficient utilisation of the available rainfall.

Thus it is quite clear that the marked association outlined above is dependent on localized variations of several environmental factors, which in turn are related to the existence of a perennial species which has established itself in the first place. This example is clear cut since the environment is so extreme that it approaches the level below which plants cannot maintain themselves, and a slight (advantageous) shift of the environment produces a marked and a readily visible response in the vegetation. Similar relationships will hold in more temperate regions of the world, but here they are much less obvious and little is known about them at present.

The fact that vegetation modifies and changes the environment to a considerable extent is largely responsible for plant succession in general (see Chapter 3). Numerous examples of environmental modification by vegetation are available (e.g. Fig. 3.6) and quantitative data are given by Crocker and Major (1955) relating the accumulation of organic carbon to the age of a stand of *Alnus crispa* (Fig. 5.4). This

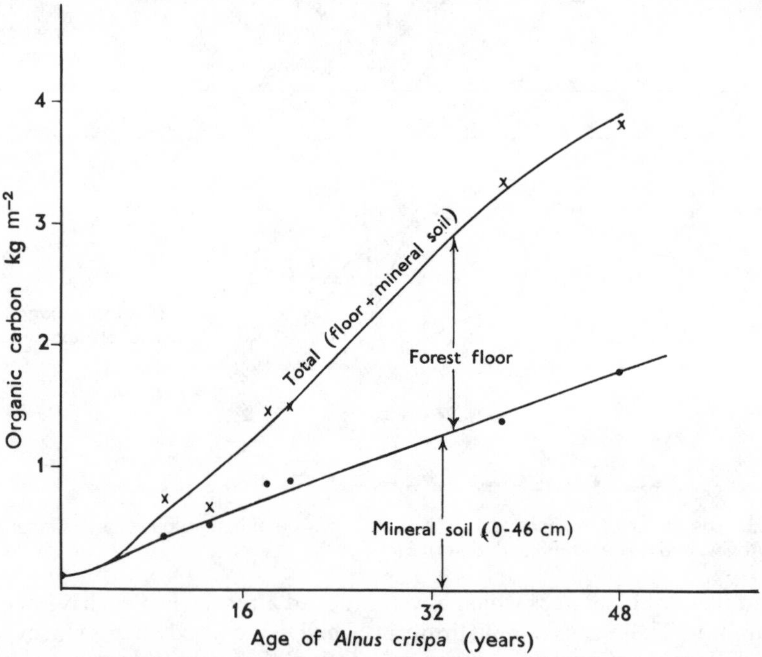

Fig. 5.4 Accumulation of organic carbon in the mineral soil and forest floor under *Alnus crispa* in Alaska. (From Crocker and Major, 1955; courtesy of *J. Ecol.*)

accumulation of carbon in a 50-year period and the resultant changes in the vegetation is in fact a characteristic and well marked succession (see Chapter 3). Moreover, this example illustrates the steady build up of humus material in relation to vegetational cover. The rate of build up of humus is of course also dependent to a considerable extent on climate and drainage. Thus in *Astragalus* the nitrogen accumulation will be relatively small, but sufficient to have an effect on the associated vegetation, which leads to the marked positive correlation of species. It is a reasonable assumption that similar effects are present in temperate herbaceous vegetation, but the relatively infrequent occurrence of strong positive association between species in this type of vegetation would suggest that such modifications of the environment are often overshadowed by environmental variations resulting from other factors.

Production of toxic substances by plants

As far back as 1832 De Candolle suggested the importance of root excretions and since 1900 considerable attention has been directed towards the detection of substances in plants which are toxic to other species. The demonstration of direct chemical interaction between species is surprisingly difficult, and much of the earlier work designed to prove an interaction between two species based on the exudation of a toxic substance can be criticized on the ground of experimental design or may be explained in terms of competition for some environmental factor.

Bedford and Pickering (1919) examined in some detail the vigour of apple trees in relation to the grass cover of an orchard. Observation showed a very marked difference between the height and fruit production of trees grown in tilled soil compared with trees of similar age grown in plots with a complete grass cover. For example a plantation of Bramley's Seedling apples established for 23 years was selected and half of it laid down to grass. In the first year the grass did not establish well but even then the season's crop was reduced by 5% (compared with the crop from the tilled section). In the next year the crop was markedly affected as fruit production was reduced by 89% following the complete establishment of the grass. Similar results were obtained in several other experiments, all of which strongly suggested the action of a toxic substance produced by the grass. This possibility was examined further experimentally. Young apple trees were grown in pots while a 'surface crop' of grass was suspended above in a perforated container. Water was supplied through the surface crop which leached any toxic substance into the pot below. The resultant inhibition by the surface crop was very marked.

The results at first sight seem conclusive but the field observations can be explained in part by competition for nitrogen and water. The inhibition of the trees by grass cover can be almost completely eliminated by addition of nitrogenous fertilizers, together with irrigation of the orchard. The pot experiments are quite convincing, however, and one is forced to conclude that the presence of a toxic substance is probable, but under field conditions it is either destroyed by bacterial action or is absorbed by soil particles. It should, however, be borne in mind that the removal of an inhibitory effect by the addition of nitrogen and water does not necessarily prove that a toxic substance is *not* operative. It could be that the toxic substance produced by the grass for example reduces the root development of the apple trees and accordingly limits their ability to take up nitrogen and water. With the availability of abundant nitrogen and water the limiting effect of a reduced root system is minimal and the apparent toxic effect disappears. In other words the mechanism of competition for nitrogen and water could be through the action of the toxin.

The relationship between apple trees and surface crop illustrates clearly the variety of suggestions that can be offered as an explanation of an observed result. The experiments have to be designed with considerable care, and the above example does emphasize the attendant difficulties in this field of research. More recent work has revealed numerous examples of inhibition of one species by another; these have been usefully reviewed by Rice (1974; 1979). Allelopathy is defined as any biochemical interaction between two plants, or a plant and micro-organisms which can be either beneficial or detrimental to their growth. Allelopathy is thus completely separate from competition which involves the *removal* or reduction of some factor from the environment that is utilized by another competing individual. However, the distinction between these two types of interactions, at a mechanistic level, is often difficult to achieve and Muller (1969) has suggested that the term 'interference', that includes both allelopathy and competition, should be reserved for the general case where there is an undefined but clear-cut 'influence' of one plant on another.

Over the last 15 years there has been a considerable level of interest in allelopathy providing numerous examples of both interspecific and intraspecific interactions and several distinct concepts are now considered to be fundamentally important.

1. The interactive effects can be directly evident either as a suppression of germination, or as a reduction of growth and performance of mature individuals.

2. Where the interaction is evident in mature plants it can be mediated directly by one plant on another or it can operate indirectly as a secondary effect. Thus the inhibition of, for example, an organism in the rhizosphere can then reduce the availability of some soil factor resulting in a subsequent restraint to the growth of a third species.

3. The source of the active inhibitory substance can be from roots as an exudate, from leaves or buds as a leachate, or as breakdown products from raw humus. In some instances the allelopathic substance can be transferred in vapour form from the source to the site of interaction. A wide range of organic compounds are involved and these have been reviewed in detail by Rice (1974) and Horsley (1977).

4. Suggested mechanisms of action were widely variable ranging from effects on cell elongation, membrane permeability, rates of mineral uptake to indirect and direct effects on plant photosynthesis.

Friedman *et al.* (1977) provide a classical example of sagebrush allelopathy in the Negev desert, Israel, comparable with similar examples documented in California by Muller and co-workers (see below). Experimental manipulation of established individuals or particularly *Artemisia herba-alba* by either severe pruning or complete removal show a significant increase in the numbers as well as the dry matter production of associated annual plants growing in the experimental plots (Table 5.2).

Table 5.2 The numbers of dry matter yields of annual plants per 0.25 m² in variously treated plots: March 1975 (mean of five plots ± standard error). (Data from Friedmann *et al.*, 1977, courtesy *J. Ecol.*)

Treatment	North-facing Slope: Number of annual plants	North-facing Slope: Total dry matter (g)	South-facing Slope: Number of annual plants	South-facing Slope: Total dry matter (g)
(*i*) All perennials removed	7.5 ± 3.0 **	6.5 ± 2.2 **	19.1 ± 5.6 *	12.2 ± 2.4
(*ii*) All perennials pruned	3.2 ± 1.0 *	1.1 ± 3.2	17.7 ± 6.0 *	9.3 ± 2.4
(*iii*) Soil between perennials hoed	2.7 ± 0.8	1.8 ± 0.5 *	11.0 ± 3.6	5.9 ± 2.6
(*iv*) All perennials removed and soil between them hoed	8.7 ± 3.2 **	5.9 ± 2.8 **	12.5 ± 5.8	8.3 ± 2.6
(*v*) Control	1.2 ± 0.4	0.6 ± 0.2	11.9 ± 2.6	10.8 ± 4.4

Significance level of difference from control: * $P < 0.05$; ** $P < 0.01$.

Similarly tomato seeds as a test species are markedly inhibited by aqueous leachates obtained directly from twigs of *Artemisia* as well as by volatiles from shoots placed separately in a small beaker inside the experimental vessel (Fig. 5.5). Similarly Chou and Muller (1972) working with *Arctostaphylos glandulosa* var. *zacaensis*, one of the

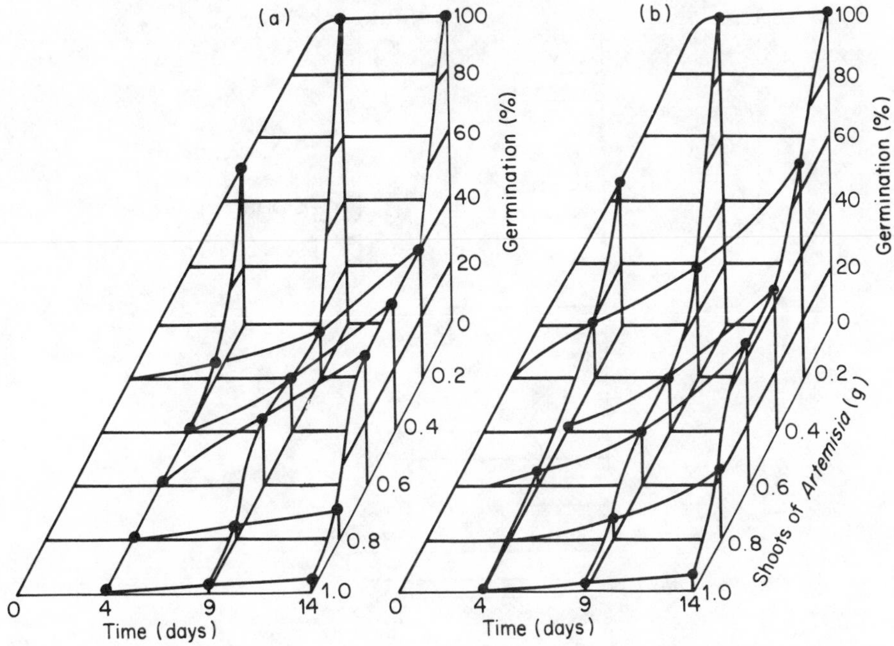

Fig. 5.5 Germination with time of tomato seeds in the presence of increasing amounts of *Artemisia herba-alba* shoots: (**a**) immersion method; (**b**) volatiles method. (From Friedman *et al.*, 1977; courtesy of the *J. Ecol.*)

dominant shrub species of chaparral vegetation in California, showed that aqueous extracts from leaves strongly inhibited the subsequent elongation of the radicle following germination in a range of herbaceous plants normally associated with chaparral vegetation, particularly after fires (Fig. 5.6). Fire is a regular feature of chaparral vegetation and after the destruction of the shrub layer and presumably the source of the allelopathic substances, there is an immediate flush of germination of herbaceous species from the seed pool in the soil. This growth of herbs is gradually suppressed as the shrubs re-sprout from underground perennating organs not killed by the fire; again re-establishing the source of allelopathic leachate. However more recently Kaminsky (1981) has established that for some species at least the inhibition of seed germination is largely associated with the soil and not necessarily the shrub layer itself and has proposed soil micro-organisms as the source of the active toxin. The evidence is derived from a number of experiments designed to interrupt or promote the activity of soil fungi or bacteria, and shows that indeed the soil toxicity appears to be the result of soil micro-organisms. These organisms produce substances that are in sufficient quantities in the soil to account for significant reductions in seed germination and seedling growth. The role of fire in fact is to suppress soil microbial activity bringing to an end, temporarily, the inhibition of seed germination.

Thus, drying and rewetting soil collected from under mature shrubs, which promotes a burst of microbial activity and presumably an exudation of toxin, also represses seed germination and root elongation of lettuce when it is used as a test organism (Fig. 5.7). Furthermore 72 hours of incubation following rewetting is necessary for the toxic effects to develop, presumably in response to the build-up of the microbial population (Fig. 5.8). Treatment with chloroform vapour which strongly inhibits microbial metabolism, immediately after the potential promotion of activity by drying and rewetting the soil, largely removed the usual level of inhibition of seed germination

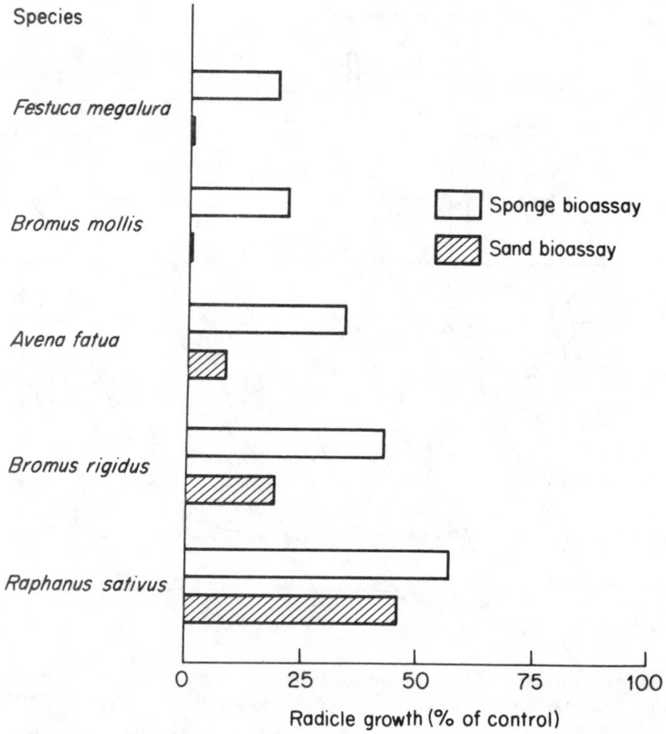

Fig. 5.6 The toxic effect of 5% aqueous extract of freshly fallen leaves of *Arctostaphylos* upon the growth of various species in sand and sponge bioassays. Values are expressed as the per cent of distilled water controls. Results of tests are significantly different from the control at 0.1% of confidence, using the *t*-test. (From Chou and Muller, 1972; courtesy of *American Midland Naturalist*.)

induced by soil filtrates (Fig. 5.9). Equally an additional incubation period after removal of the chloroform vapour, restored microbial metabolic activity and the subsequent normal levels of germination inhibition.

Although *all* cases of allelopathy in chaparral vegetation need not invoke soil microbial toxicity, these data are particularly important in that they do demonstrate both active levels of toxicity in natural soils as well as the major potential provided by soil micro-organisms for the control of herbaceous growth and performance. The role of the allelopathy in natural conditions has often been questioned when the experimental evidence for the presence of toxic substances was developed from aqueous or sand cultures. Accordingly the level of inhibition reported by Kaminsky (1981) does substantiate the realistic role of allelopathy in the replacement of one species by another under natural field conditions.

There has been considerable interest in the potential importance of allelopathy as a major factor in old-field succession (Rice, 1974, 1979; Sarma, 1974; Grant and Clebsch, 1975; Jackson and Willemsen, 1976; Anaya and Del Amo, 1978). Evidence suggests that direct inhibition may be involved in some instances, but Rice (1974) and Kapustka and Rice (1976) have both emphasized that the inhibition of soil biological nitrogen-fixation by the species dominating the first stages of succession is of primary importance. Similarly Jobidon and Thibault (1982) have shown that aqueous extracts of leaves, buds, and leaf litter of *Populus balsamifera* actively inhibits growth of *Alnus crispa* seedlings as well as the rate of nitrogen-fixation in the *Alnus* root nodules (Table 5.3). Nitrification has also been shown to be sensitive to allelopathic influences, particularly in forest ecosystems (Lodhi, 1979;

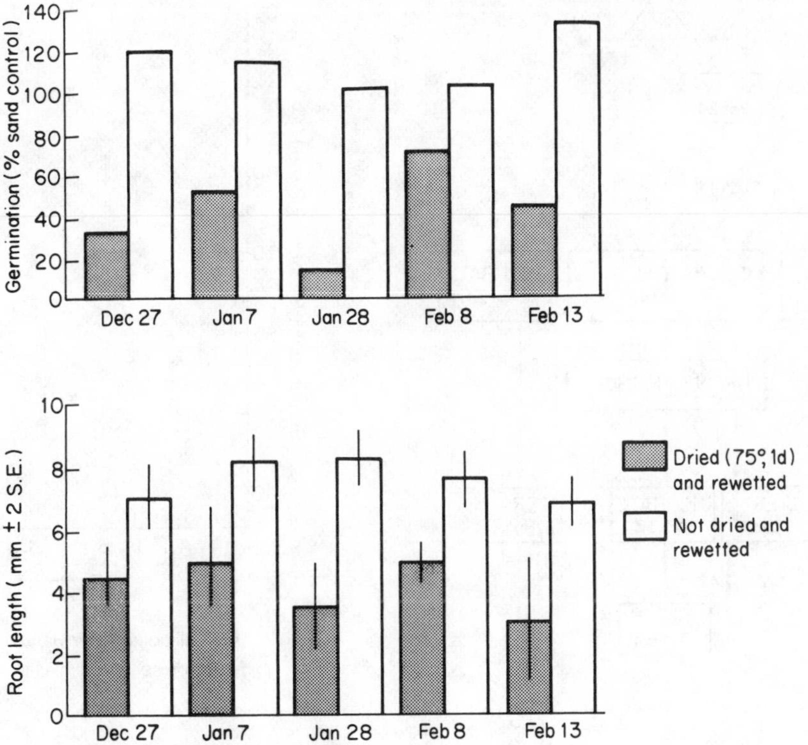

Fig. 5.7 Effects of soil drying and rewetting on the germination and growth of lettuce. (From Kaminsky, 1981; courtesy of *Ecological Monographs*.)

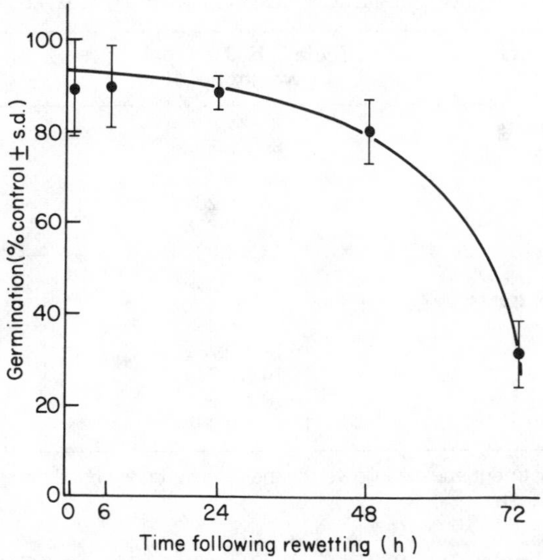

Fig. 5.8 Effect of time following rewetting of dried soil on the production of germination inhibition. (From Kaminsky, 1981; courtesy of *Ecological Monographs*.)

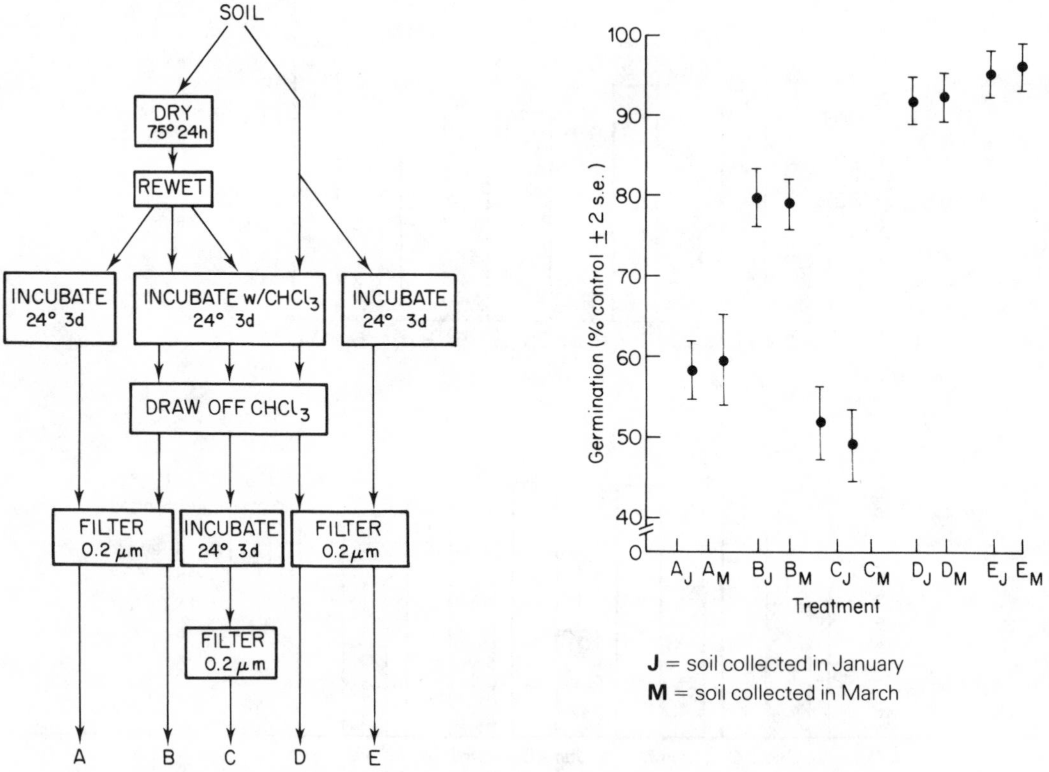

Fig. 5.9 Schematic layout of the procedures prior to, and after chloroform fumigation of experimental soil samples with the subsequent results of seed germination tests using soil leachates from the different combinations of treatments. (From Kaminsky, 1981; courtesy of *Ecological Monographs*.)

Table 5.3 Effects of different parts of *Populus balsamifera* on nitrogenase activity (mole C_2H_4 3 h^{-1} root dry weight^{-1} × 10^{-11}) (±SE). (Data from Jobidon and Thibault, 1982; courtesy *Amer. J. Bot.*)

Treatments (%, w/v)	**Mole C_2H_4 3 h^{-1} root dry weight^{-1} × 10^{-11a}**	**% of inhibition**
Bud extracts		
2	4.36 ± 0.59*a*	61.72
1	4.12 ± 0.49*a*	63.88
0.5	5.26 ± 0.26*ab*	53.89
0.1	6.54 ± 0.63*b*	42.67
Control	11.41 ± 0.99*c*	0.00
Leaf litter extracts		
2	4.25 ± 0.96*a*	62.74
1	6.19 ± 0.78*a*	45.74
0.5	4.14 ± 0.54*a*	63.71
0.1	5.20 ± 0.55*a*	54.42
Control	11.41 ± 0.99*b*	0.00

a = Within each treatment, means followed by the same letter are not different at 0.05 level, Duncan's new multiple range test.

Lodhi and Killingbeck, 1980), and Thibault *et al.* (1982) for example show nitrification is almost completely inhibited by aqueous extracts of *Abies balsamea* needles. The data were expressed either as residual ammonium in the experimental

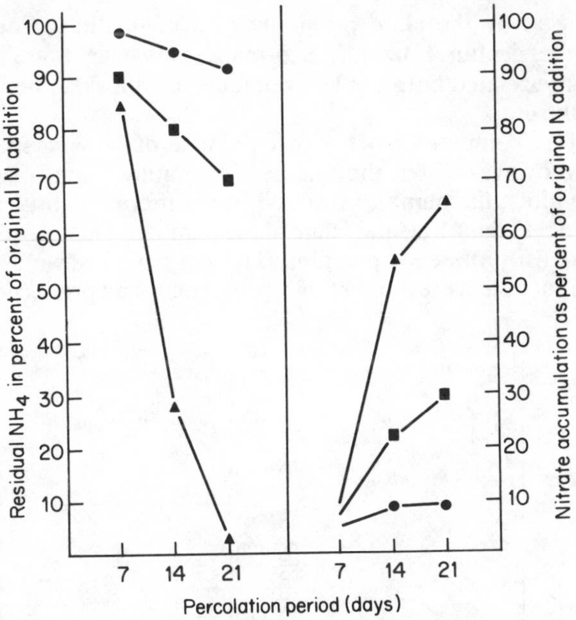

Fig. 5.10 Effect of aqueous extracts (2%, w/v) of *A. balsamea* needles (●——●) and *P. balsamifera* dormant buds (■——■) on rates of nitrification in forest soils, controls (▲——▲). Statistical t values for treatments and controls indicate significant difference at $P < 0.05$. (From Thibault *et al.*, 1982; courtesy of *American Journal of Botany*.)

soil column after treatment with ammonium sulphate or as accumulation of nitrate (Fig. 5.10), and particularly emphasize the difficulties of exactly documenting reaction as the causal mechanism in plant succession (see also discussion in Chapter 3). It also remains to be seen whether allelopathy also plays an important role in population demographic changes.

Competition leading to negative and positive correlation or association

The term 'competition' is used in a wide variety of contexts with a slightly different meaning and it has been suggested that 'interference' (Harper, 1961) be used as a general term to describe the loss of vigour and productivity of an individual due to the close proximity of another. In this context, however, the term is used to describe the struggle between individuals for some enviromental factor such as light or nitrogen, and the final effect is not one of complete elimination of one individual by another but a lowering of the performance of one of the competing members. This is usually manifested in a reduction in the numbers of a plant in relation to another species and leads to a negative correlation between the two.

Much of the detailed work on competition has been carried out under controlled conditions, since in the field it is often difficult to distinguish factors involved in competition from the other possible causes of a negative correlation. In a consideration of the effects of competition it is important to realize at which point of the life cycle competition exerts the most pressure, whether it is at the seedling stage or whether the effect is over a longer period of time. The effect of marked competition at seedling stages is usually of greater importance in relation to the actual numbers of plants which survive to maturity. Reduction in numbers of plants is more easily observed in ecological situations than the less obvious effect of reduced performance caused by the close proximity of a competitor. Conversely, in an agricultural or horticultural field, performance and yield are of paramount importance and are related both to plant numbers as well as to the general level of performance of the crop.

An example of the competition between individuals of the same species is given by Harper (1960) who investigated the numbers of mature plants of four species of poppy in sample plots; the numbers resulted from different densities of sown seed. The results (Fig. 5.11) show a clear decline in the number of mature plants relative to the rise in the density of seeds per plot. The density of sown seed produces a corresponding reduction in the number of capsules per plant produced by the mature

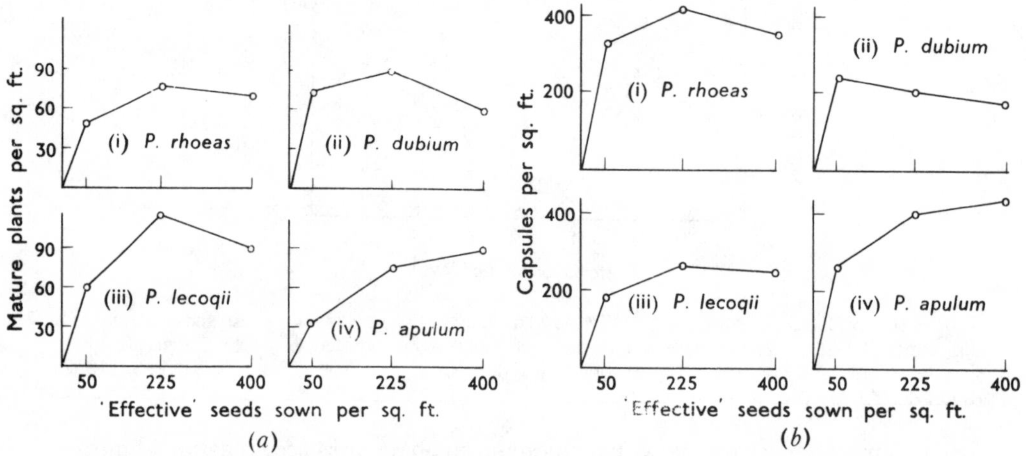

Fig. 5.11 Competition between individuals of the same species, as expressed by number of mature plants per square foot (*a*), and capsules per square foot (*b*), in plots sown with different densities of seed. (From Harper, 1960; courtesy of Blackwell Sci. Pub.)

plants. Similarly Donald (1951) (see also Harper, 1961) gives data relating the plant weight of *Trifolium subterraneum* to the density of sowing at successive dates (Fig. 5.12). At the time of sowing individual plant weight is of course not dependent on density, but as the plants develop those at the highest density compete for available nutrients and light and show a reduction of plant weight. This effect is seen earliest at high densities but is apparent at progressively lower densities as time goes on and the plants grow larger. These examples show quite conclusively the dependence of plant numbers and their performance on initial density of seeds and on the density of mature individuals respectively.

Data showing the effect of different species on the survival of *Juncus effusus* seedlings at different levels of soil fertility are given by Lazenby (1955). Five combinations of species were used: *Juncus* growing alone (J), *Juncus, Molinia caerulea,* and *Trifolium repens* (JTM), *Juncus* and *Trifolium* (JT), *Juncus, Trifolium* and *Lolium perenne* (JTL) and *Juncus, Trifolium* and *Agrostis tenuis* (JTA). The results are summarized in Fig. 5.13 and all show a marked reduction of both the numbers of plants as well as the performance of *Juncus* in all the competitor combination used, and at all nutritional levels. Lazenby suggests that the important effects are on the early seedling stages of *Juncus* since its seeds require light before they will germinate. Accordingly those species which form a very close sward inhibit germination to a considerable extent.

The examples discussed above are all taken from experiments performed under fairly well controlled environmental conditions, but they all demonstrate the sort of effect which can be expected in natural vegetation. Well documented examples from

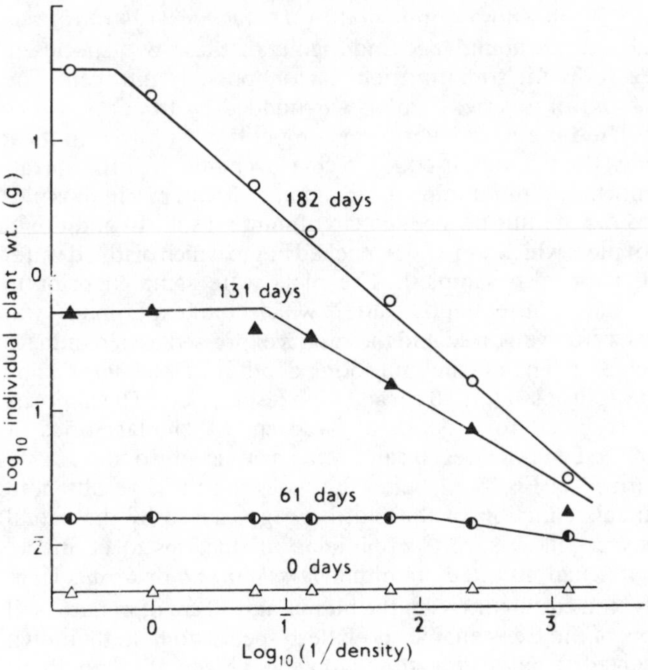

Fig. 5.12 Reduction of individual plant weight by competition between individuals of *Trifolium subterraneum* at different densities and at different times from sowing. (From Harper 1961; courtesy of Cambridge Univ. Press.)

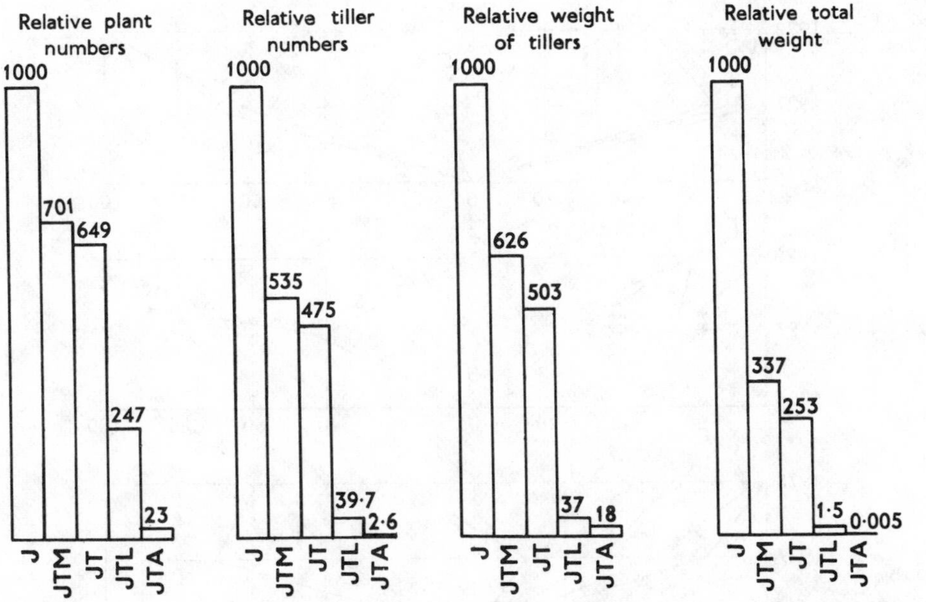

Fig. 5.13 Relative plant numbers, tiller numbers, weight per tiller and total weight of *Juncus effusus* developed with different companion species during 18 months' growth. (From Lazenby, 1955; courtesy of *J. Ecol.*)

natural vegetation are few and perhaps the most informative data are those relating to the interactions between a tree canopy and the ground vegetation. It is sometimes assumed that the paucity of vegetation beneath trees is a direct result of the low light intensity that prevails at ground level for most of the growing season. This is by no means always correct. Watt and Fraser (1933) examined this relationship in a *Pinus sylvestris* woodland with a ground flora dominated by *Deschampsia flexuosa* and *Oxalis acetosella*, and related the abundance and vigour of these two species to competition with the tree roots for some nutrient factor, possibly nitrogen. The experimental layout consisted of a series of plots surrounded by trenches, which severed tree roots and isolated the herb layer from competition with tree roots at different depths. Five trench depths were used 2.5–5 cm (raw humus), 8–10 cm (raw humus cut to its entire depth), raw humus plus 10 cm of mineral soil, raw humus plus 20 cm of mineral soil, and raw humus plus 33–45 cm of mineral soil. In addition a control plot, and a series of plots which were not trenched but to which distilled water or nitrogen were added, were also sampled. The plots were sampled prior to trenching and for two years subsequently, after which the experiment was discontinued. The samples were oven dried and the results expressed as percentages of the original crop in each plot. The maximum rooting depths of *Deschampsia* and *Oxalis* were observed to be approximately 60 cm and 8 cm respectively. The majority of the *Pinus* roots were confined to a depth of 8–20 cm. A similar series of experiments were also conducted under beech canopy in an adjacent wood.

The results are summarized in Fig. 5.14 below. In assessing these results note must be taken of the annual reduction of the plot herbage caused by the actual harvesting. However, they do allow several preliminary deductions to be made. Firstly, the addition of water had no effect on either *Deschampsia* or *Oxalis* (Fig. 5.14, a2 and b2). Secondly, the cutting through the humus and 10 cm of mineral soil and through the main roots of the trees enables both herb species to hold their own and probably in the absence of harvesting they would have actually benefited.

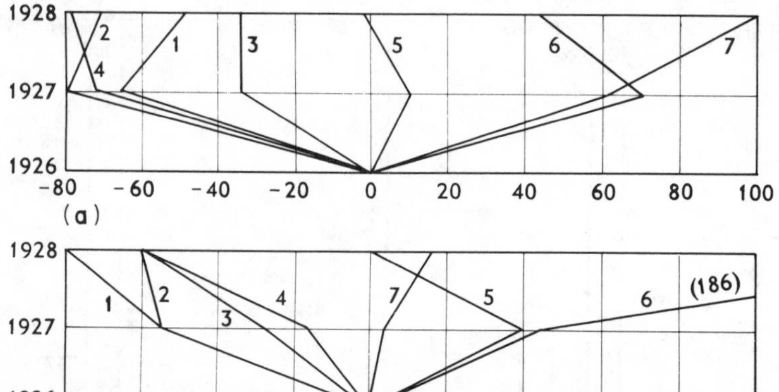

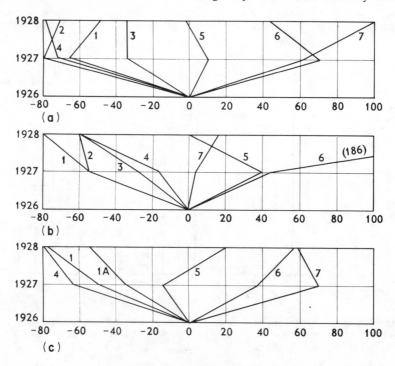

Fig. 5.14 Percentage increase or decrease in the weights of the original field layer of *Deschampsia flexuosa* under pine (**a**), *Oxalis acetosella* under pine (**b**), and *O. acetosella* under beech (**c**). The treatments were: 1. Control. 2. Distilled water added. 3. Raw humus cut 2.5–5 cm. 4. Raw humus cut to surface mineral soil (8–10 cm). 5. Raw humus plus 10 cm mineral soil cut. 6. Raw humus plus 20 cm mineral soil cut. 7. Raw humas plus 33–45 cm mineral soil cut. (From Watt and Fraser, 1933; courtesy of *J. Ecol.*)

Trenching to greater depths shows a general increase in crop weight (Fig. 5.14 (6) and (7)). The addition of nitrogenous fertilizers gave mixed results, *Oxalis* showed no change when the rates of application were small but was completely eliminated by the larger dosage. However, *Deschampsia* benefited from small doses although the largest dose had an adverse effect. Thus it appears that nitrogen may be one of the important factors for *Deschampsia* at least, which is limited by tree root competition. Other possible effects of the trenching which should also be noted are that the subsequent death and decay of the severed tree roots would increase the level of nutrient in the soil to a slight extent and also that the decay of the roots would allow greater rates of oxygen diffusion along 'channels' once filled by the roots. The increase in vigour of the herbaceous vegetation is sufficiently large to suggest an increased benefit which is greater than these two possibilities would provide. It is remarkable that the most marked effect on both herb species is at the maximum depth of trenching even though *Oxalis*, for example, has a shallow rooting system.

The inhibition of a stand of trees by the ground vegetation is perhaps, at first sight, surprising, but is a well-known factor in the establishment of certain tree crops and offers a further example of what has been interpreted as competition. One of the more valuable crops planted by the Forestry Commission is Sitka spruce (*Picea sitchensis*), as its yield of timber is high. Early trial plantings showed that spruce planted in *Calluna*, though establishing fairly well, was immediately 'checked' in its growth. Addition of basic slag, as a source of nitrogen and phosphorus improved growth initially, but did not prevent the final onset of 'checking'. The experimental plantings included some mixed stands of Sitka spruce and Scots pine and it was noted that the growth of spruce was greatly assisted by the presence of pine. Weatherall (1957) gives data comparing two stands of equal age (Table 5.4) which show quite clearly an increased growth after 10 years which is even greater after a longer period of time (Table 5.5). This use of a 'nurse' crop is now standard practice. Japanese larch is more commonly used than pine, and the apparently beneficial effect of one tree species on another is related to the presence of *Calluna vulgaris* as a dominant species

Table 5.4 Comparison of Sitka spruce pure and in mixture with Scots pine at 10 years of age. (From Weatherell, 1957; courtesy of *Quat. J. Forestry.*)

	Pure plantation (Sitka spruce)		**Mixed plantation (Sitka spruce and Scots pine)**	
	Mean height spruce (cm)	Mean growth 1937 (cm)	Mean height spruce (cm)	Mean growth 1937 (cm)
Not manured	55	3.6	73	10.4
Basic slag applied shortly after planting	79	8.6	98	16.8

Table 5.5 Comparison of Sitka spruce pure and in mixture with Scots pine at 12 years of age. (From Weatherell, 1957; courtesy of *Quat. J. Forestry.*)

	Pure plantation spruce		**Mixed plantation spruce and pine**	
	Mean height (cm)	Mean growth 1939 (cm)	Mean height (cm)	Mean growth 1939 (cm)
Not manured	61	4.3	98	16.3
Basic slag applied after planting	95	7.1	128	17.8

of the ground layer. Observations show that the roots of spruce growing adjacent to Japanese larch tend to be very much more developed under the larch than in the area dominated by *Calluna* and tend to run in the larch needle litter, only rarely descending into the mineral soil. Japanese larch very effectively suppresses *Calluna* and concurrently with the removal of this competition spruce will grow at a greater rate than a pure stand which only slowly suppresses *Calluna*. It seems likely that the roots of larch and spruce occupy different levels in the soil and accordingly any competition between them is minimized. The effect of Japanese larch as a nurse crop is well marked (Table 5.6), the height of spruce being doubled over a period of 13 years. The beneficial effects of basic slag suggest that nitrogen (the addition of calcium accelerating nitrification) or phosphorus or both are possible requisites in the competitive interaction between spruce and *Calluna*; however, other factors may well be involved.

Table 5.6 Height in metres of various species mixtures after 13 years. (From Weatherell, 1957; courtesy of *Quat. J. Forestry*.)

Crop	Height of spruce (m)	
	Complete ploughing	Single furrow ploughing
Three rows of spruce alternating with two rows of Japanese larch	4.02	3.78
Sitka spruce plus broom	4.05	2.74
Two rows of spruce alternating with two rows Scots pine	2.87	2.59
Alternating rows of spruce and pine	3.20	3.23
Two rows of spruce alternating with single row Scots pine	2.29	2.59
Pure stand spruce	1.83	1.83

Thus Handley (1963) interprets the interaction between *Calluna* and *Picea abies* as being due to the inhibition of ectomycorrhizae which are essential to the growth of spruce. Robinson (1972) demonstrated subsequently that indeed run-off from roots of living *Calluna vulgaris* on the raw humus formed by this species contained a factor toxic to several mycorrhizal fungi in culture. As a result the interpretation of Weatherell (1957) that 'root-competition' is responsible for the checking of growth in spruce is probably incorrect. An allelopathic interaction operating on the root ectomycorrhizal systems that are particularly important for phosphorous uptake, is potentially a more likely mechanistic explanation.

The relationship between the ground layer and tree layer outlined above demonstrates a presumed competition between the two life forms. None of the results is conclusive and the same difficulty exists in attempting to relate the observed effect with a causal mechanism. That is to say, it seems a reasonable assumption that the observed increase in vigour in these instances is directly related to the removal of competition, but it can be argued that an identical effect would be produced by the removal of the source of an inhibitory substance. This is the converse of the problem outlined above (p. 83); when the results of controlled experiments are extrapolated into field conditions the definition of causality is immediately open to several explanations unless the data are obtained by extremely critical sampling of both vegetation and environment coupled to an extensive factorial experimental design. Due to the time factor, such sampling presents considerable difficulty, even if one assumes that soil analyses can be done to the required order of accuracy; this latter difficulty at the moment appears insurmountable.

6

Plant Population Dynamics

Studies of animal populations involving a census of their numbers have occupied a central position in animal ecology for a considerable period of time. Plant ecology however, has been largely dominated by investigations into the spatial configurations of species in relation to their environment. At a small scale such configurations are termed pattern (Chapter 7) and at larger scales, plant associations, which can be characterized by their 'unique' species spectrum (Chapters 9 and 10). Until relatively recently very little attention has been focused on plant populations and the control of the *numbers* of individuals involved. This lack of information partly stems from the pre-occupation with spatial configurations and partly from the difficulties inherent in studying changes in natural populations of plants. In animal populations changes can be fairly readily documented by changes in numbers. In plant populations, changes in numbers of individuals do occur, but more frequently a population change is initially reflected by much less obvious criteria. Changes in dry weight, growth morphology, number of seeds per individual, all indicate pressure on the population from environmental changes or by increased interference from other species. A change in actual numbers of mature individuals is a much more gradual process, especially where perennial species are involved. Understandably what information is available is largely derived from experimental studies on populations of annual weed species, which with their short life cycle are more amenable to an experimental approach. The evidence available for perennial species suggests that at least the numbers of the population established by seed are controlled in the same way as annual species.

Harper (1977) has extensively reviewed in depth the development of plant population dynamics whilst Silvertown (1982) has provided an excellent yet more concise coverage of the topic. Only examples of the more salient points are covered here and these are summarized as a simple flow diagram for two hypothetical species A and B (Fig. 6.1). Each stage of the life cycle which interacts with or is affected by either an environmental parameter or another species is shown. Equally the overall importance of season, biotic factors, and genetical make-up of the potential population is indicated.

Events in the soil

The number of individuals reaching maturity in a population at any given point in time is clearly not only a function of how many seeds are available, but also of what physiological state they are in. Many plant species produce seeds which do not germinate immediately but which lie dormant until a particular sequence of either intrinsic events, extrinsic events, or both, trigger germination. Harper (1977) conveniently classifies dormancy into three categories: innate dormancy, which is a specific property of a species and requires a particular treatment to break it; induced dormancy which can develop when a non-dormant seed is exposed to a particular (harsh) environmental situation; and enforced dormancy developed in response to external conditions but removed immediately when the conditions are suitably modified.

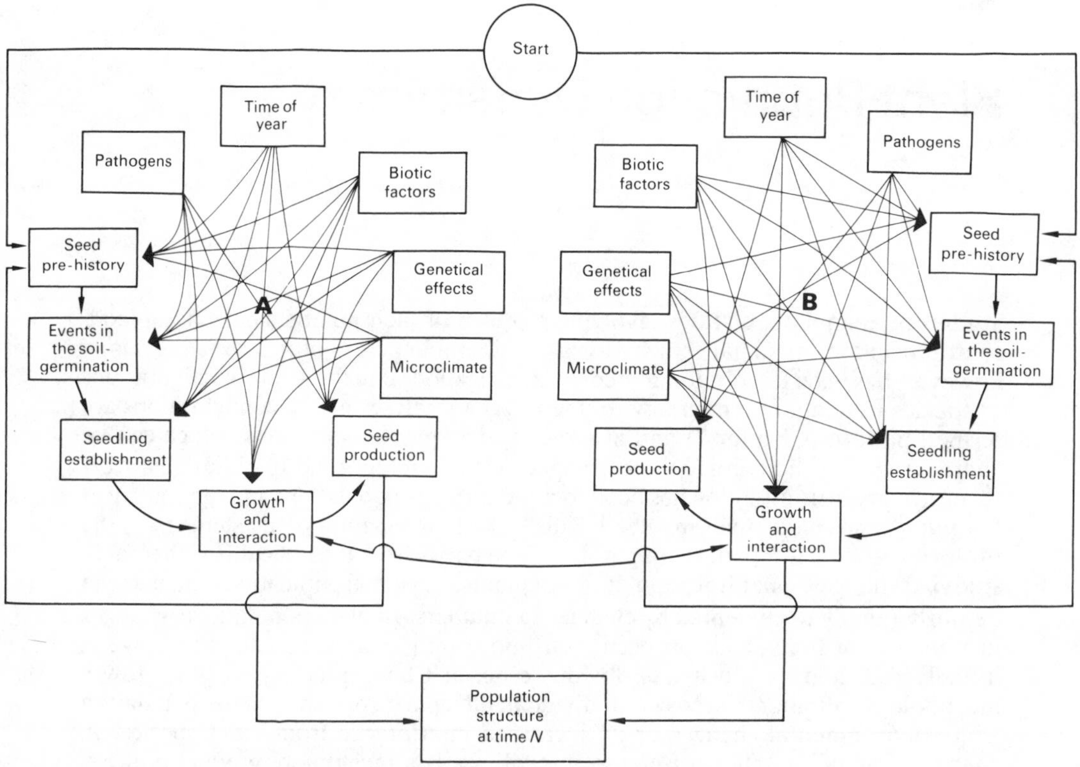

Fig. 6.1 Schematic representation as a flow diagram of the seed germination, establishment, interference, and seed production of two hypothetical species A and B.

Increasing O_2 tension around the seed embryo by pricking or abrading the seed coat, giving a fluctuating temperature or low temperature treatment, light treatment, or an increased level of nitrate availability, have all been shown to induce germination (Stokes, 1965; Thurston, 1959; Toole *et al.*, 1956; Vegis, 1963). All of these parameters are likely to be operative in all plant systems. Seed coats decay with time allowing an increased oxygen supply to the embryo and equally temperature fluctuates seasonally. Since nitrate concentration also varies throughout the season due to the temperature dependence of nitrogen-fixing organisms, the control of the actual number of seeds available for germination at time *N* is extremely complex. Popay and Roberts (1970*a* and *b*) examined the emergence of populations of the weed species *Senecio vulgaris* and *Capsella bursa-pastoris* over a two-year period and

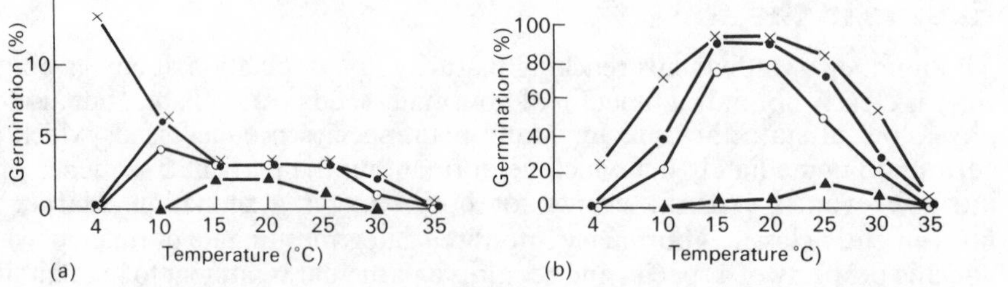

Fig. 6.2 Effect of different constant temperatures on the germination of seeds of (**a**) *Capsella* and (**b**) *Senecio* when exposed to light. ▲, after two days in *Senecio* or after five days in *Capsella*; ○ after 10 days, ●, after 14 days; ×, after 28 days. (From Popay and Roberts, 1970*a*; courtesy of *J. Ecol.*)

correlated in part the observed flushes of germination with the breaking of dormancy.

The germination requirements of *Senecio* are a light treatment (which is essential for germination), and a temperature of 10–15°C (Fig. 6.2). In addition dry storage at temperatures up to 35°C further increases germination (Fig. 6.3). Complete

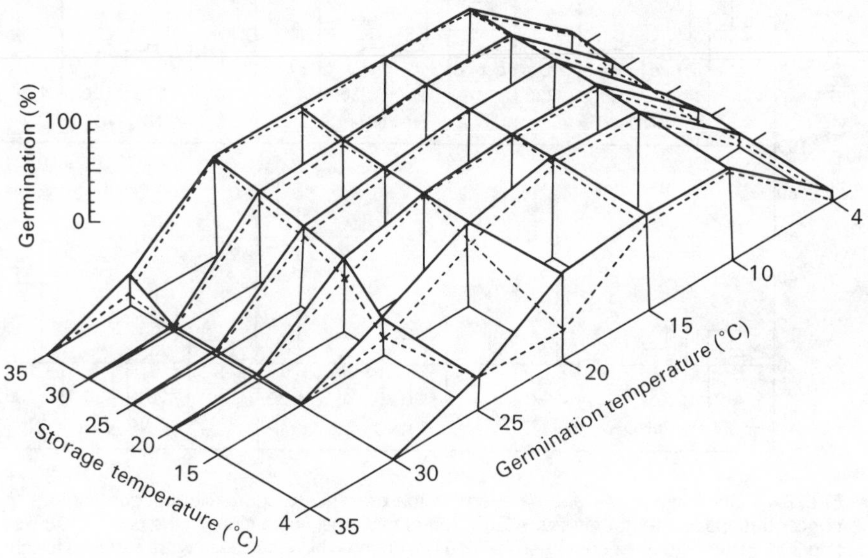

Fig. 6.3 The effect of dry storage at different temperatures on the germination of *Senecio* seeds over a range of temperatures. (———) Germination after 8 weeks' storage; (– – –) after 4 weeks' storage. The seeds were exposed to diffuse light for a few minutes each day. (From Popay and Roberts 1970*a*; courtesy of *J. Ecol.*)

suppression of germination is produced by darkness, low temperatures, low oxygen supply and high CO_2 concentration. Conversely *Capsella* requires as an essential factor a low-temperature treatment (below 10°C) as well as light. Dry storage has little or no effect. Popay and Roberts also showed a marked interaction between per cent germination and concentation of nitrate whilst a fluctuating temperature régime was employed in the experimental lay-out. Low concentrations of nitrate solution (10^{-1} or 10^{-2} mol dm^{-3}) coupled with a diurnal temperature fluctuation of 19 h at 4°C and 5 h at 30°C gave a very marked response (Fig. 6.4).

The resultant ecological implications are clear-cut: The germination behaviour of freshly collected *Senecio* seed varies tremendously throughout the year (Fig. 6.5). There is little dormancy in seeds collected in the summer, whilst conversely, material collected in the spring showed only 30 per cent germination at the optimum temperature of 10°C. Thus the dry storage treatment (see above) seems to be effectively accomplished during July and early August, whilst the seeds are ripening on the plant, allowing subsequently a wide temperature range for successful germination. Their results indicate a burst of germination in *Senecio* as the soil temperature rises above 10°C in January and again in February (Fig. 6.6). There is also a marked emergence of a further population of both species in June to August which is correlated with maximal diurnal soil temperature fluctuation and simultaneously an increase in soil nitrate level.

Other sources of variation in germination of seed populations are known. Black (1955) has shown that depth of sowing affects the pre-emergence weight changes and cotyledon area in *Trifolium subterraneum*. The latter has considerable

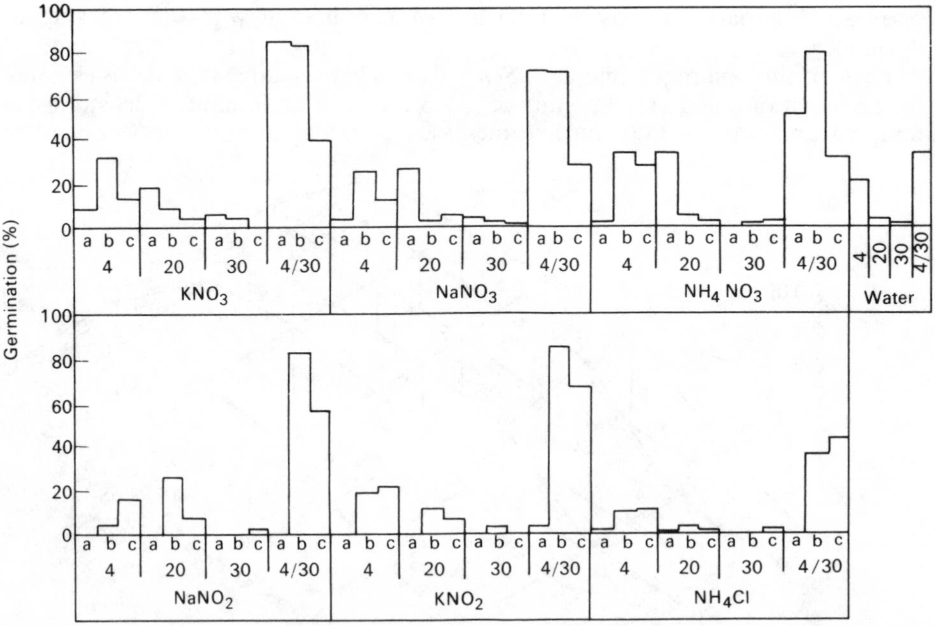

Fig. 6.4 Germination of *Capsella* seeds in the presence of inorganic nitrogen sources at different concentrations and temperatures. 4/30°C refers to diurnal fluctuations in the ratio of 19/5h between 4° and 30°C. The tests were continued for 28 days, during which the seeds were exposed to light for a few minutes each day. (**a**) 10^{-1} M, (**b**) 10^{-2} M, (**c**) 10^{-3} M solution. (From Popay and Roberts, 1970*a*; courtesy of *J. Ecol.*)

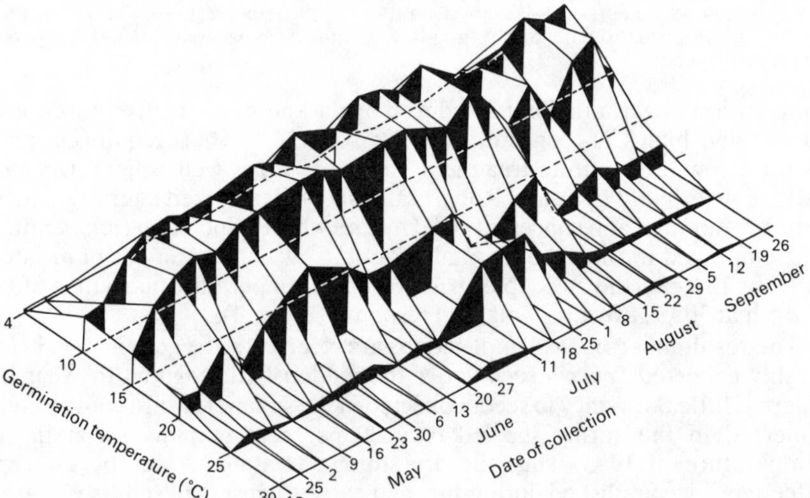

Fig. 6.5 Germination behaviour of *Senecio* seeds freshly collected at different times of the year from the same habitat. The seeds were exposed to diffuse light for a few minutes each day during the 32-day germination tests. (From Popay and Roberts, 1970*b*; courtesy of *J. Ecol.*)

effect on the subsequent performance of the plant (see below). This parameter is only likely to affect those plant populations where there is a regular soil disturbance factor, notably populations of weed species. The possibility of a biotic interaction and the burying of seeds by such a vector cannot be ignored, however, in natural

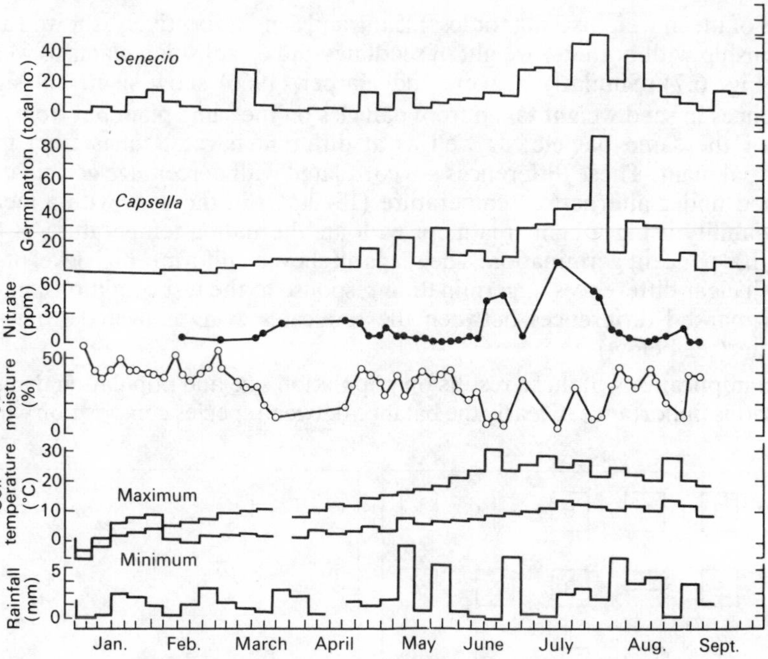

Fig. 6.6 Weekly emergence of seedlings at the experimental site in 1967. The values for the experiments on natural seed populations are the total numbers of seedlings emerging each week. Nitrate concentration is expressed in parts per million (p.p.m.) of nitrate in solution in the soil water. (From Popay and Roberts, 1970*b*; courtesy of *J. Ecol.*)

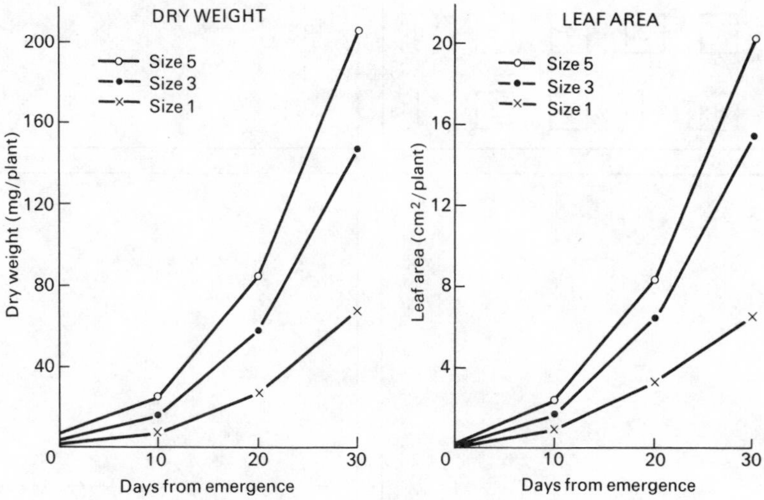

Fig. 6.7 The influence of seed size on increase of dry weight and leaf area in subterranean clover. (From Black, 1956; courtesy of *Aust. J.Agric. Res.*)

populations. Watt (1919) showed, for example, that the viability of acorns depended on their percent moisture content being maintained by trampling into the soil by grazing animals. Other similar relationships may also exist.

The size of a seed can have considerable control over its subsequent germination pattern and growth. Black (1956) using subterranean clover and three seed size classes of mean weights 3 mg (or less), 5 mg and 8 mg respectively, showed a marked relationship with both dry weight of seedlings and establishing plants as well as leaf area (Fig. 6.7). Similarly, Cavers and Harper (1966) show slight but significant differences in seed weight taken from panicles on the same plant but from different parts of the same panicle, as well as at different harvest times from the same individual plant. These differences are correlated with percentage germination data obtained under alternating temperature (15–30°C) in the dark with a measure of seed viability using a light treatment with an alternating temperature of 10–20°C giving 100 per cent germination. The seeds all show a uniformly high level of viability yet with clear differences in germination response to the test conditions, in addition to the marked differences between the replicates A–D collected from separate locations (see below).

The implications of these results on population size and population dynamics are of obvious importance. Clearly the balance between species composition will depend

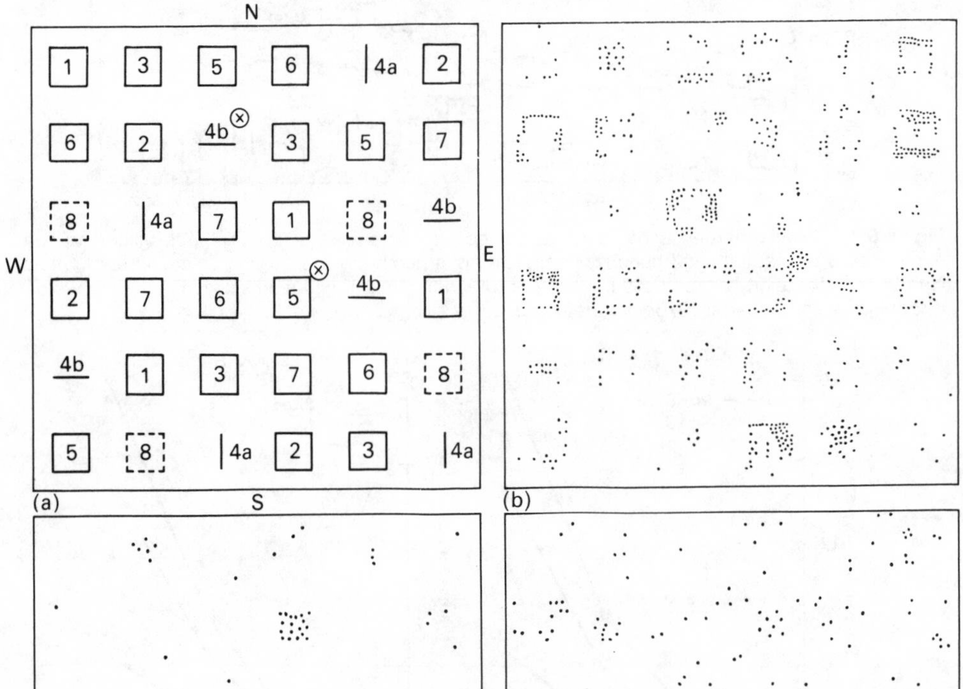

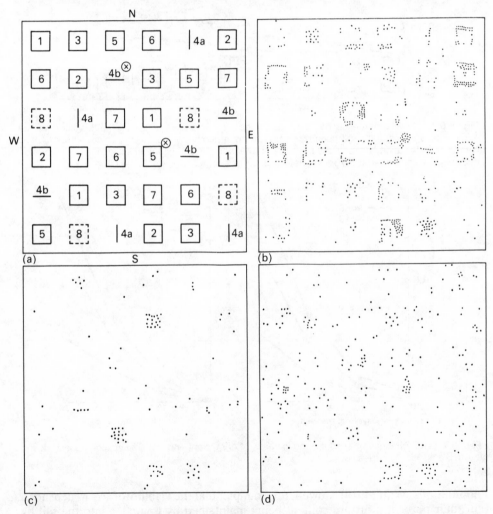

Fig. 6.8 (**a**) Plan showing the distribution of various types of objects (1–8) placed on a soil surface sown with seeds of *Plantago* species. ⊗ = worm casts. (**b**) The distribution of seedlings of *P. lanceolata* in relation to objects and depressions on the soil surface. (**c**) As for (b) with seedlings of *P. media*. (**d**) Seedlings of *P. major*. (From Harper *et al.*, 1965; courtesy of *J. Ecol.*)

in part on the climatic régime of the preceeding months interacting with the particular dormancy/germination characteristics of the species components. Different sites having slightly different microclimates will as a result carry slightly different species populations. For perennial species, originally having established by seed, these differences could be perpetuated or even intensified by subsequent interactions (see below).

Harper *et al.* (1965) have further demonstrated the importance of both soil surface texture and small modifications of soil microtopography on germination of *Plantago* and *Bromus* species. Sheets of glass were used to make square depressions 1.25 cm and 2.5 cm deep, or were left on the soil surface. Sheets of plate-glass were inserted vertically into the soil leaving different heights projecting above the soil surface and aligned either N–S or E–W. Thin-walled wooden boxes without top or bottom were also inserted to different depths in the soil. The results (Fig. 6.8) show a marked positive correlation of position of germination with the different objects and surface microtopography. Additional experiments using soil clods of different grades, as a

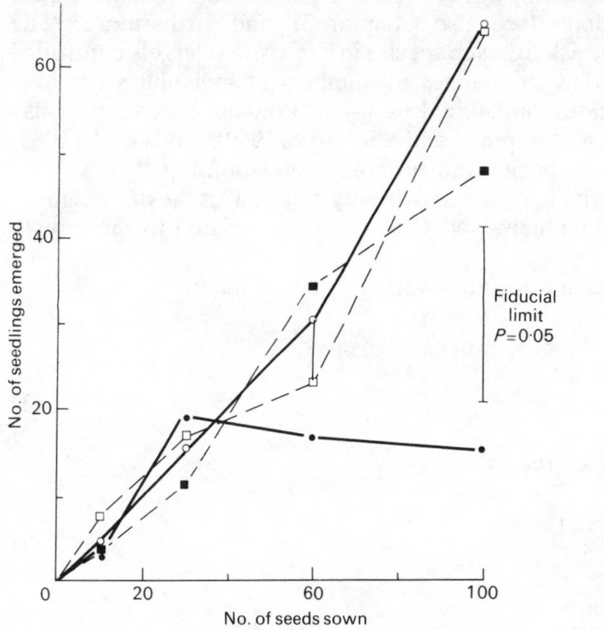

Fig. 6.9 The relationship between the numbers of seeds of *Bromus* species sown and the numbers of seedlings established on soils of varied microtopography. (□) Soil microtopographical variance = 189.5 mm²; (○) = 166.5 mm²; (■) = 60.6 mm²; (●) = 2.8 mm². (From Harper *et al.*, 1965; courtesy of *J. Ecol.*)

surface layer on top of a soil-sand mixture, again demonstrated a marked relationship between surface irregularity (expressed as microtopographical variance) and numbers of seeds germinating (Fig. 6.9). Furthermore, *Bromus madritensis* established much more effectively on rough-surfaced soils than *B. rigidus*. Seedlings of Kale and *Chenopodium album* showed a similar response to surface roughness establishing much better on rough surfaces than on smooth. These roughness effects could be cancelled out by covering with polythene covers, and Harper *et al.* interpret this result as indicating seed-water relationships to be the factor affected by surface roughness, with temperature playing a lesser part. However, in the light of the findings of Popay and Roberts (1970*a*, *b*) (see above) the relationship may be much more complex.

In addition to the control of seed dormancy and germination by the environment, there exists a considerable number of soil pathogens capable of reducing the seed population prior to germination (Wernham, 1951). Mortality of seeds prior to germination and emergence, as a result of exposure to unexpected low temperatures, is a further source of loss of viable seeds. Harper *et al.* (1955) show clearly that maize grains (4 varieties used) are unable to tolerate the cold and wet soil conditions of late autumn through to spring in Britain. This factor alone, over and above any requirement of a longer growing season, limits the extension of maize into temperate regions. Similar considerations will apply to other 'exotic' species and raises the opposite problem of how in fact the seeds of indigenous species do successfully overwinter. Clearly other factors are involved in the soil about which we have as yet, no detailed information.

Establishment and growth

Harper and McNaughton (1962) showed that the number of mature plants of *Papaver* species, sown in pure and mixed populations does not relate to the number of seeds sown, above a certain density of seed. Self-thinning occurs, minimizing the interference between individuals (see also Chapter 5), and furthermore, 'The density-induced establishment risk for each species in a mixture is largely controlled by its own density.' In addition to changes in numbers of individuals reaching maturity, the dry weight of mature individuals is also profoundly affected (see also Harper and Chancellor, 1959, and Aspinall and Milthorpe, 1959). Yoda *et al.* (1963) made a survey of a number of species and proposed a relationship $W = C.p^{-3/2}$ between dry weight of the surviving plants and density, where W is the dry weight, p is the density of surviving individuals, and C is a constant related to the growth architecture of the species.

Thus if a plant has a linear dimension L, covering an area S then:

$$S \propto L^2 \quad (1)$$

and $$W \propto L^3 \quad (2)$$

From 1 and 2 $$S \propto W^{2/3} \quad (3)$$

Since density is also related to area

$$S \propto \frac{1}{p} \quad (4)$$

From 3 and 4 $$W^{2/3} \propto \frac{1}{p} \quad (5)$$

$$W^{-2/3} \propto p$$
$$W \propto p^{-3/2}$$
$$W = C.p^{-3/2} \quad (6)$$

The full significance of this 3/2 power law was recognized by White and Harper (1970) who extended its application to a wide range of both species and species mixtures (Fig. 6.10), and showed it to be of general application. Root and shoot components of plant yield appear to follow the same law and equally the yield tables employed by foresters also closely coincide. In addition they tested and confirmed the assumption that it is the smallest plants in a population which are the first to thin, leaving the larger ones to grow more rapidly. This is an important relationship and enables the subsequent pathway of a population to be predicted from initial events. The cause of the thinning phenomenon is not clear. However, the factors influencing

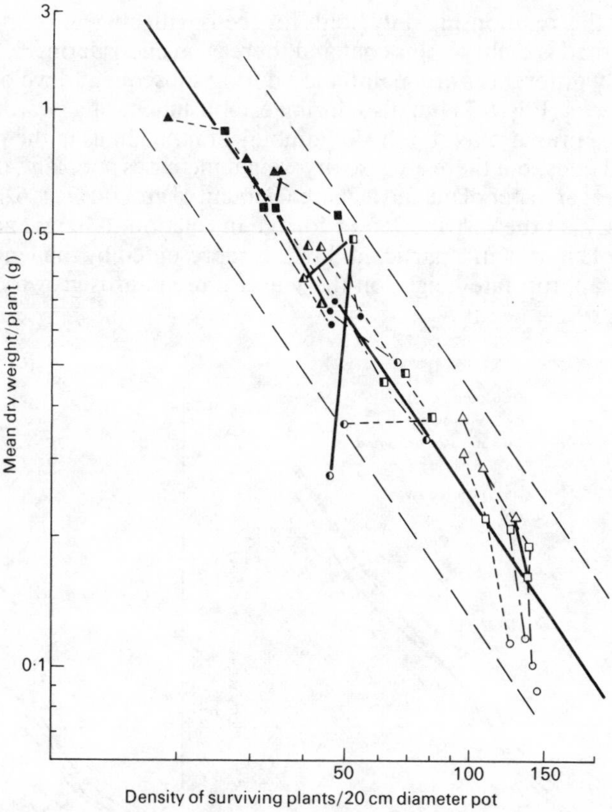

Fig. 6.10 Changes in numbers and individual plant weight with time of *Brassica napus* (———) and *Raphanus sativus* (. . . . .) in pure stands and in mixtures; the mixtures consisted of 37% *Brassica*, 63% *Raphanus* (–·–·–) and 66% *Brassica*, 34% *Raphanus* (– – –) 2 weeks after sowing, before changes in numbers occurred. Three fertility levels (○, low; □, medium; △, high) were imposed on the populations. Harvests were taken at 6½ (○), 13 (⊙) and 17 (●) weeks after sowing. Confidence limits are drawn at $P = 0.05$. (From White and Harper, 1970; courtesy of *J. Ecol.*)

seed germination discussed above indicate some of the likely parameters and emphasize the irregularity of emergence likely from a population of seeds. There will be an immediate disparity in age imparting morphological differences which will be rapidly intensified by effects of seed size and soil microtopography. In turn this will influence and alter the canopy nanoclimate generating still further differentials within the population over a period of time.

Growth and interaction

Black (1955, 1956, 1958, 1960) in an elegant series of papers examined the population numbers of subterranean clover, and how they changed. They are especially interesting since they demonstrate one of the mechanisms of interference pressure between individuals.

The influence of seed size on the area of the cotyledons has been discussed briefly before (p. 99, Fig. 6.7) and similarly the interaction with depth of sowing. Black (1955, 1956) showed that seed size determined the depth from which successful seedling emergence can take place and also the initial area of cotyledons. It appears that once successful emergence has been achieved the area of the cotyledons determines the extent of the difference in the early stages of growth. It should be noted that this relationship only holds for seeds with epigeal germination and where the whole seed is embryo plus coat and there is no endosperm.

The early differences are maintained during subsequent development not only in the first weeks (Fig. 6.7) but also during establishment of a sward when there is any interference pressure. Not only do numbers of individuals in the population alter at differential rates, but there are also dry weight increases per plant, dry weight increase per area, leaf area per plant and leaf area per unit of ground (Figs 6.11 and 6.12). Black (1958) analysed the distribution of foliage in relation to height above ground (Fig. 6.13). The leaf area in a particular layer is represented by the length of a horizontal line at the appropriate height on the *y*-axis. Pure swards grown from two different

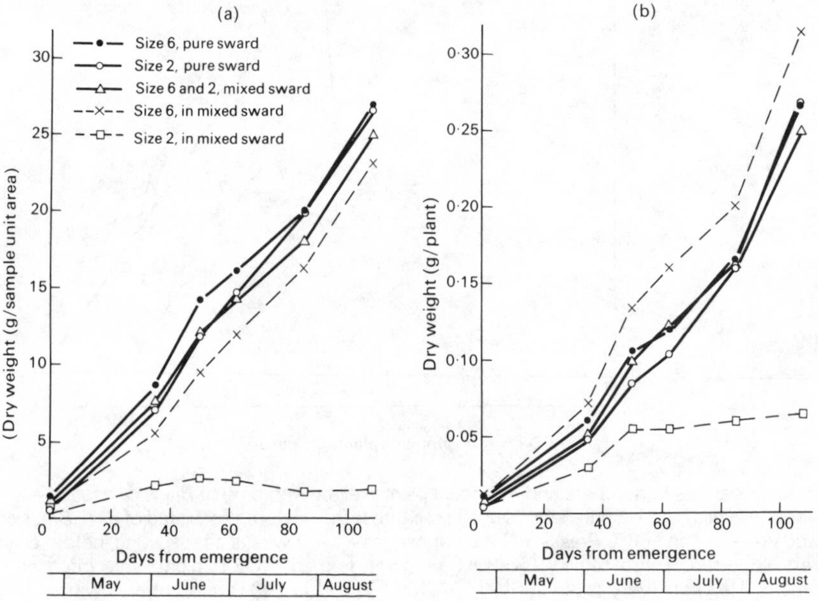

Fig. 6.11 Dry weight increases, per square link (**a**), and per plant (**b**). (From Black, 1958; courtesy of *Aust. J. Agric. Res.*)

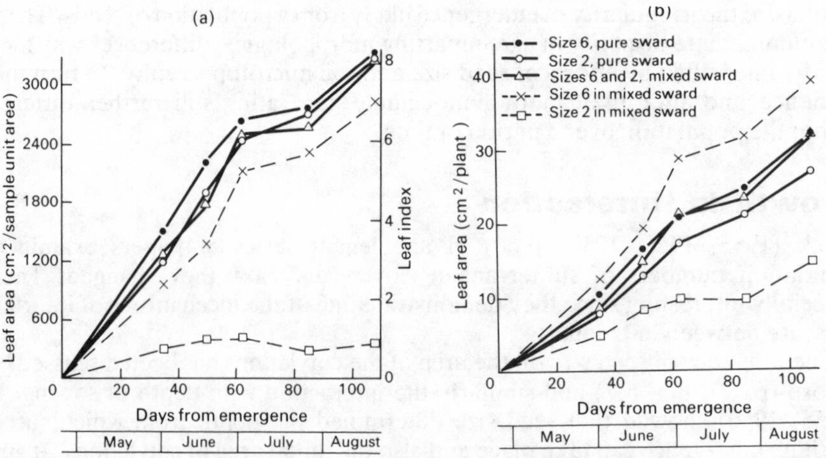

Fig. 6.12 Increases in leaf area, per square link (**a**), and per plant (**b**). (From Black, 1958; courtesy of *Aust. J. Agric. Res.*)

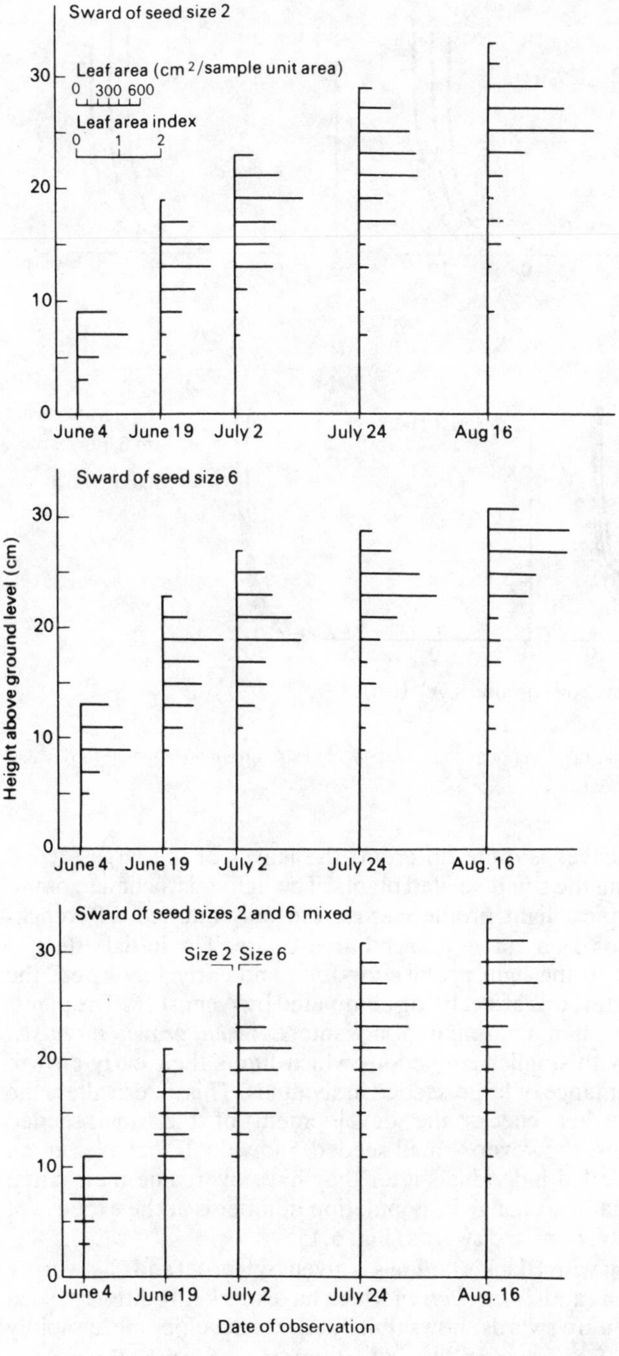

Fig. 6.13 Leaf area at each 2-cm layer of large-seeded small-seeded, and mixed swards. (From Black, 1958; courtesy of *Aust. J.Agric. Res.*)

seed sizes show closely similar results, both producing the bulk of their leaves between 25 and 30 cm above ground and by the final harvest (August 16) also having little difference in dry weight yield or leaf area (Figs 6.11 and 6.12). The sward from mixed seed sizes is markedly different. The leaves of the small-seeded plants are shown to the left and the large-seeded plants to the right of the vertical axis. The

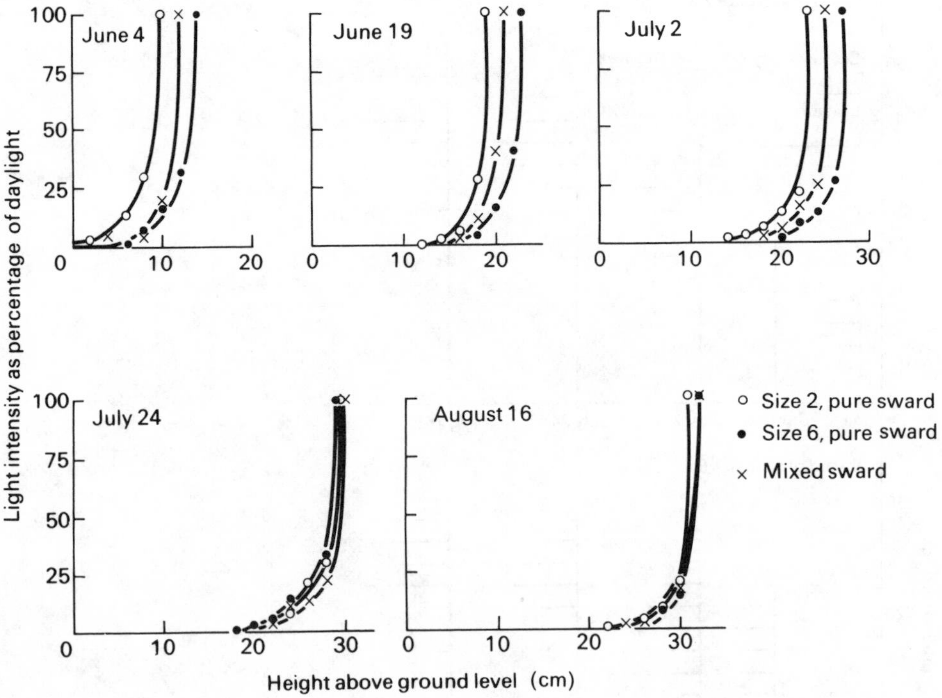

Fig. 6.14 Light intensity relative to full daylight at each 2-cm layer in subterranean clover. (From Black, 1958; courtesy of *Aust. J. Agric. Res.*)

spatial distribution of the leaves is quite different, the leaves of the large-seeded plants completely dominating the small-seeded plants. The light microclimate of the sward (Fig. 6.14) shows typical light profile curves, the only marked differences being the relative position of each curve at each harvest date. The initial effect of cotyledon area can be seen in the light profiles for June and early July where the small-seeded plants are shorter, this effect being eliminated by August 16. The plants from small seeds are thus at a continual disadvantage *when grown in mixed population*: They emerge with smaller cotyledons which limits their early growth thus allowing the rapid dominance of large-seeded individuals. This in turn alters the light microclimate and further checks the development of the small-seeded individuals. In pure swards, however, small-seeded individuals behave in an identical fashion to large-seeded individuals after they have overcome their initial slow start. It is significant that the change of population number is at the expense of small-seeded individuals only, in mixed swards (Fig. 6.15).

An interesting comparison with Black's findings is given by Harper and Clatworthy (1963), using *Trifolium repens* and *T. fragiferum* which have markedly different seed sizes (Fig. 6.16). Growth of pure swards shows that *T. repens* develops more rapidly in the first 14 weeks than *T. fragiferum* and has a higher leaf area index (the ratio of leaf area per unit of ground covered – L.A.I.) (Fig. 6.17); subsequently, *T. fragiferum* develops a higher L.A.I. which levels out around 16–18 weeks, whilst that of *T. repens* falls off reflecting its growth habit (the results were partly affected by rotting of some leaves after lodging). In mixed swards, a similar sequence of events occurs with *T. repens* maintaining a slightly higher L.A.I. and both species coming into equilibrium with each other at 15 weeks. Similarly the vertical distribution of leaf

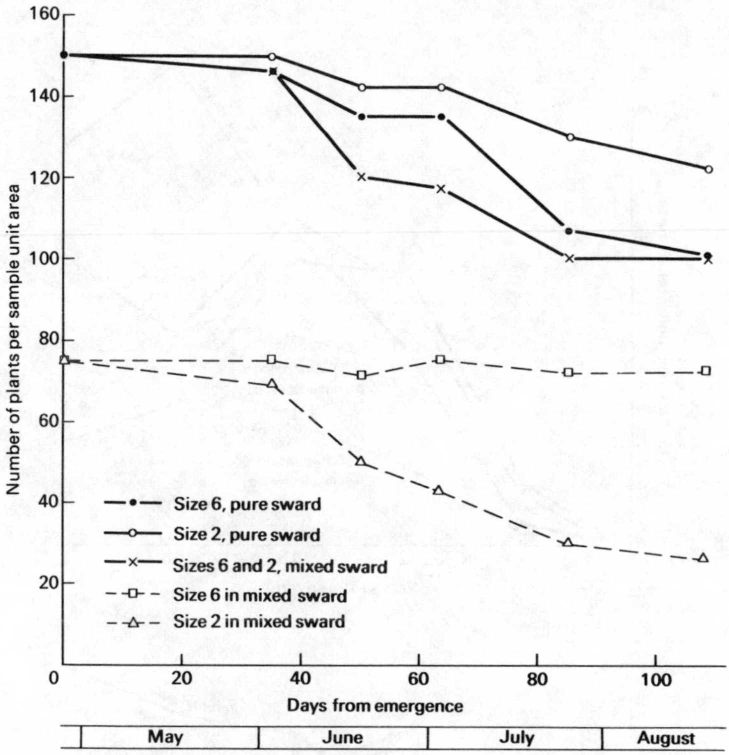

Fig. 6.15 Changes in plant numbers for large-seeded, small-seeded, and mixed swards, and for both components of the mixed sward. (From Black, 1958; courtesy of *Aust. J. Agric. Res.*)

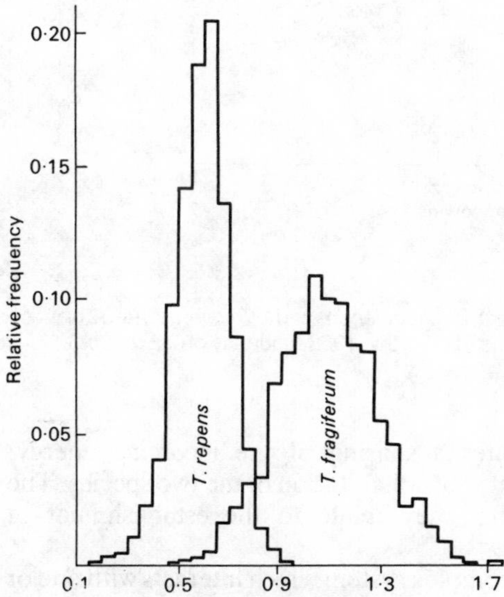

Fig. 6.16 Frequency distribution of seed sizes of *Trifolium repens* and *T. fragiferum*. (From Harper and Clatworthy, 1963; courtesy of *J. Exp. Bot.*)

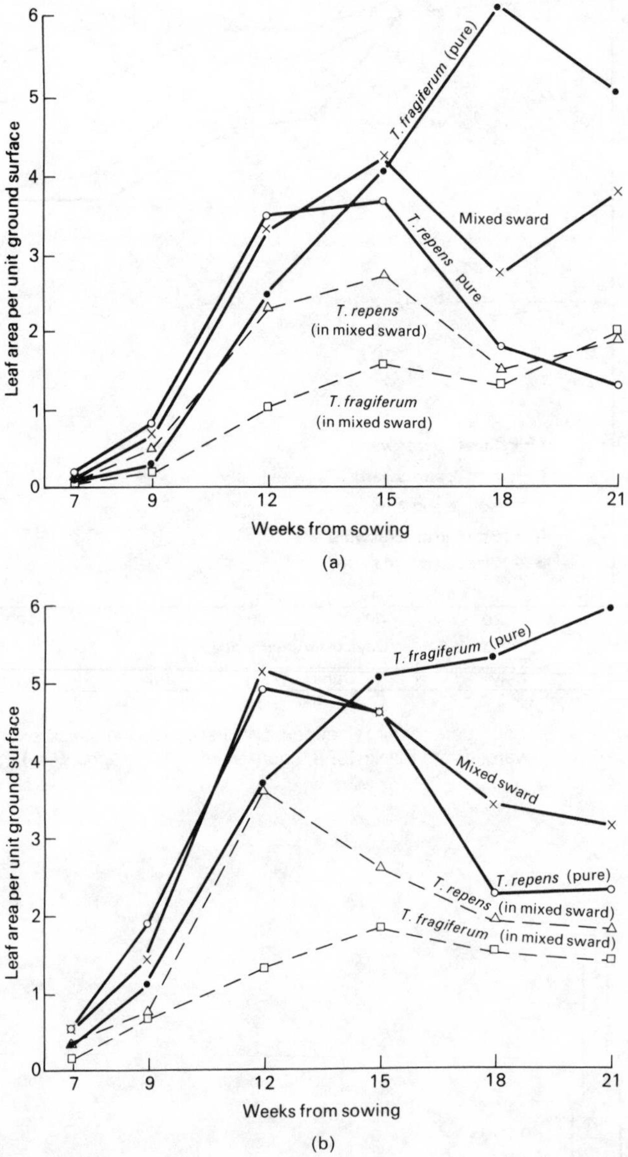

Fig. 6.17 The development of leaf area in swards of *Trifolium repens* and *T. fragiferum*. (**a**) 32 plant per 900 cm², (**b**) 64 plants per 900 cm². (From Harper and Clatworthy, 1963; courtesy of *J. Exp. Bot.*)

area in relation to light profiles indicates a sharing of the incoming energy, substantiating the trend of a satisfactory level of cohabitation of the two species. The results are particularly interesting in that they relate to the establishment of perennial species.

Thus seed size is merely one factor in a complex system which interacts with one or several other factors to decide the final outcome of population balance. Black's data demonstrate a situation where the initial slight advantage of one seedling over another becomes exaggerated by other factors in the system. Conversely, Harper's data show a system where the reverse is true and potential deleterious effects of seed size are eliminated by other factors in the system, again emphasizing the extremely complex control mechanisms of population numbers.

A provisional summary of the factors that have so far been identified as involved during the initial establishment and early growth phases of a species is given in Fig. 6.18. It should be recognized that the interaction of the 'environment', which may also include intrinsic properties of each species in addition to climate, occurs not only during establishment and early growth, but is a continuous process that also operates on mature annuals or perennials. Such interactions are often termed 'competition' and have in part been considered in the previous chapter (Chapter 5). More detailed case studies however are considered below. The second important and fundamental point to recognize in Fig. 6.18, is that only a fragment of the potential species diversity in an area of vegetation is included in the diagram. This level of simplification is also usually inherent in the field studies so far completed.

Depletion and survivorship curves in natural populations

A population of individuals is in a state of continual flux as the unit constituents die, survive or reproduce. There has been some discussion as to the exact form a unit constituent of the population takes. Thus it has been argued that the number of embryos within a population of seeds, for example, that reaches maturity constitutes the primary population in any one year. A proportion of these may be destroyed by pathogens, grazed by herbivores, or over-stressed by macroclimate. Subsequently there is potentially a further sizeable decline of the seedling population in the soil, before even any seedlings can be identified and any further population changes documented. However, because of the formidable, if not impossible, nature of following such pre-germination events with accuracy, it has been standard practice to accept the population unit as an emergent seedling, where establishment by seed is important, or where a vegetative unit is involved, some measure of the stolon/rhizome number provides the established population. These two forms of 'recruits' are termed *genets* (Kays and Harper, 1974; Harper, 1977) and *ramets* respectively. A genet is a single seedling or, in many perennial plants, an extensive clone of individual rhizomes which then can be subdivided into ramet units. Normally genets are expressed by N whereas η is reserved for the number of ramets per genet, and thus $N\eta$ gives the total number of propagules (Harper, 1977) present in the whole population.

There are several ways of expressing the changes in numbers of individuals of a species with time. In Fig. 6.19 the changes in number of individuals of *Ranunculus acris* (Sarukhán and Harper, 1973) are expressed over a two year experimental period in different ways. Thus there is a 'net' population at any point in time which cycles regularly up and down as individuals are recruited from seed or as seedlings and mature individuals are eliminated from the population. However, this net population can also be examined as a sequence of new recruits added to the existing population or as a sequence of deaths and this information is best presented as cumulative gains and losses. Two points then clearly emerge: (1) although there is little difference between the absolute numbers of individuals at the end of each of the two years there has been a tremendous flux of individuals over the experimental period; (2) there is a very marked period of synchronous recruitment and mortality, with an almost constant level of survivorship throughout the intervening periods of the year. Examination of net numbers together with cumulative gains and losses does

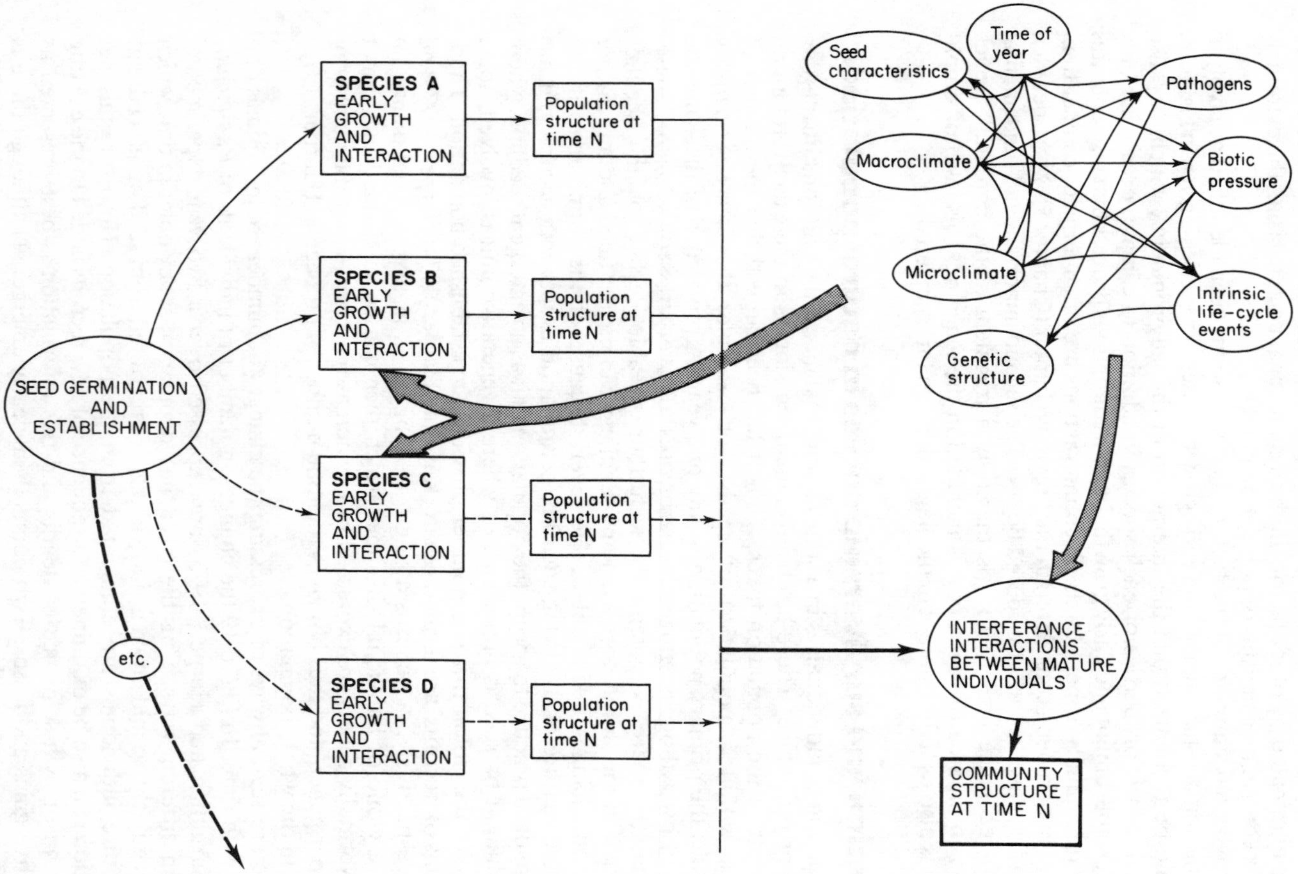

Fig. 6.18 Simplified schematic representation of the potential complexity involved in an understanding of plant population dynamics.

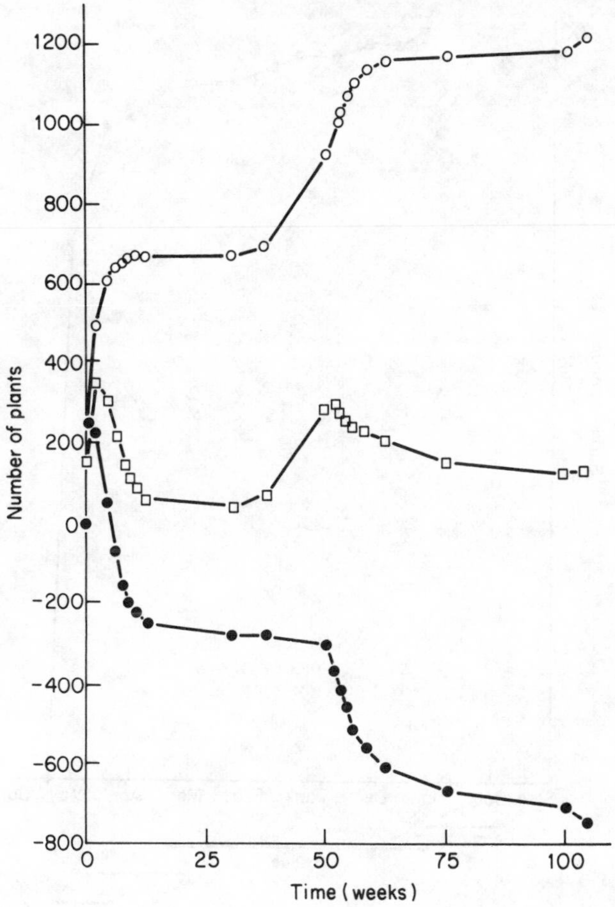

Fig. 6.19 The population flux of *Ranunculus acris* within a plot of 1 m² in permanent grassland. ○ = cumulative gains; ● = cumulative losses; □ = actual population present. (From Sarukhán and Harper, 1973; courtesy of *J. Ecol.*)

not however give any information of events which may correlate with specific age groups of individuals. At the start of an experimental period the age structure of the population is clearly unknown. However, if each individual is tagged, subsequent annual recruitment to the population can then be documented together with the subsequent mortality. The term *cohort* has been applied extensively to the group of individuals that are recruited in any one year, i.e. one generation of recruits. Detailed analysis of each group of cohorts can then provide an accurate *survivorship curve* on an annual basis. Such curves are *age-specific* (see for example Angevine, 1983) and contrast with *depletion curves* which follow the numbers of plants of *all ages* that survive.

A survivorship curve for an annual species *Cerastium atrovirens* is given in Fig. 6.20. Established individuals over winter as vegetative rosettes with an almost constant but low mortality rate until late March when flowering and anthesis occurs. During this period greater demands are made on the energy resources of the environment as a result of this growth and development but at the same time the seedlings also

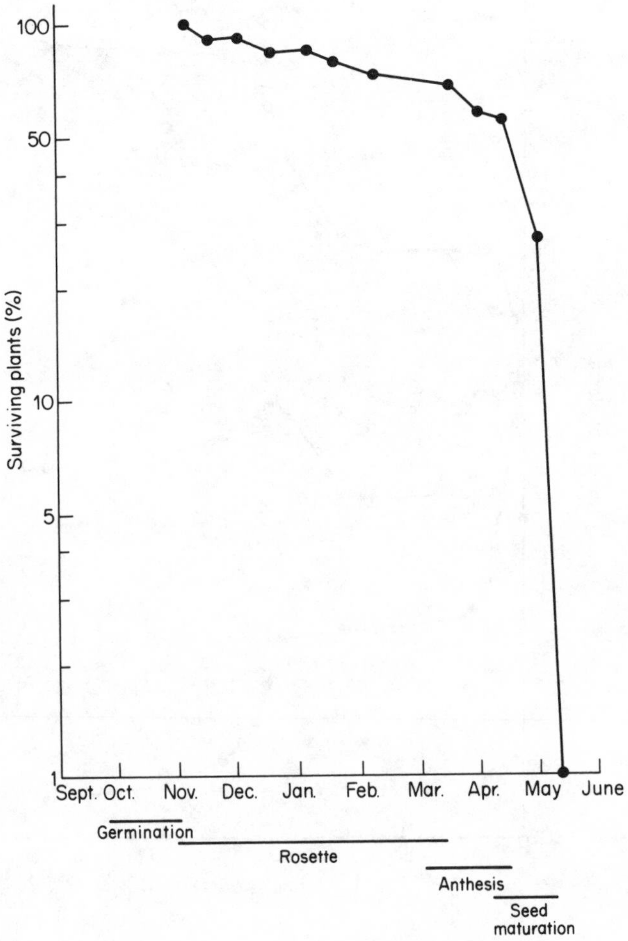

Fig. 6.20 Survivorship curve for *Cerastium atrovirens* on fixed dunes at Aberffraw, North Wales. (From Harper, 1977; courtesy of Academic Press.)

experience interference both from each other as well as from other associated species. This results in a slight increase in the rate of mortality. Harper also provides survivorship curves for other annual species (Fig. 6.21) and it is evident that increased levels of mortality are often (although not always) associated with specific seasonal events particularly during the initial establishment period and/or during reproductive phases. Similarly *Ranunculus acris* (Fig. 6.19) clearly has a rate of mortality more or less restricted to the establishment phase.

Perennial species also show a higher risk of mortality associated with seedling establishment as well as with the season of most active growth. However, if the reproductive unit is vegetative, usually the mortality risk associated with the establishment phase is little different from that experienced by fully established mature plants. However, over several years, established ramets and genets typically show a log-linear decline in numbers of survivors with a slight superimposed seasonal pattern of increased and decreased mortality. Thus it appears that a more or less constant interference pressure controls survivorship of cohorts over time, rather

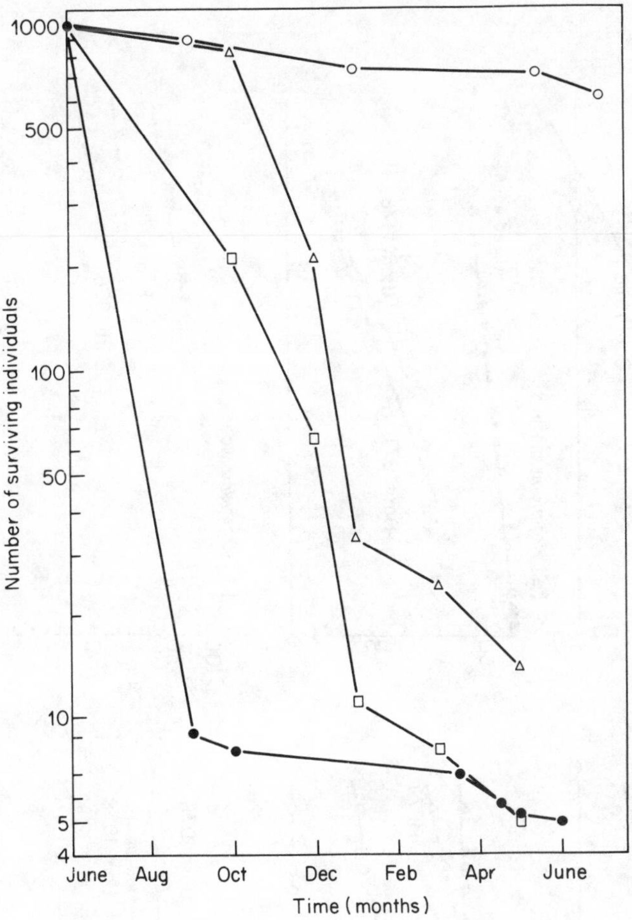

Fig. 6.21 Survivorship curves for natural populations of four species of annual plants. ○ = *Vulpia membranacea* on fixed dunes in N. Wales; = ● *Spergula vernalis* on fixed dunes in the Torun Basin, Poland; □ = *Minuartia uniflora* and △ = *Sedum smallii* on granitic outcrops, Georgia, USA. (From Harper, 1977; courtesy of Academic Press.)

than year to year variations of macro- or microclimate. In Fig. 6.22, the survivorship curves of six populations of *Fragaria vesca* and *F. virginiana* provided by Angevine (1983) show clearly that although recruitment varies from year to year and from population to population, rates of mortality within each group of cohorts are remarkably log-linear although they can vary significantly from year to year as well as from population to population.

Despite the prolific work that has appeared on plant demography over the last decade, part of which has been well summarized by Harper (1977), depressingly few general concepts have emerged. Equally unfortunately a good proportion of these data are relentlessly descriptive with very little attempt to relate the findings to potential mechanisms of change. Despite the potential complexity of even the simplest system further progress in an understanding of plant population dynamics, hopefully in the next decade, will very much depend on an identification of the specific mechanisms involved in even a few selected cases.

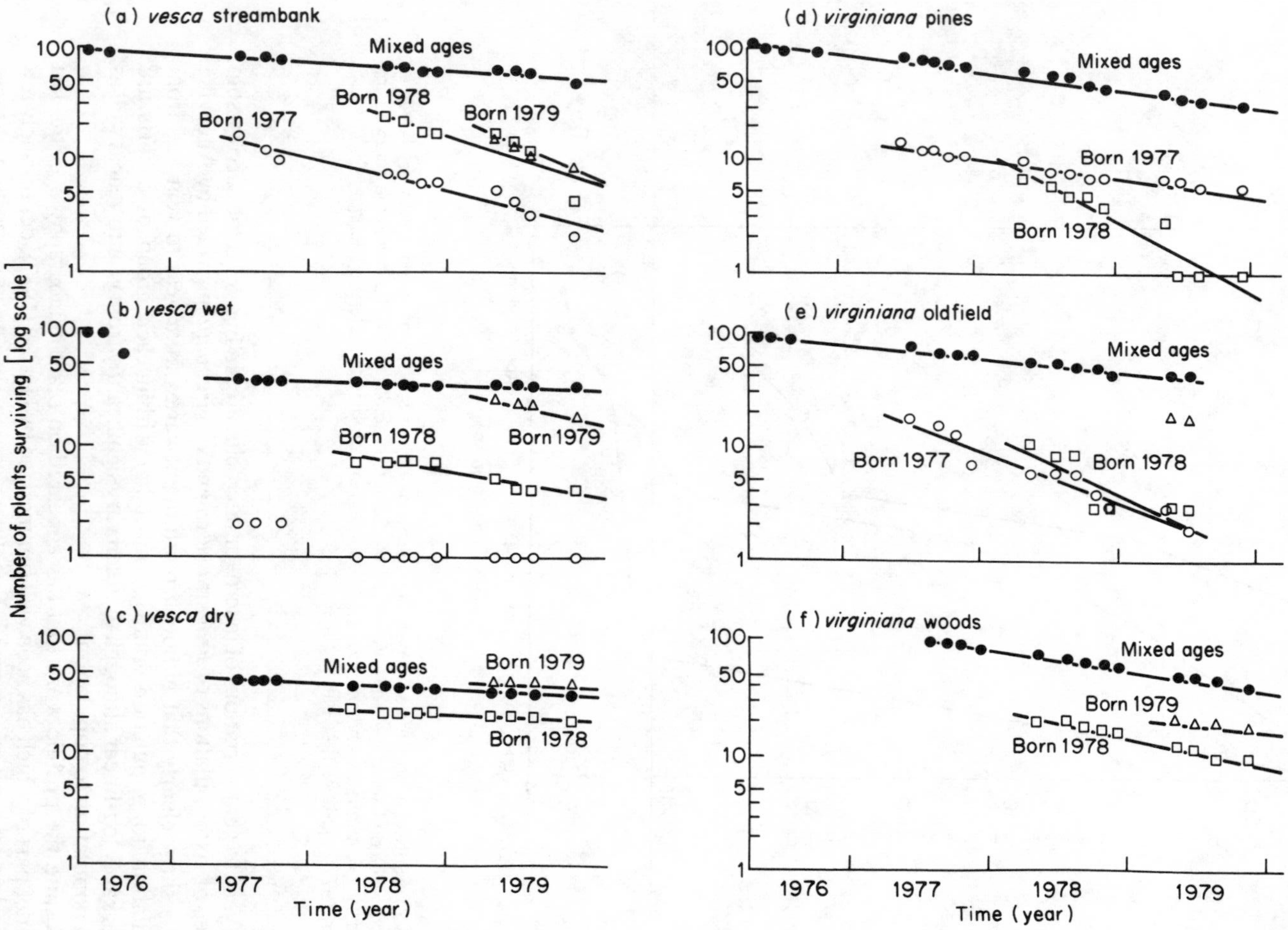

Fig. 6.22 Age-specific survivorship curves of six populations of (**a**) –(**c**) *Fragaria vesca* and (**d**) –(**f**) *F. virginiana* in Tompkins County, New York, U.S.A. over 4 years. Straight lines represent fitted least-squares regressions. Symbols: ● = plants of mixed age, born in 1976 or previously; ○ = born in 1977; □ = born in 1978; △ = born in 1979. (From Angevine, 1983; courtesy of *J.Ecol.*)

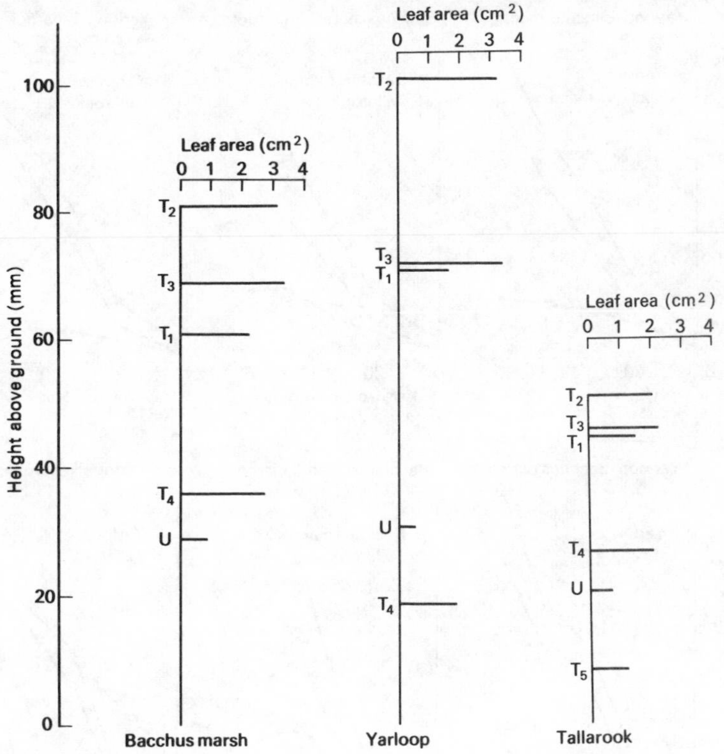

Fig. 6.23 Leaf development in three strains of subterranean clover grown under reduced light intensity. (From Black, 1960; courtesy of *Aust. J. Agric. Res.*)

Genetic control, seed production and total system interaction

Overall control of the growth pattern and behaviour is controlled by the genetic make-up of the individuals, and in many instances occurs at a level where phenotypic differences are not necessarily evident. Black (1960) showed that three strains of subterranean clover with leaf development at different heights above ground (Fig. 6.23) produced different light profiles resulting in marked interference between individuals in mixed populations (Fig. 6.24). Cavers and Harper (1966) using replicates of individuals of *Rumex crispus* and *R. obtusifolius* from six different locations showed an extremely variable response to two contrasting germination conditions (Fig. 6.25). Not only were there marked differences between locations but different individuals from the same site showed as much variation (Fig. 6.26). Although this polymorphism can be related to nanoclimate of different parts of the inflorescence, or microclimate of different parts of the site, it is highly probable that there remains a residuum of variability directly related to genetic control. These two cases suggest for some species at least, that predictions as to population behaviour pattern will have to allow for the distribution of the gene population and its effect on seed germination and seedling development. There is now available a wide range ot well-documented examples of ecotypic variation which has been very usefully collated by Heslop-Harrison (1964). In addition to relatively gross morphological

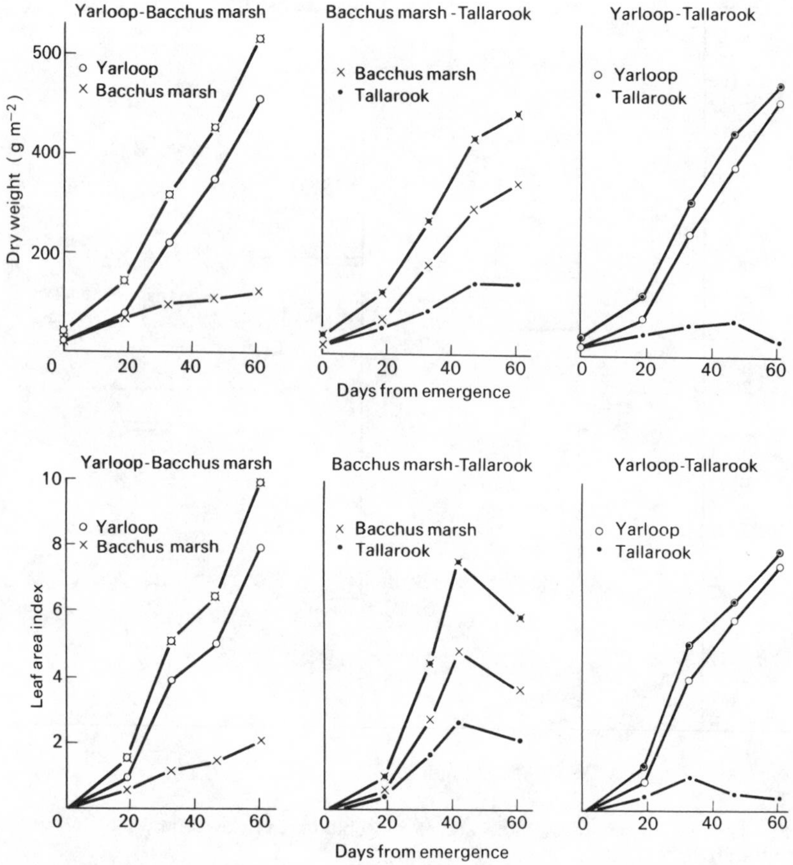

Fig. 6.24 Changes in dry weight and leaf area index of the components of mixed swards. (From Black, 1960; courtesy of *Aust. J. Agric. Res.*)

differences (leaf lengths, hairiness, etc.) there is a wide range of physiological ecotypes. McKell *et al.* (1960) compared two races of *Dactylis glomerata* and showed very different responses to differing conditions of soil moisture stress, which correlated with their known distribution. Björkman and Holmgren (1963) in a study of photosynthetic efficiency in the sun and shade ecotypes of *Solidago virgaurea* found a more efficient use of weak light in shade ecotypes than in those from open habitats. Conversely, clones from open habitats utilized intense light more efficiently. Similarly Mooney and Billings (1961) for *Oxyria digyna* showed that plants of northern populations had higher respiration rates at all temperatures than plants of southern alpine populations. Numerous additional examples are also available in the literature. Although many of these ecotypic differences relate to geographically separated populations, similar though less extreme variability will occur in adjacent populations, where environmental pressure has been exerted on the gene population. Equally although the documented examples above do not relate directly to seed characteristics, they are all indirectly related and will control to a greater or lesser extent seed numbers, or quantity of stored metabolites and hence seed size.

The evidence from Cavers and Harper (1966) also indicates a marked seasonal effect on seed weight, harvested at three different times from individual plants of *Rumex crispus* and *R. obtusifolius*. This results in a corresponding variation of

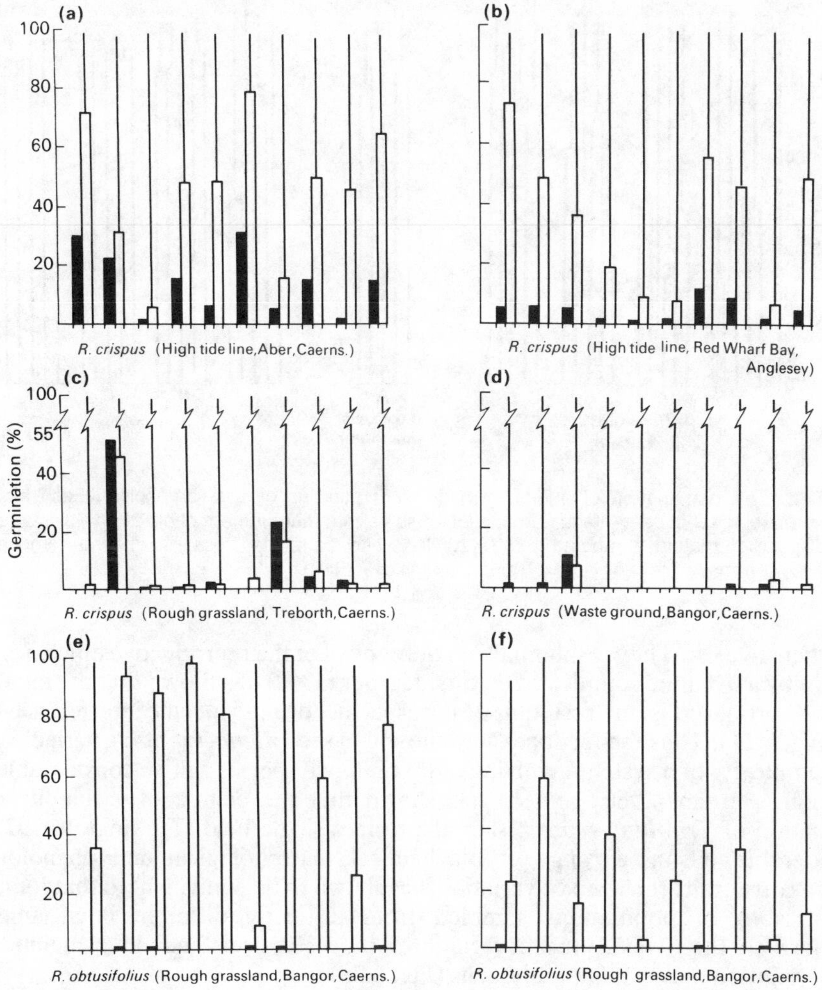

Fig. 6.25 The percentage germination of seeds from ten different plants in each of six different habitats tested under contrasting germination conditions. Vertical lines: Light, alternating temperature; Black bars: darkness, constant temperature (20°C); White bars: (**a–d**) darkness, alternating temperature (10–20°C), (**e**) and (**f**) darkness, constant temperature (20°C) interrupted by one brief exposure to daylight. (From Cavers and Harper, 1966; courtesy of *J. Ecol.*)

percentage germination (Fig. 6.27a and b). There is a marked dynamic situation existing in the fruiting panicle. Variations of temperature and light induced by the detailed structure of the panicle and its particular orientation to the sun will not remain constant throughout the year. Both the continued development of the panicle itself, as well as the progressive change in angle of inclination of the sun's rays will profoundly affect nanoclimate at any one sample point. Where there is a temperature requirement for after-ripening of the seed, a further level of heterogeneity will thus be imposed on the seed population. In addition, seed weight which is known to subsequently affect the performance of the resultant seedlings is controlled by translocation of reserve metabolites up from the leaves. The availability of these metabolites is also a function of time of year.

The interactions outlined above relate only to documented examples and are not

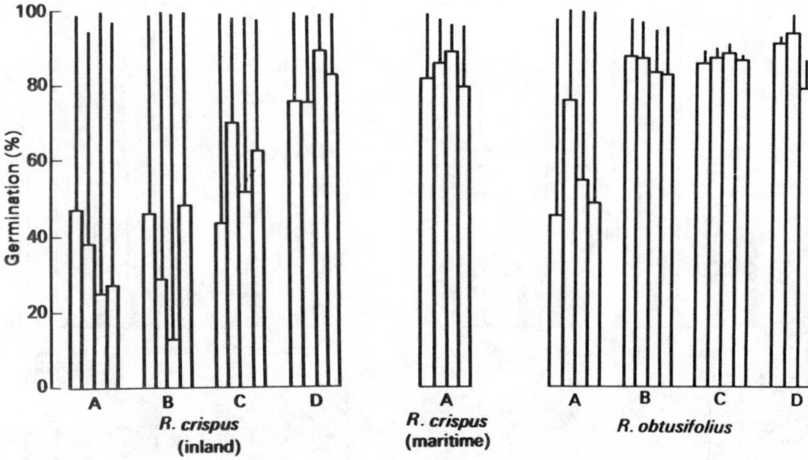

Fig. 6.26 The germination of seeds from different panicles of the same plant. Each block of histograms represents percentage germination of seed from four different panicles on a plant. Vertical lines: light, alternating temperature (10–20°C); white bars: darkness, alternating temperature (15–30°C). (From Cavers and Harper, 1966; courtesy of *J. Ecol.*)

an exhaustive list. They result finally in the control of the reproductive efficiency of a species by a dynamic complex of factors. Harper (1960) has shown that the numbers of seeds produced is one resultant of intraspecific, or interspecific interference (see also Fig. 5.11). Thus the feedback into the soil of seed *number* is also a variable, and the complexity of a system involving more than two species will be considerable. In addition, Burden (1980*a* and *b*) has shown that the degree of variability of a population of *Trifolium repens*, sampled from a single field (Turkington, 1975) is considerable. Using a range of biochemical, morphological and phenological comparative traits the level of variation was shown to be equivalent to that found in comparisons of populations sampled from distinctly different environments. Furthermore there was signficant variation in the resistance of individual members of the population to two foliar pathogens (Fig. 6.28).

Thus the level of complexity in even a 'simple' system is formidable and this has prompted a number of ecologists to provide a rather bleak assessment of current and future progress (Grime, 1984; Harper, 1982; Woolhouse, 1981). This has prompted Grime (1979; 1984) and others to examine alternatives to the rather descriptive approaches that have very much dominated community and population ecology. Thus information on the contrasting field distributions of a range of plants and animals reveals the importance of sets of traits which in turn has focussed attention on particular strategies, the most widely known being that of *r*- and *K*- selection (MacArthur and Wilson, 1967; Pianka, 1983). The term strategy, when used in an ecological context, has attracted some criticism, but in agreement with Grime (1984), we retain the concept here as a grouping of similar or analagous genetic characteristics that recur repeatedly among species or populations and underlie the similarities of their ecology.

Despite the apparent ready and embracing acceptance of *r*- and *K*- selection it is quite incomplete in terms of the full range of plant ecology (see Grime, 1979 particularly), and the alternative triangular-model approach proposed by Grime (1974; 1979; 1984) offers a more satisfactory alternative. The model involves three primary strategies characteristic of competitors (C), ruderal species (R) and those species that tolerate stress (S) respectively (Fig. 6.29). This approach provides a

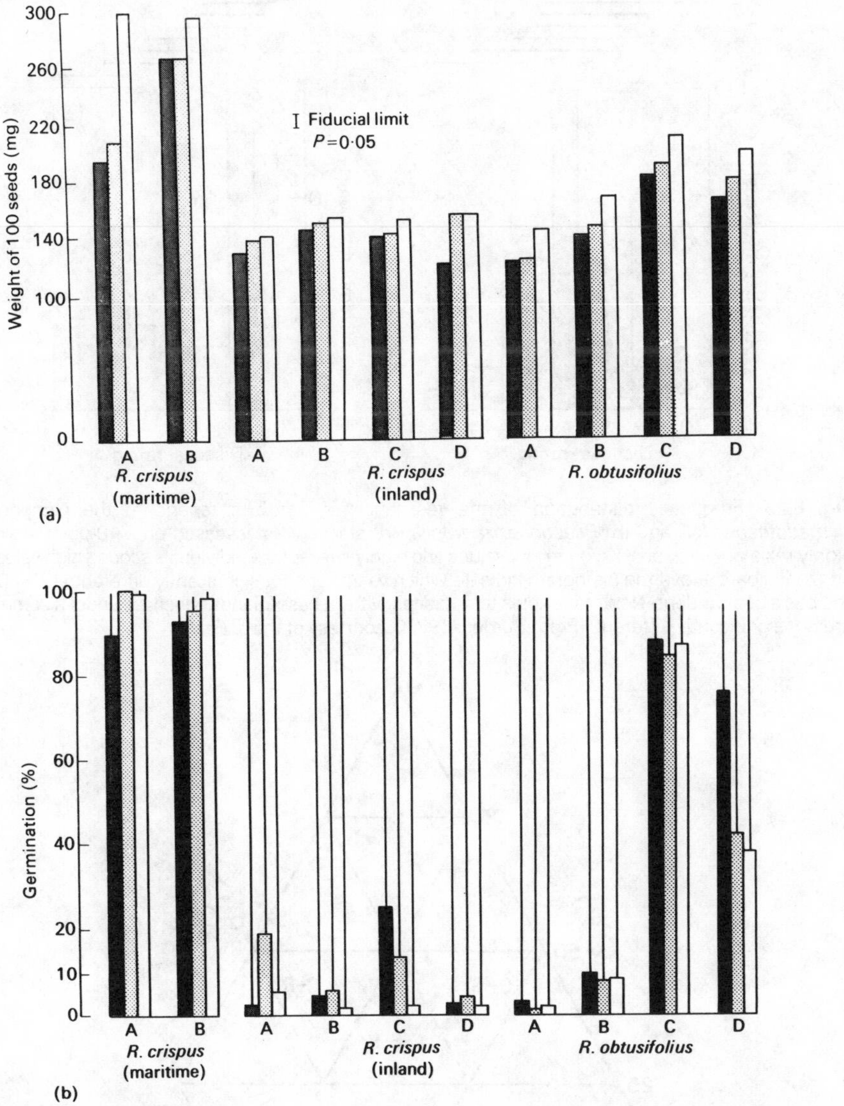

Fig. 6.27 (**a**) Variations in seed weight harvested at three different times from individual plats (A–D). Black bars, eary harvest; stippled bars, medium harvest; white bars, late harvest. (**b**) Percentage germination of seeds harvested at different times from plants (A–D) as above. Histograms record germination under darkness and alternating temperature, vertical lines under light and alternating temperature. (From Cavers and Harper, 1966; courtesy of *J. Ecol.*)

framework in which information from a range of ecological studies can be related together. The central assertion is that intensity of competition declines progressively under increased levels of stress or disturbance and as a result the equilibrium between C, S and D is a major determinant of vegetational structure and composition. Only at the three apexes will vegetation be exclusively composed of species with identical strategies and conversely an understanding of the diversity of the strategies within vegetation will also provide an appreciation of the mechanisms that control dominance, co-existence and the relative abundances of the constituent species.

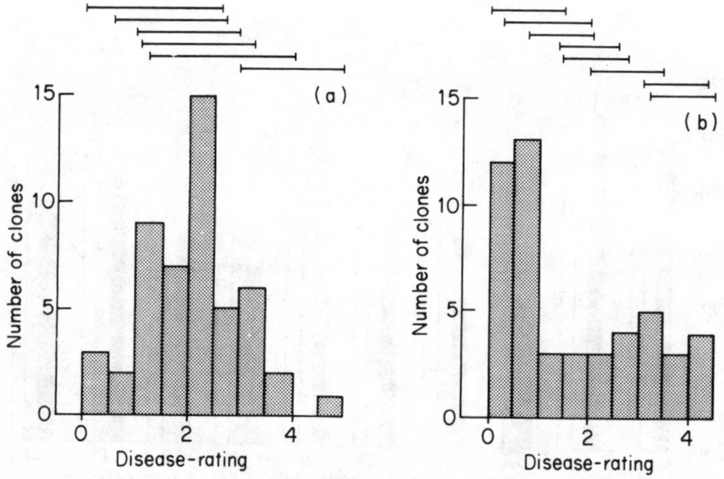

Fig. 6.28 Frequency distribution of the resistance of *Trifolium repens* to the pathogens (**a**) *Cymadothea trifolii* and (**b**) *Pseudopeziza trifolii*. Resistance was assessed on a 10-point scale, with highly resistant individuals scoring low values and highly susceptible individuals scoring high values. The horizontal bars above the histogram indicate which values are not significantly different at $P = 0.01$ on the basis of a Student–Newman–Keuls test; clones with disease-ratings which fall under a common bar are not significantly different. (From Burden, 1980*b*, courtesy of the *J. Ecol.*)

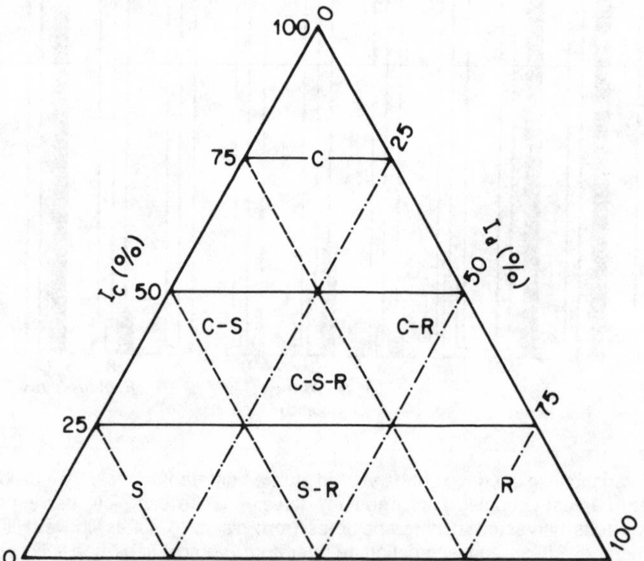

Fig. 6.29 A triangular ordination of the strategies developed by plants in relation to competition, stress and disturbance. I_c, relative importance of competition (———); I_s, relative importance of stress (– – –); I_d, relative importance of disturbance (–·–·–). (From Grime, 1979; courtesy of John Wiley and Sons.)

7

The Poisson Series and the Detection of Non-randomness

In addition to the detection of non-randomness in species distributions by correlation methods discussed in the last chapter, a number of approaches have been made to the non-random distribution of individuals of the same species. This, in part, can be accounted for in similar terms as correlation between species, but several other methods of detecting departure from randomness are available.

Historical

The earliest accounts of the non-random nature of the distribution of plants in a community are those of Gleason (1920) and Svedberg (1922) who independently showed that several species were markedly non-random. Svedberg's approach to the problem has since become one of the standard methods of detecting non-randomness in vegetation and consisted of relating the observed number of individuals per quadrat to the expected number derived from the Poisson series e^{-m}, $m\, e^{-m}$, $m^2/2!\, e^{-m}$, $m^3/3!\, e^{-m}$, $m^4/4!\, e^{-m}$, . . ., where m is the mean density of individuals. Successive figures of the series give the probability of quadrats containing 0, 1, 2, 3, 4, . . . individuals respectively, and the expected number of quadrats thus falling into each of these classes can be readily calculated. The nature of the non-randomness was termed as 'overdispersed' when individuals tended to be clumped together, and 'underdispersed' when individuals were scattered very evenly over the area. Overdispersion is thus characterized by large numbers of both empty quadrats and quadrats containing a large number of individuals, and similarly underdispersion by the majority of quadrats containing an intermediate number of individuals. Svedberg showed that dispersal of a number of species agreed satisfactorily with the Poisson series which indicated random dispersion, but also that several species showed either 'overdispersion' or 'underdispersion'. The terms 'overdispersed' and 'underdispersed' refer to the distribution curve of the data and *not* to the pattern of individuals on the ground. This has led to some confusion and it has been suggested by Greig-Smith (1957) that the terms 'contagious' and 'regular' should replace 'overdispersion' and 'underdispersion'. In a Poisson series the variance is equal to the mean and thus the ratio of these two values is equal to 1. Svedberg used this ratio as a measure of randomness; when the values were greater than 1 the distribution was assumed to be contagious, when less than 1 it was assumed to be regular.

Tests of significance

The use of variance:mean as an index of contagion in vegetation has since been used by a number of workers (Clapham, 1936, Archibald, 1948, Dice, 1952, etc.) usually employing a significance test for the difference between the observed and expected variance:mean ratio (Blackman 1942) or a χ^2-test to compare the terms of the Poisson series with the observed data (Blackman 1935). In addition to the test of goodness of fit, and variance:mean ratio (sometimes termed the Coefficient of Dispersion (Blackman 1942) or relative variance (Clapham, 1936)) several additional tests for non-randomness have been devised which are adequately described by Greig-Smith (1957) (see also Moore, 1953; Ashby, 1935; David and Moore 1954; Whitford, 1949) and are not discussed here. However, as Evans (1952) has pointed out the variance:mean ratio may give a widely different estimate of non-randomness from a χ^2-test of goodness of fit and it may be necessary to use more than one test for non-randomness. Evans gives the following hypothetical case with a mean and variance of 1.00 and hence a variance:mean ratio of 1 which indicates a random distribution within the population:

Table 7.1 The observed and expected number of quadrats containing 0, 1, 2, . . . individuals in a hypothetical case, where the variance and mean of the population equal 1.

Number per quadrat	Observed number of quadrats	Expected number of quadrats
0	20	36.79
1	76	36.79
2	—	18.40
3	—	6.13
4	—	1.53
5	5	0.31
>5	—	0.05

Obviously there is a considerable discrepancy between the observed and expected data ($\chi^2 = 63.24$, $p \ll 0.1$ per cent) and in general the χ^2-test of goodness of fit is a more reliable indication of non-randomness than the variance:mean ratio.

The method of testing departure from randomness is illustrated below. The data are taken from two communities, one consisting entirely of random individuals (Fig. 7.1) and one with hypothetical offspring grouped around the 'parents' (Fig. 7.2). The data have been obtained from randomly placed quadrats located by pairs of co-ordinates,

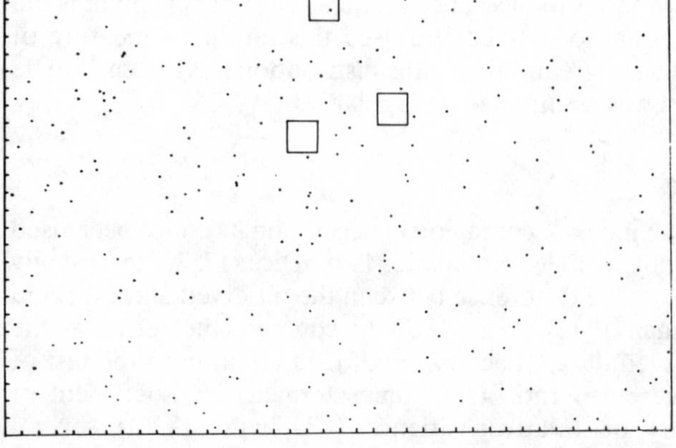

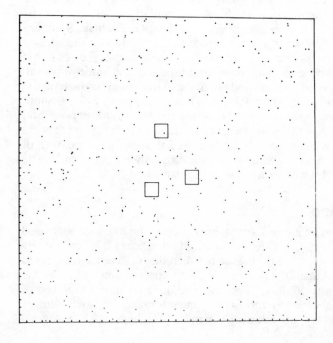

Fig. 7.1 A random 'community' with sample quadrats positioned by pairs of random co-ordinates.

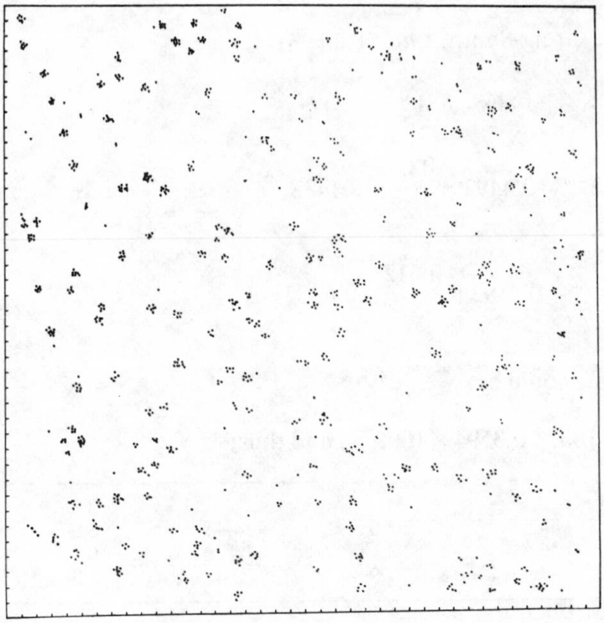

Fig. 7.2 A contagious distribution of individuals.

taken from tables of random numbers. Thus the first quadrat is located by co-ordinates 240.200 (the bottom-left hand corner of the quadrat), the second by co-ordinates 163.187 and third 179.241 and so on (a procedure identical with that used to position the 'individuals' (Fig. 7.1) in the 'community'.

Random population (Fig. 7.1)

The data for 100 random samples are given below:

Table 7.2 The observed number of quadrats containing 0, 1, 2, . . . individuals taken from a random population.

Number of individuals in each quadrat (*a*)	0	1	2	3
Frequency of occurrence in 100 quadrats (*f*)	46	34	14	6

(*a*) χ^2 *goodness of fit*

$$\text{Mean density of the population, } m = \frac{\Sigma af}{100} = 0.8$$

Thus from the series e^{-m}, $m\,e^{-m}$, $m^2/2!\,e^{-m}$, $m^3/3!\,e^{-m}$, . . ., the expected number of quadrats containing 0, 1, 2, . . . individuals can be calculated:

$$e^{-m} = e^{-0.8} = 0.4493$$
(see Greig-Smith, 1983; Table 3)

$$m\,e^{-m} = 0.4493 \times 0.8 = 0.3594$$

$$\frac{m^2}{2!} e^{-m} = 0.4493 \times \frac{0.64}{2} = 0.1438$$

$$\frac{m^3}{3!} e^{-m} = 0.4493 \times \frac{0.512}{6} = 0.0383$$

Thus the expected distribution should be:

0.4493×100, 0.3594×100, . . . and thus:

Number of individuals per quadrat	0	1	2	3
Expected frequency	44.9	35.9	14.4	3.8
Observed frequency	46	34	14	6
Difference	1.1	1.9	0.4	2.2

The goodness of fit is calculated as the sum of the differences squared divided by the expected frequency:

$$= \frac{(1.1)^2}{44.9} + \frac{(1.9)^2}{35.9} + \frac{(0.4)^2}{14.4} + \frac{(2.2)^2}{3.8}$$

$$= 0.0269 + 0.1006 + 0.0111 + 1.2737$$

$$= 1.4123$$

The χ^2-table is entered with 2 degrees of freedom (2 less than the number of terms used to calculate the χ^2 total) which shows a value of 1.386 when $p = 0.5$. Thus the chance of this difference between the two sets of data arising fortuitously is 50:50 and we can regard the observed data as showing a very good fit with the expected series and the population sampled was randomly distributed. (When the chance of a difference between two sets of figures arising completely fortuitously, falls as low as 0.05 (1 in 20 or a 5 per cent level of significance) it is usually considered that such a level of odds necessitates some other hypothesis, and the difference can be safely regarded as 'real' rather than 'accidental').

(*b*) *Variance:Mean ratio*

From the data above:

$$N = 100, \qquad x = \frac{\Sigma(x)}{N} = \frac{80}{100} = 0.8$$

$$\Sigma(x)^2 = 144$$

$$(\Sigma x)^2 = 6400$$

$$\therefore \quad \frac{(\Sigma x)^2}{N} = 64$$

The variance of the population is given by

$$\frac{\Sigma(x)^2 - \dfrac{(\Sigma x)^2}{N}}{N-1} = \frac{144-64}{99} = \frac{80}{99} = 0.8080$$

$$\text{Thus the variance:mean ratio} = \frac{0.8080}{0.8000} = 1.01$$

The variance:mean ratio shows the population to have apparently some degree of contagion, and this difference from the expected ratio of 1 must be tested by a *t*-test:

Standard error of the variance:mean ratio is given by

$$\sqrt{[2/(N-1)]} = \sqrt{\tfrac{2}{99}} = 0.1421$$

$$t = \frac{\text{Observed} - \text{Expected}}{\text{Standard Error}}$$

$$= \frac{0.01}{0.1421}$$

$$t = 0.0704, p < 0.9$$

Again this difference is not significant, for such a difference could arise by chance very frequently, and the distribution of the population can be regarded as random.

Contagious population (Fig. 7.2)

The data for 100 random samples is given below, the sampling procedure being identical to that used above:

Table 7.3 The observed number of quadrats containing 0, 1, 2, . . . individuals taken from a contagious population.

Number of individuals in each quadrat	0	1	2	3	4	5	6	⩾7
Frequency of occurrence in 100 quadrats	47	6	5	8	5	6	7	16

(*a*) χ^2 *goodness of fit*

Mean density of the population, $m = 2.44$

The Poisson series with this mean value is given by the probabilities:

$$e^{-m} = e^{-2 \cdot 44} = 0.0872$$

$$m\,e^{-m} = 0.2128$$

$$\frac{m^2\,e^{-m}}{2!} = 0.2596$$

$$\frac{m^3}{3!}\,e^{-m} = \frac{2.44^3}{6}\,0.0872 = 0.2111$$

$$\frac{m^4}{4!}\,e^{-m} = \frac{2.44^4}{24}\,0.0872 = 0.1288$$

$$\frac{m^5}{5!}\,e^{-m} = \frac{2.44^5}{120}\,0.0872 = 0.0628$$

$$\frac{m^6}{6!}\,e^{-m} = \frac{2.44^6}{720}\,0.0872 = 0.0256$$

$$\frac{m^7}{7!}\,e^{-m} = \frac{2.44^7}{5040}\,0.0872 = 0.0089$$

Number of individuals per quadrat	0	1	2	3	4	5	6	⩾7
Expected frequency	8.7	21.3	26.0	21.1	12.9	6.3	2.6	0.9
Observed frequency	47	6	5	8	5	6	7	16
Difference	38.3	15.3	21.0	13.1	7.9	0.3	4.4	15.1

$$\chi^2 = \frac{(38.3)^2}{8.7} + \frac{(15.3)^2}{21.3} + \cdots + \frac{(15.1)^2}{0.9}$$

$$= 470.334, \quad p < 0.001$$

The difference between the observed and expected numbers of occurrences is highly significant, the chances of this difference arising accidentally are very much greater than 1000:1.

(*b*) *Variance:mean ratio*

$$N = 100, \qquad m = \bar{x} = 2.44$$

$$\Sigma(x)^2 = 1364.0$$

$$(\Sigma x)^2 = 59\,536.0$$

$$\frac{(\Sigma x)^2}{N} = 595.36$$

$$\frac{\Sigma(x^2) - \dfrac{(\Sigma x)^2}{N}}{N-1} = \frac{768.64}{99} = 7.7640$$

$$\therefore \text{ Variance:mean ratio} = \frac{7.7640}{2.44} = 3.182$$

Standard error = 0.1421

$$\therefore \qquad t = \frac{3.182}{0.1421}$$

$$= 22.3927 \text{ with 99 degrees of freedom}, \quad p < 0.001$$

Again the probability of this difference arising by chance is very much less than 1000:1.

Various criticisms of the variance: mean ratio have been made, in addition to the case outlined by Evans (1952) described previously. Thus Jones (1955–6) suggests the interpretation of variance: mean ratios is very unreliable when the mean density of individuals is very high or very low, and Skellam (1952) has criticized this approach on the grounds that the success of the ratio as an indicator of non-randomness is dependent on the size of the quadrat used for sampling. This latter criticism is, in fact, applicable to all of the methods used in the detection of non-randomness of vegetation. This effect of size of quadrat can be very useful in gaining information on the scale of the non-randomness present and is discussed below. In general the χ^2-test for goodness of fit gives a reliable indication of the occurrence of non-randomness relative to the choice of quadrat size, provided very abundant and very rare species are not included in the investigation. Both these tests and others which have appeared are not applicable to distributions where an individual is not an obvious entity, and work on grassland for instance has accordingly been severely handicapped (see below).

Several attempts at deriving a method for detecting non-randomness in the distribution of tree species have been made. These have involved measures of point-to-plant and plant-to-plant distances. Basically all the methods employing distance measures between individuals or between random points and individuals depend on the relationship between these measures in a random population. Thus when individuals are distributed at random, the ratio of mean distance from a random point to the nearest individual, to the mean distance between randomly selected individuals to its nearest neighbour, should be unity. This ratio will be greater than one when individuals are contagiously distributed and less than one when they are evenly distributed (corrected for the relative density of individuals). The significance of departure from expectation can be tested in a variety of ways.

Contagious distributions

Following the demonstration that the observed distribution of individuals in a plant community did not fit a Poisson series, a considerable effort was made to find some mathematical series to which field data of this nature could be satisfactorily fitted. The type of function used in all cases involves parameters relating the distribution of individuals to random central points, around each of which a number of 'offspring' is scattered. Thus contagious distribution of individuals is related to the most obvious and likely causal factor – that of vegetative spread or heavy seeds from a parent individual, the parents being distributed at random. Archibald (1948; 1950) found a satisfactory fit for many (though not all) species, to Neyman's contagious distribution and also to a similar distribution, Thomas's Double Poisson. Similarly, Barnes and Stanbury (1951) found a satisfactory fit to Neyman's and Thomas's distributions using data taken from uniform deposits of china clay residues in the process of being colonized. On the other hand, Thomson (1952) found only one species out of three tested that fitted these distributions at all satisfactorily. Several other distributions have been suggested, based on the premise that contagion in vegetation is largely due to the morphology of the individual or the efficiency of seed dispersal mechanisms, and operates at one (small) scale only. Unfortunately the value of this approach is minimized by two considerations: (*i*) later work has shown that contagion in vegetation is due to a multitude of factors and is present on numerous scales in any one site; (*ii*) the mathematical parameters employed to define the distribution and generate the series have no meaning ecologically, or at least it is impossible to relate known ecological factors to these parameters. Thus whilst the approach is of considerable academic interest to the mathematician it is of little use to the ecologist and it would appear that the complexity and variation in the distribution of individuals is of such an order that the formulation of a mathematical model is virtually impossible and it is necessary to fall back on rather simpler empirical approaches.

The effect of quadrat size on the detection of non-randomness

One of the criticisms made of the variance:mean ratio and goodness of fit as tests of non-randomness in vegetation, has been the dependence of its detection on quadrat size (Skellam, 1952). On theoretical grounds it can be readily shown that in any contagious population the use of a Poisson series and tests of departure from it will show both random, contagious and regular distribution as the size of quadrat is steadily increased. This is illustrated below (Fig. 7.3) where sampling with qudrats *A*

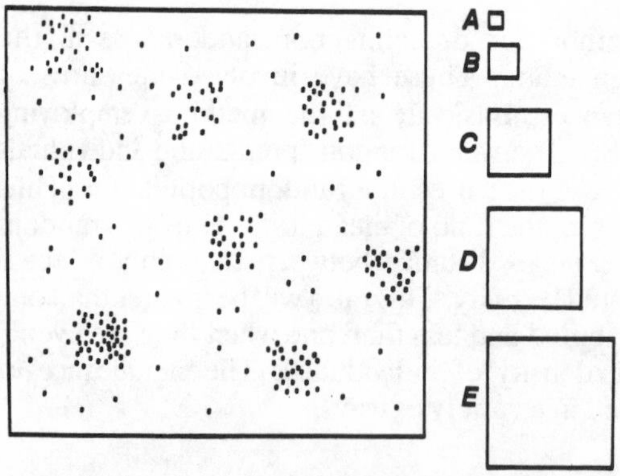

Fig. 7.3 The relationship between quadrat size and variance (see text).

or *B* the population would probably show slight contagion, sampling with quadrat *C* very marked contagion (each quadrat would contain very many or very few). With quadrats *D* and *E* the distribution would appear to tend towards a regular distribution with all quadrats containing approximately the same number of individuals. Thus the most marked demonstration of contagion would be with the quadrat with an area approximately equal to the area of the clump.

Table 7.4 Density data from Experiment 5 (Mosaic of irregularly shaped areas), showing the change of contagion detected with increase of quadrat size. (From Greig-Smith 1952*a*; courtesy of *Ann. Bot., Lond.*)

Quadrat size (cm)	**Mean**	**Mean per 100 cm²**	**Variance mean ratio**	**t**	**p**	χ^2	**n**	**P**
10	0.07	0.07	0.9394	0.43	0.6–0.7	–	–	–
15	0.18	0.08	1.0527	0.37	0.7–0.8	–	–	–
20	0.29	0.07	0.9958	0.03	0.9	–	–	–
25	0.39	0.06	1.1342	0.94	0.3–0.4	1.56	1	0.2–0.3
30	0.68	0.08	1.3928	2.76	0.001–0.01	4.00	1	0.02–0.05
35	0.79	0.06	1.5674	3.99	0.001	2.29	1	0.1–0.2
40	0.97	0.06	1.6340	4.46	0.001	26.14	2	0.001

Greig-Smith (1952*a*) gives data obtained from a series of artificial layouts of coloured discs, and some of the data are presented above (Table 7.4), which illustrate the apparent change of distribution of individuals in a population with increase in size of the sampling unit. For quadrat sizes 10, 15, 20 and 25 cm there is marked indication of randomness; at 30 cm the relative variance (variance:mean ratio) increases sharply to a significant level which is maintained in the 35 and 40 cm quadrats, indicating the contagious nature of the artificial population. (The relationship between quadrat size and variance can be readily appreciated from the hypothetical scheme above (Fig. 7.3). With the small quadrat *A* roughly equal proportions of the quadrats will contain high, intermediate and low densities of individuals. With quadrat *C* on the average high or low values will predominate and the variance accordingly will be high. Finally, a very large quadrat (*E*) will contain roughly equal numbers of individuals and the variance of the data will be very low.) The χ^2 goodness of fit gives a closely comparable picture but it does not reach a significant level of departure from expectation until the 40 cm sampling quadrat. The approximate area of the 'mosaic' employed of high and low densities in this particular layout corresponds more or less with the 35–45 cm quadrat.

The importance of the relationship between quadrat size and detection of contagion lies in the fact that the actual scale at which clumping occurs can be detected merely by resampling the population several times with different sizes of quadrats. Since non-randomness may thus be demonstrated in vegetation where a close visual inspection does not reveal any obvious sign of clumping, this is an important step forward. The mere demonstration of non-randomness in vegetation is of very limited interest but once information is available as to the scale or scales at which this non-randomness occurs, it becomes possible to relate some environmental factor or factors to scales of the detected non-randomness.

The analysis of a contiguous grid of quadrats and the detection of pattern

The development of the analysis of a grid of quadrats is a logical step following the demonstration of the dependence of the detection of non-randomness on quadrat size. Instead of throwing a range of quadrat sizes over an area, a grid of contiguous quadrats is laid out and enumerated, the increasing 'quadrat' sizes then being built up by blocking adjacent quadrats in pairs, fours, eights, etc. An analysis of variance of the data is then carried out, the variance being partitioned between the different block sizes (see Appendix 6). In the graph relating the mean square (variance) to block size the different scales of pattern appear as peaks at a block size corresponding to the mean area of 'clump'. As was pointed out above, this approach is normally used where no visual contagion (pattern) is detected and the term 'clump' is, more strictly speaking, an area where a species is at a slightly higher or lower density than in the surrounding area. This slight difference in density is not apparent on a visual inspection.

An example of such an analysis is given below with data taken from an artificial layout of clumps 2 units square grouped into a second scale of pattern at 64 units (Fig. 7.4). The graph of mean square against block size (Fig. 7.5) shows very clearly the two scales of pattern present in the layout as a double peak.

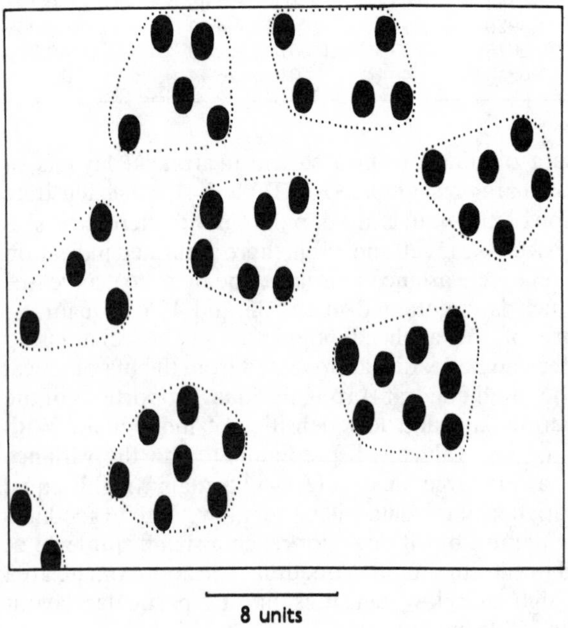

Fig. 7.4 An artificial 'community' with two scales of pattern (see text).

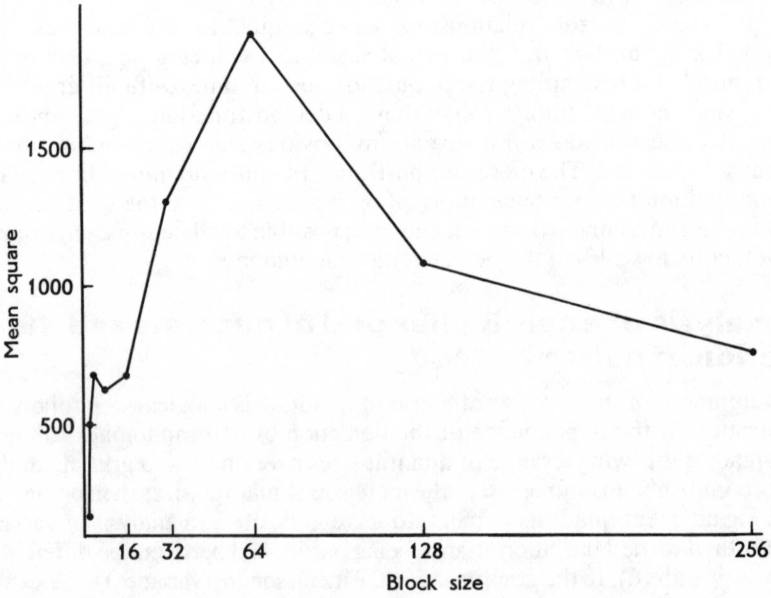

Fig. 7.5 Mean square (variance)/block size graph with a double peak representing two scales of pattern (see text).

The original method entailed using density as a measure of abundance but the approach was modified and extended (Kershaw, 1957) to enable percentage frequency or cover data to be used when the technique was being applied to grassland communities, etc. Frequency can be very readily utilized in this type of analysis, the basic unit of the grid being a sub-divided quadrat instead of a simple quadrat. Presence or absence is recorded for each sub-division of the basic grid unit (25 sub-divisions are usually sufficient) and a percentage frequency figure calculated for each quadrat in turn for the whole grid. The frequency data are then blocked up as before and analysed, the graph mean square/block size peaking at the mean area of the 'clump'. The conversion of the technique to cover data necessitates a change of approach in that instead of detecting the mean *area* of a 'clump' a line of cover readings is taken and analysed giving the mean *dimension* of a scale or scales of pattern. A Perspex frame has been designed to enable contiguous cover readings to be taken at intervals as small as 1 cm along a transect (Fig. 7.6). Thin needles are

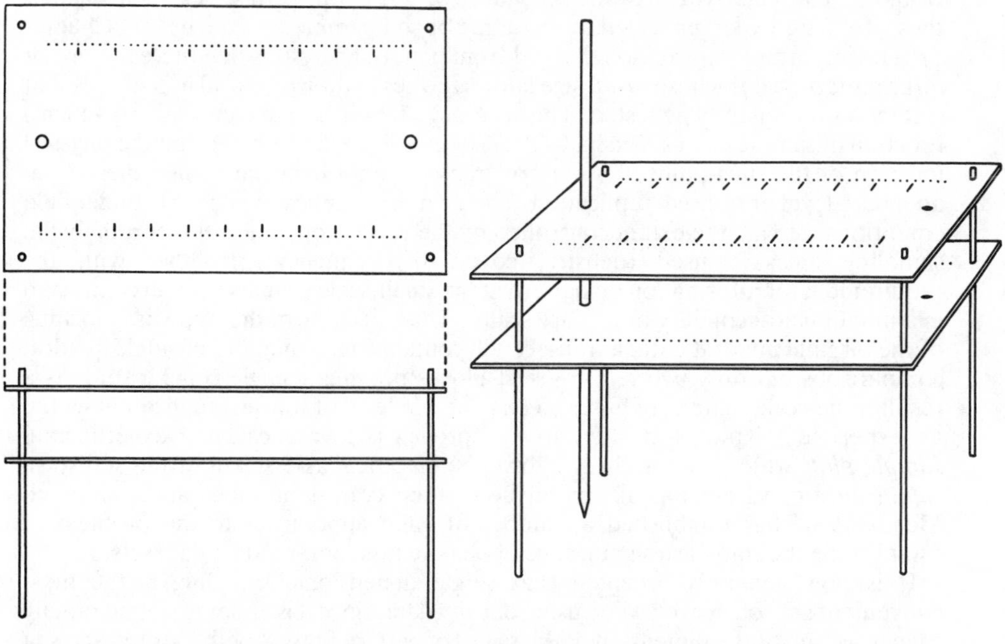

Fig. 7.6 A point quadrat for taking contiguous cover readings along a transect. (From Kershaw, 1958; courtesy of *J. Ecol.*)

lowered, or optical measure taken (by sighting through the two sets of holes on to the ground layer) and recorded, down one edge of the frame. The frame is then pivoted and the readings continued along the next row of holes that have thus been brought into line. Each set of five adjacent readings are grouped together and expressed as a percentage cover reading. The cover readings are then blocked up and analysed, the peaks in the graph representing the mean *dimension* of a scale of pattern.

Originally it was thought necessary to transform the data before analysing it but it appears that this is not essential (Greig-Smith, 1961*a*) unless tests of significance are to be made. The significance of any particular peak represents a difficulty that was initially overcome by a subjective approach. Under non-random conditions the use of a variance ratio test is invalidated and it was therefore necessary to fall back on the consistent appearance of a peak in a set of replicates. Thus if in a series of replicate samples two scales of pattern are detected at the same or closely similar block sizes in all the analyses, they can be taken as certain departures from randomness at these two scales. Thompson (1958) considered the statistical implications of the method and also suggested repeated sampling of a population would distinguish chance effects from steady trends inherent in the community. The general reliability of this approach has been tested by trials on artificial layouts of different types (Kershaw, 1957). In general as long as the basic unit is smaller than the smallest scale of pattern to be detected, the abundance of the species being investigated is not too low and particularly if some degree of replication is obtained, the results are reliably consistent (see also Greig-Smith, 1961*a*).

However, subsequent to this early work there has been an unfortunate degree of misinterpretation of the purpose of pattern analysis, coupled with attempts to measure *exactly* the particular scales of pattern present (Usher, 1969; 1975). The original concept recognised that in typical field situations there would probably be a *naturally large degree* of variation in the actual size of the pattern scales present. This would be particularly true of the secondary or tertiary pattern scales, controlled as they often are by larger scaled topographic or soil parameters (see below, Chapter 8). The important information derived from pattern analysis is that there are *two* or *three* pattern scales whose *mean* size allows the generation of causal hypotheses that can then subsequently be tested. The *accuracy* of this mean of course is very much a function of sample size as Kershaw (1957) demonstrated, and as a result the physical location of the start-point of a pattern transect is not important when there is an adequate level of transect replication. The degree of replication possible under field conditions is to a large extent controlled by the vegetation type and accordingly the sampling measure used (density, cover, or frequency), together with the requirement for information on pattern at the small-scale primary level in contrast to examination at secondary or tertiary scales. Thus, because of the step-wise blocking of the original quadrat data, a transect will contain a large number of single quadrat comparisons but only two half-transect blocks proving a single comparison. As a result large-scale pattern requires a considerable level of transect replication within an experimental plot. An alternative approach is to replicate the experimental *sample plots* which then usefully allows the empirical assessment of the statistical *significance* of variance peaks by virtue of their consistent appearance, although Mead (1974) has established a number of valid approaches to the problem of establishing the statistical significance of peaks within non-random data sets.

It is now generally accepted that single dimensional sampling is the most convenient approach even when using density data, since it is easier to obtain rapidly a clear picture of the dimension of the scales of pattern present with four transects of quadrats than it is to add more rows of transects to build up a complete grid. Numerous examples are available of various patterns in vegetation that have developed in response to certain factors and it is clear that the technique is more than sufficiently sensitive to detect very faint pattern.

The use of this approach to non-randomness in vegetation does enable a detailed approach to be made to the causal factors underlying the pattern (see below). When a pattern is very slight it is likely that one or relatively few environmental factors will be controlling it. Conversely, if a pattern is very marked, it is probable that a number of environmental factors are involved. In the former case it is often possible to make an approach to the causal factor involved, in the second instance the factors are often so numerous as to defeat any attempt at elucidating their effects and interactions. Similarly where a pattern is clearly visible without recourse to detailed statistical analysis, the complexity of the environmental background of such patterns is too great to enable a clear insight into the detailed mechanism. Thus a study of pattern is of considerable use in studies of interactions between the environment and vegetation and also between individual and individual. Pattern can similarly be expressed as a level of association between species (see an earlier section) and in fact the trend of association between species is often of considerable use in gaining an understanding of the causal factors of the pattern. Thus once two species are positively or negatively related in a community, the population will be non-random and the pattern thus present can be expressed either as an analysis of variance or as a correlation coefficient.

The relationship between quadrat size and the trend of detected association discussed in an earlier section is very marked in the variance analysis. Correlation coefficients similarly can be calculated for different block sizes and the changes of trend of association related to the scales of pattern of the species. A single approach to this aspect of non-randomness can be made by considering the covariance between two species and analysing it in a similar way to variance (Kershaw 1960*a*). Thus if two species A and B are not related in any way then Var A + Var B = Var $(A + B)$. Conversely if they are associated in some way, the observed variance will be different from the expected variance. This is conveniently illustrated by considering two species A and B having two scales of pattern as shown in Fig. 7.7 A and B. If they are

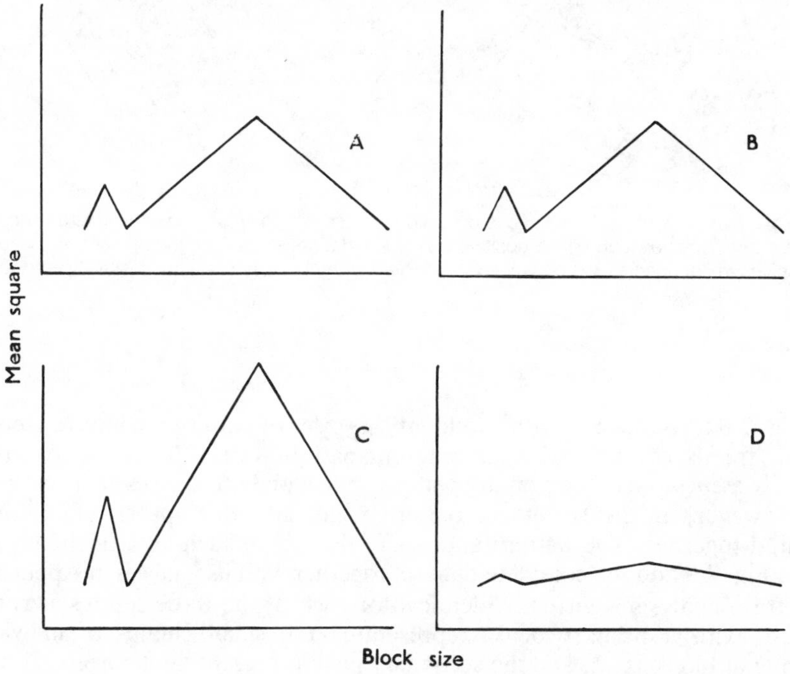

Fig. 7.7 *A* and *B* represent hypothetical patterns of two species when analysed separately. If analysed jointly their pattern will be as *C* if they are positively associated and as *D* if negatively associated.

associated together positively then the analysis of the joint data of A and B (i.e. the original data of A added to the original data of B) will show an increase of variance over and above that expected on theoretical considerations (Fig. 7.7C). Similarly if A and B are negatively related, the joint analysis of A and B will show a decrease of variance compared with that expected (Fig. 7.7D). The difference between the observed mean square/block size graph for AB and that expected (i.e. the mean square/block size graph of A plus that of B) is a measure of the co-variance between A and B at different block sizes (see also Appendix 6).

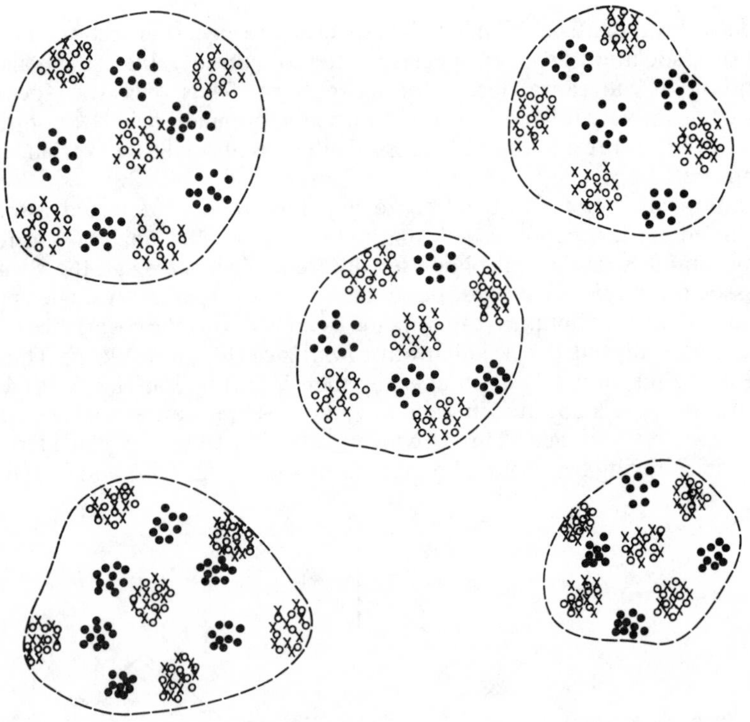

Fig. 7.8 Artificial layout of three speces (x, o, and ●) each with a small primary and large secondary scale of pattern. Species x and o are positively associated together at both the primary and secondary pattern scale, whilst species ● is negatively associated at the primary scale but positively associated at the larger secondary scale.

In Fig. 7.8 three species with two identical scales of pattern are interrelated with different trends of association at different pattern scales. Thus species 1(●) is negatively associated at the primary pattern scale with both species 2(o) and species 3(x). However, at the secondary pattern scale, all three species are *positively* associated together. The pattern analysis of data from such an artificial layout is given in Fig. 7.9 and the complete data set together with its analysis in Appendix 6. The pattern analysis is virtually identical for each of the three species as is to be expected, with a primary peak, representing the small clumps of individuals appearing at blocksize 2, and the secondary peak representing the aggregations of clumps, at block size 16. The covariance analysis (Fig. 7.9) similar shows the extremely clear-cut relationships present, with species 2 and 3 markedly positively correlated at both primary and secondary scales, whilst species 1 and 2, and 1 and 3, are negatively correlated at block size 2 but positively correlated at the secondary pattern scale at blocksize 16 (see also Appendix 6).

An example from actual field data is given below, showing the relationship between *Carex bigelowii* and *Festuca rubra* in *Rhacomitrium* heath of Iceland (Kershaw, 1962*c*) (Fig. 7.10). The micro-topography of the area shows two scales of pattern at block size 8 and 32 (equivalent to 80 cm and 3.2 m). The density and performance data (expressed as mean leaf length per quadrat of *Carex*) show an identical pattern which immediately suggests a correlation with the micro-topography of the area. The variance analysis of the density data for *Festuca rubra* is not as clear cut, apparently

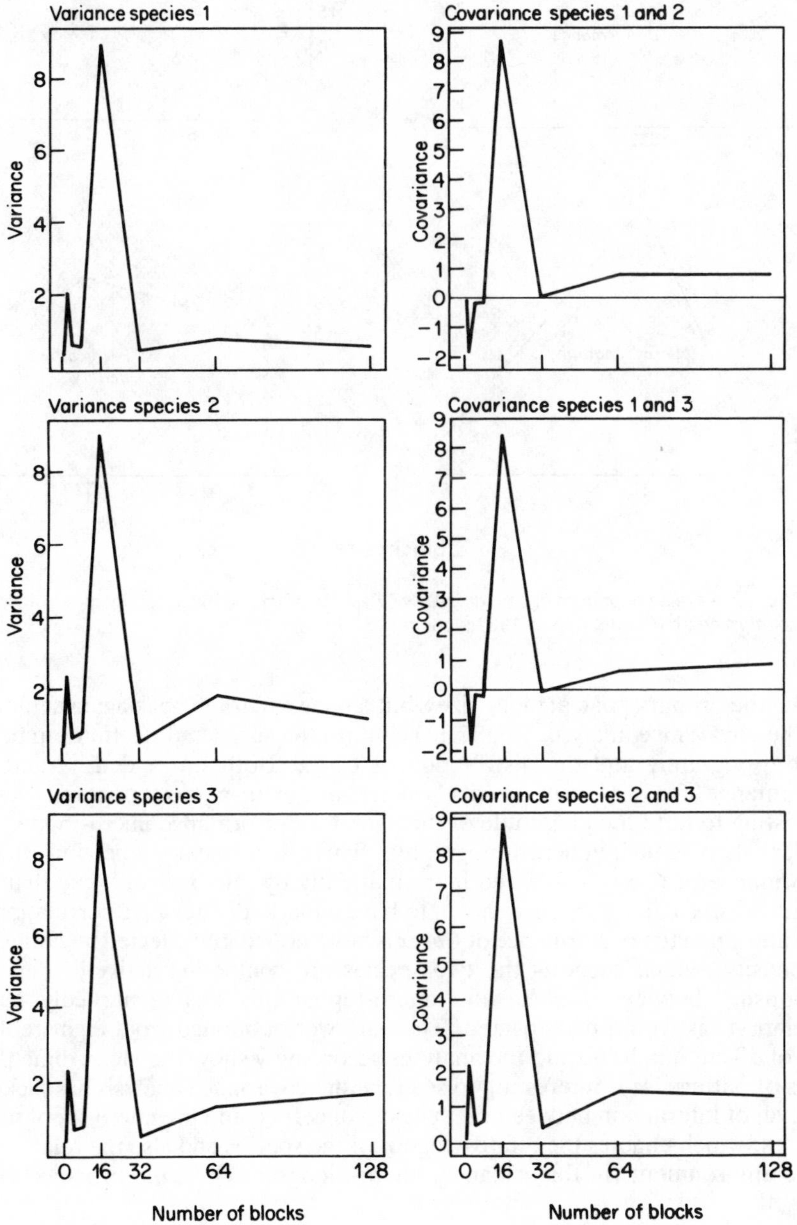

Fig. 7.9 Pattern and covariance analysis of the artificial data set provided in Appendix 6 and illustrated in Fig. 7.8 above.

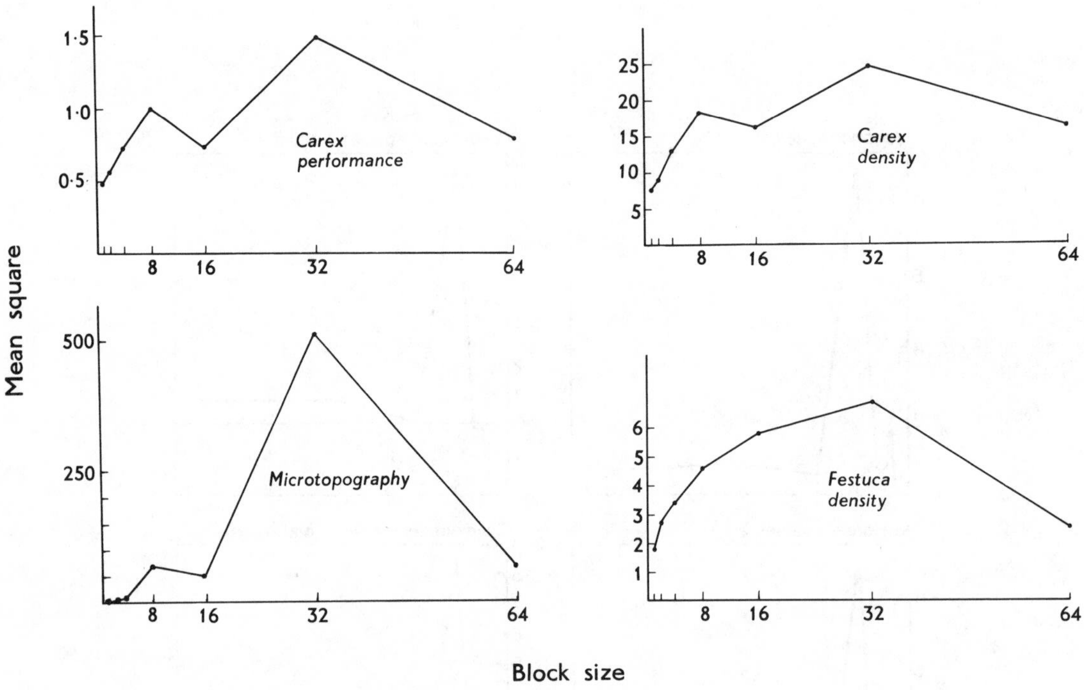

Fig. 7.10 The scales of pattern in *Carex bigelowii, Festuca rubra* and micro-topography in an Icelandic *Rhacomitrium* heath (From Kershaw 1962*c*; courtesy of *J.Ecol.*)

lacking the primary peak at block size 8 but having a marked peak again at block size 32. The covariance analyses (Fig. 7.11) confirm the suspected relationship between micro-topography and the distribution of *Carex*. Both *Carex* density and *Carex* performance are obviously positively correlated with micro-topography and it is interesting to note the amplitude of the peak for performance/micro-topography is greater than density/micro-topography. This is a clear indication that the performance of *Carex* is affected more markedly by micro-topography than is the density. *Festuca* and *Carex* are shown to have a markedly negative correlation (Fig. 7.11) and again the performance of *Carex* is more noticeably affected by *Festuca* than the density, which suggests the two species are competing actively. Finally the relationship between *Festuca* and micro-topography has a markedly negative correlation, as would be expected. The data were obtained from eight replicated lines of 10 cm quadrats, and the analyses accordingly show the mean dimension of scales of pattern. It is interesting to see how the covariance analysis has picked out the level of interaction between the species concerned and the micro-topography of the area, which enables the relative vigour of the species and also the relative effect of the environment (in this instance, micro-topography) on the two species to be assessed.

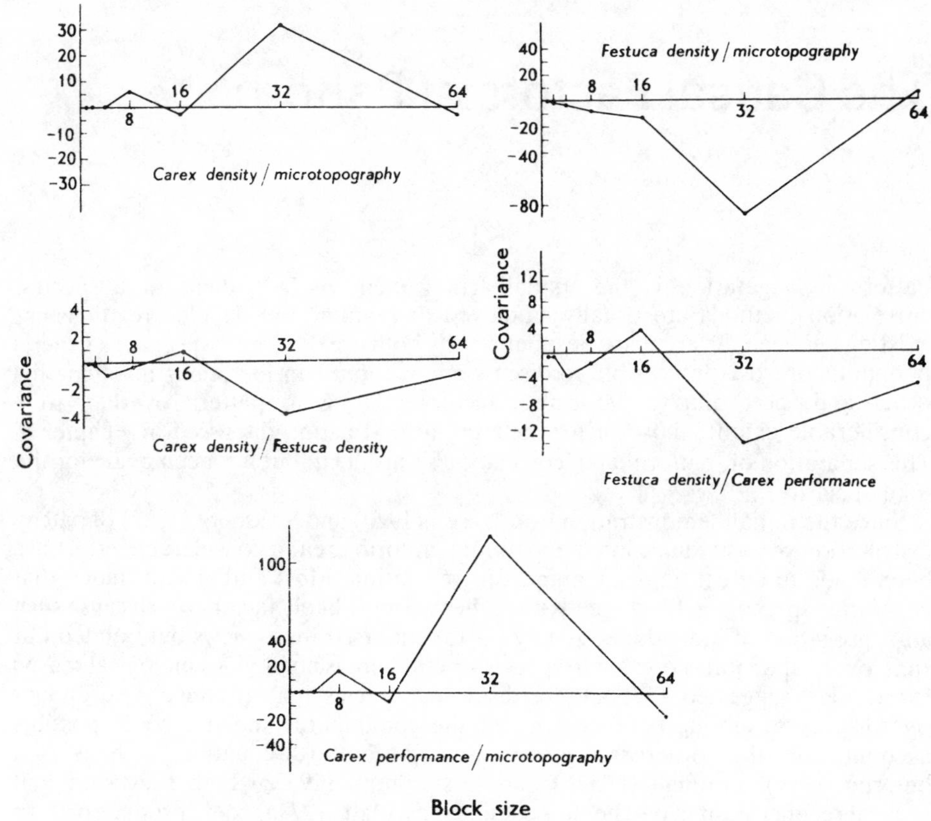

Fig. 7.11 The relationships between *Carex* density, *Carex* performance, *Festuca* density and microtopography, expressed as covariance (see text). (From Kershaw, 1962*c*; courtesy of *J. Ecol.*)

8

The Causal Factors of Pattern

Pattern in vegetation is the spatial arrangement of individuals of a species; correlation methods are usually employed to examine the detailed relationships existing between different species, but clearly both are different aspects of a general phenomenon. It is impossible to discuss one without considerations involving the other and, accordingly, the causal factors relating to pattern overlap, to a considerable extent, those factors relating to correlation discussed in Chapter 5. This separation of 'pattern' and 'correlation' is artificial, but has been made for the sake of clarity.

Since the initial demonstration by Gleason (1920) and Svedberg (1922) of patchy distribution of individuals in an apparently uniform area, a considerable effort has been made to detect non-randomness in vegetation. Most authors concluded that vegetative spread and heavy seeds were the two most likely factors which cause such an aggregation of individuals. Ashby (1948) endorsed these views but pointed out that 'overdispersion is present in species where there is no obvious biological reason for it'. He suggested that an overdispersed species may produce a secondary overdispersion among other species of the community, and this could possibly account for the otherwise inexplicable pattern (discounting obvious soil heterogeneity). Goodall (1952*b*) and Greig-Smith (1952*a*) both suggested that cyclical regeneration in patches as postulated by Watt (1974*a*) could produce pattern in vegetation, but there still remained numerous cases where an explanation was difficult. Following the development of techniques of pattern analysis (Greig-Smith 1952*a*, 1961*a*; Kershaw, 1957, 1960*a*), it is apparent that a very wide range of pattern scales are present in vegetation. Some of these patterns probably reflect a correspondingly wide range of variation of some factors of the environment. Sometimes it has proved impossible to understand all the factors responsible for a certain pattern and often a scale of pattern is related to one of the more obvious or more easily measured of a group of closely interrelated environmental factors. In addition to the original summary of Kershaw (1963), Greig-Smith (1979) has recently provided an excellent updated summary of the range of pattern scales occurring and their causality.

Morphological pattern

Where a species has a densely tussocked form, or where the basal leaves form a rosette, then the smallest pattern present in an analysis will be a measure of the mean size of an individual. In these cases it is clear that the morphology of a species will produce small scale pattern and, further more, this scale of pattern will be imposed on other species present. An example of apparent spatial exclusion is given by Kershaw (1958) for *Agrostis tenuis*, which showed a marked scale of pattern at 3.2 m. *Dactylis glomerata* and *Lolium perenne* also showed this same scale of pattern but were negatively associated with *Agrostis*, and thus appeared to be spatially excluded. In fact the distribution of these species was controlled in this area mainly by the depth of soil, *Agrostis* growing more densely on the shallower areas of soil, and the other species on the areas of deeper soil. Thus pattern which would appear at first sight to be due to spatial exclusion of one species by another, may in fact be controlled by independent factors of the environment.

Where the species under consideration has an extensive rhizome system, several scales of pattern are usually evident as the result of the characteristic system of branching and leaf-production of that species. Phillips (1954*a*) showed for *Eriophorum angustifolium* three scales of pattern that could be attributed to morphology (Fig. 8.1). The morphology of *Eriophorum* gives a characteristic grouping

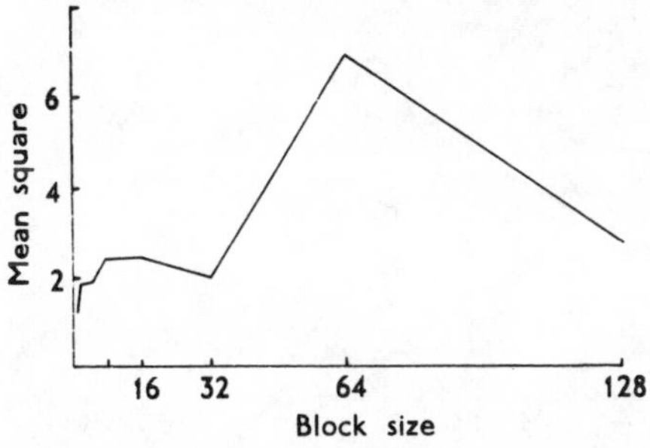

Fig. 8.1 These scales of pattern in *Eriophorum angustifolium* represented by three peaks in the graph mean square/block size. (From Phillips, 1954*a*; courtesy of *J.Ecol.*)

of aerial shoots (Fig. 8.2) separated by a length of rhizome which is equivalent to a year's increment of growth. The double primary peak is due to the grouping of shoots round the parent shoot from backwardly directed short rhizomes giving a peak at block size 2 (200 cm^2) and a larger scale of grouping due to longer rhizomes directed

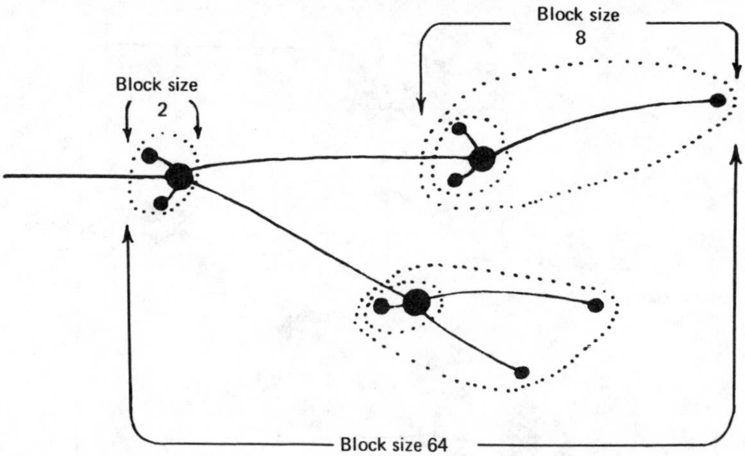

Fig. 8.2 Diagrammatic representation of the three scales of morphological pattern in *Eriophorum angustifolium*.

forwards and giving a peak at block size 8 (mean area 800 cm^2). The large peak at block size 64 reflects the mean size of a complete individual (mean area 0.64 m^2). A similar pattern series was detected in *Trifolium repens* (Kershaw, 1959) with three scales of pattern (Fig. 8.3) equivalent to first and second order branches and to the complete stolon system of the plant (Fig. 8.4). Numerous other examples of morphological pattern have been described (Agnew, 1961; Anderson, 1961*a*, 1961*b*; Greig-Smith, 1961*a*, 1961*b*; Kershaw, 1958, 1959; Kershaw and Tallis, 1958), and in general, scales of pattern at small block sizes may be attributed to the morphology of the species.

The pattern due to the morphology of the rhizome can give useful information as

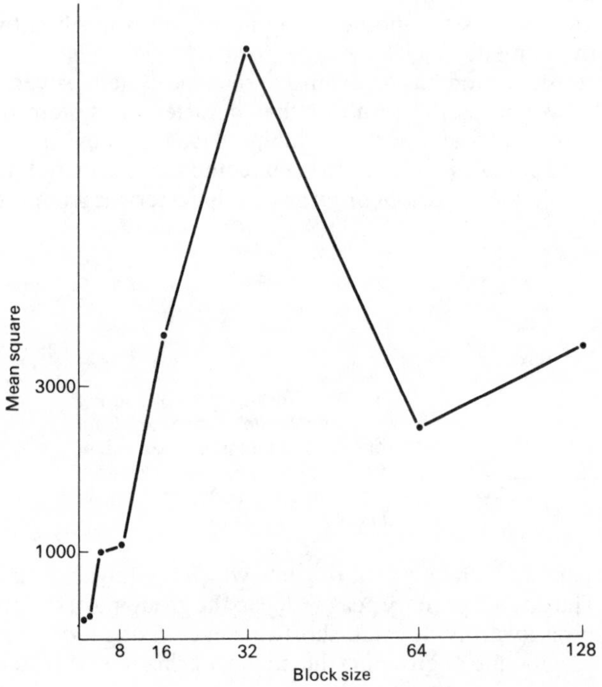

Fig. 8.3 Three scales of pattern in *Trifolium repens* shown by three peaks, at block size 1, 4 and 32, on the graph mean square/block size.

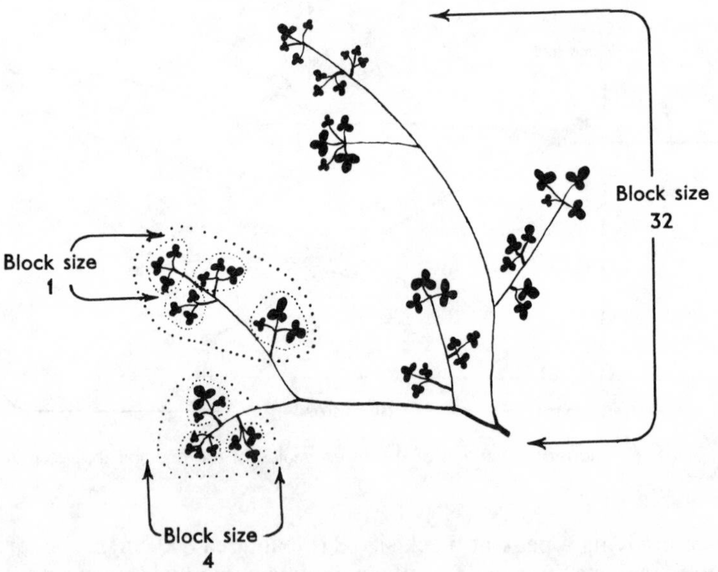

Fig. 8.4 Diagrammatic representation of the three scales of morphological pattern in *Trifolium repens*.

to the level of performance of a species in different areas. It is clear that a measure of performance such as the mean annual extension of a rhizome system gives a good indication of the vigour of a plant and one which would be markedly controlled by the environment. Thus a series of pattern analyses from different plots over the range of distribution of a species should show a progressive shift of peaks which would reflect the changing morphology. Phillips (1954*a*) demonstrated such a variation of

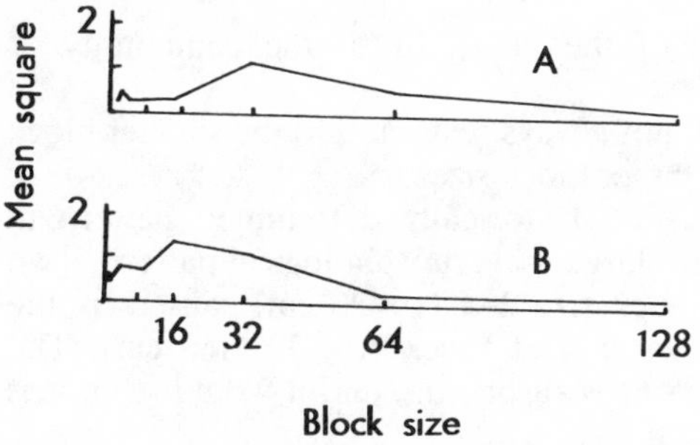

Fig. 8.5 The variation of pattern scales of *Eriophorum angustifolium* in two communities, *Calluna vulgaris, Erica cinerea* (A) and *Trichophorum caespitosum* (B). (From Phillips, 1954*a*; courtesy of *J. Ecol.*)

morphology in *Eriophorum* growing in different plant commmunities; the scales of pattern reflecting the morphology of the rhizome were very much smaller where the habitat was less favourable. (Fig. 8.5 and compare Fig. 8.1 above.) Anderson (1961*a*) and Kershaw (1959) discuss similar effects in *Vaccinium, Calluna, Pteridium* and *Trifolium* and it is apparent that valuable information can be obtained from comparisons of this sort.

The analysis of morphological pattern can also be related to the age of a rhizome system, as the pattern and size of branches vary with age and correspondingly alter the trend in a series of pattern analyses. The morphological pattern of *Agrostis tenuis* colonizing a derelict sand pit shows such a progressive change (Kershaw, 1958). The patches of sampled *Agrostis* varied in diameter from young clumps a few centimetres in diameter to large circular patches. The scales of pattern detected corresponding to the diameter of the patch are given below (Table 8.1). The three scales of pattern

Table 8.1 The scales of pattern present in analyses of tiller density from clumps of *Agrostis* of varying diameter. (From Kershaw, 1958; courtesy of *J. Ecol.*)

Clump width (cm)	Density of tillers	Peaks			
97	1.26	2	8	32	
107	1.42	2	8	64	
114	1.56	2	8	64	
130	1.74	2	8		128
137	1.72	2	8–16		128
229	1.90	1	8	32	128

(block sizes 2, 8 and 32, equivalent to 13, 52 and 208 cm^2) in the smallest clump correspond to first order branch and second order branch and to the complete rhizome system respectively. As the clumps increase in diameter the size of the complete rhizome system progressively increases, but the primary and secondary scales are simultaneously maintained. At the largest clump size investigated the maximum scale of pattern and the smaller scales of pattern are maintained and there appears a fourth scale of pattern which is caused by the primary branch systems separating away from the parent rhizome by decay of the older parts and appearing as established individuals. These 'fragments' will presumably then follow the same cycle of events.

Thus the morphological pattern, though usually expected at the smaller block sizes, if of considerable interest and can be often helpfully used to demonstrate the relationship between the morphology of the individuals and the environment, or conversely a progressive change in morphology with time. It should be pointed out that not all small-scale patterns are necessarily morphological patterns, and in fact pattern at block sizes 1–2 (mean dimensions 5–10 cm) has in several instances been attributed to other intrinsic properties of the plant, or of the community or environment (see below).

Conversely morphological patterns are not always restricted to the smaller block sizes; they can equally be detected at the larger block sizes.

Chadwick (1960) describes an interesting historically determined pattern in *Nardus stricta*, which is best described as a large-scale morphological pattern. Two scales of pattern are detected, one at block size 2–4 (10–20 cm) caused by the morphology of *Nardus*, and a secondary peak at block size 32 (160 cm). The consistency of the secondary peak (Fig. 8.6) is surprising, and it would seem it is

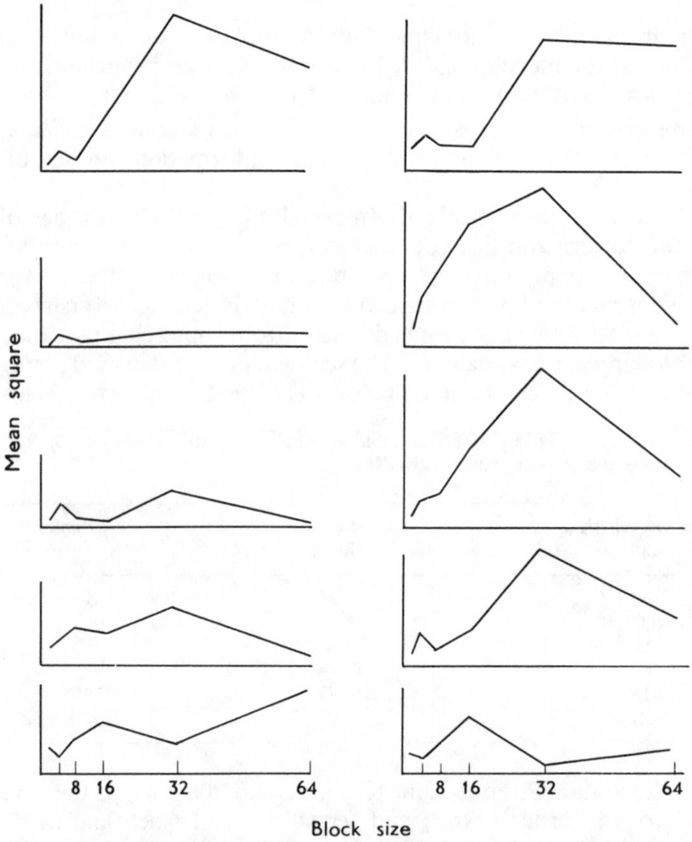

Fig. 8.6 The pattern scales of *Nardus stricta* in a number of upland grassland sites (see text). (From Chadwick, 1960; courtesy of Blackwell Sci. Pub.)

therefore not correlated with an environmental factor which would in fact fluctuate from site to site. Chadwick interprets this scale of pattern as the size of 'clone', reflecting the changes of rhizome growth since the grazing management of the Welsh mountains was changed. Thus with the advent of rationing in the first world war and the demand for smaller joints of meat the hill farmers changed over from old wethers which had effectively controlled the spread of *Nardus* by grazing, to breeding ewes which are much more selective in their grazing and do not touch *Nardus* unless more palatable species are absent. In response to this change the *Nardus* was able to spread and has actively done so ever since. Examination of the rhizome morphology shows a multidirectional growth, the old rhizome dying away behind and giving rise to a number of separate clumps each of which continues (as an individual) the centrifugal growth of the colony. The annual increment of growth is about 2 cm which over a period of 40 years would result in a scale of pattern of diameter of 160 cm equivalent in fact to the pattern at block size 32.

Austin (1968) examining the pattern in *Zerna erecta* growing on the North Downs in Kent, found very marked levels of contagion at the larger block sizes ranging from block size 32 to 128. The results were interpreted in an analogous way to Chadwick's *Nardus* data. Using an extensive series of turf dissections together with a vector analysis of the data in each block dissected, Austin demonstrated a pattern of rhizome orientation consistent with growth from centres of colonization (Fig. 8.7). The

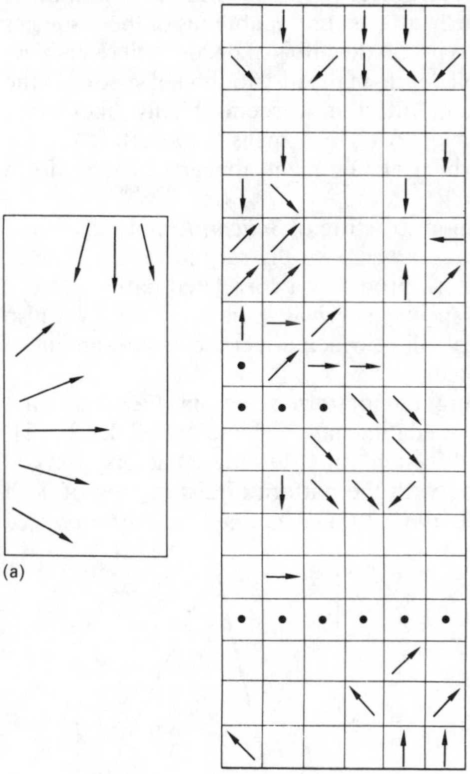

Fig. 8.7 (**a**) Proposed orientation of stolons in the turf. (**b**) Observed average orientation of stolons after vector analysis within each grid unit. (From Austin, 1968; courtesy of *J. Ecol.*)

spreading rings could be variable in size and since the length of 'intercept' of the randomized sample transects could also vary markedly, the observed variation of this pattern scale is understandable.

Environmental pattern

The effects of major discontinuities of the environment on vegetation are usually well marked and, though such reaction of the vegetation result in patterns in the true sense of the word, these major effects, often on a large scale, are obvious as a change of floristic composition. Such a pattern is readily visible without recourse to quantitative method and calls for little comment. It has become increasingly evident that extremely small variations of an environmental factor (or factors) operating over quite large areas will produce a corresponding variation in the vegetation. Such patterns usually become apparent at the medium and larger block sizes in any particular area. *Thus environmental pattern is developed in vegetation in response to a general and overall variation of one of the major environmental factors and produces a pattern of density distribution.* The first demonstration of environmentally determined pattern showed the close relationship between the pattern of *Agrostis tenuis* and soil depth in an upland grassland area which was mentioned above (Kershaw, 1958). The associated species *Dactylis, Lolium* and *Trifolium* were more abundant on the deeper area of soil while conversely *Agrostis* was more dense on the shallower soil and their pattern, though on the same scale, was in fact an inverse pattern. The mean rooting depths of these species under normal conditions suggests a possible explanation of this pattern: *Agrostis* normally produces a shallow root system whilst the other species root at much greater depths. Thus the limitation of root development in *Dactylis*, etc., seriously affects the capabilities of these species and accordingly they are only likely to survive on the areas of deeper soil (Kershaw, 1959). The effect of a small variation in soil depth on the distribution of species in the area was initially somewhat unexpected, but it now seems highly likely that numerous other soil factors will be shown to have a measurable effect on the distribution of species within an area which at first sight appears to be uniform vegetation.

Brereton (1971) examining the population structure of *Salicornia* and *Puccinellia* in salt marshes in N. Wales concluded that it was basically related to soil texture. Variation of soil texture in a spatial pattern produces a correlated pattern of soil moisture level which, in turn, controls species survival. Soil moisture level also affects soil plasticity, aeration and salinity, all of which affect the establishment of seedlings and their subsequent performance.

Anderson (1961*a* and *b*), investigating the pattern scales in *Pteridium* and *Vaccinium myrtillus*, examined a number of edaphic fators including soil depth, pH, total exchangeable bases and oxygen diffusion rate. Of these factors, oxygen diffusion rate showed a strong correlation with the pattern of distribution of both *Pteridium* (Fig. 8.8) and *Vaccinium*. Hall (1967; 1971) examined the pattern scales

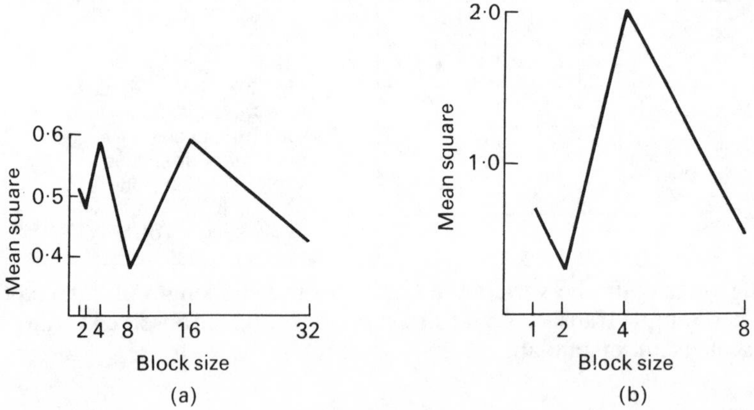

Fig. 8.8 (**a**) Analysis of *Pteridium* frond density data and (**b**) the corresponding soil oxygen diffusion data. (From Anderson, 1961; courtesy of *J. Ecol.*)

and their interactions in *Brachypodium pinnatum, Festuca rubra*, and *Bromus erecta* on the South Downs in Kent. He revealed an extremely complex situation with several marked pattern correlations with soil nitrogen, iron and boron. The pattern data for *Festuca* are shown below (Fig. 8.9) together with the marked interactions with soil iron and boron. Both these plant trace-nutrients are known to be usually of low availability in calcareous soils (Russel, 1961), and it is understandable that they are involved in the control of the pattern of at least the common chalk grass species. Hall

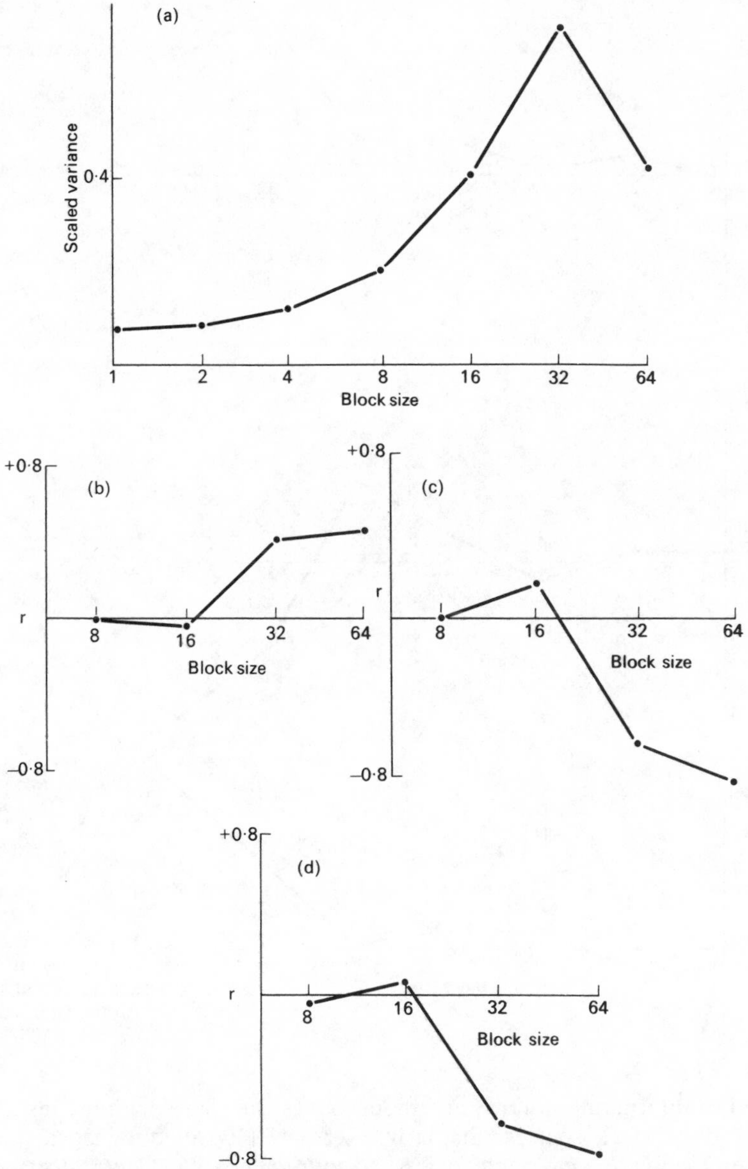

Fig. 8.9 (**a**) The scale of pattern in *Festuca* with (**b**) the correlated pattern of soil iron and (**c**) the correlated pattern of soil boron. (**d**) The correlation of pattern scales between soil iron and soil boron levels. (From Hall, 1967; Ph.D. Thesis, London.)

also shows an overall control of pattern by soil depth with a resultant species interaction which is very clear cut (Fig. 8.10). Clearly soil factors are of fundamental importance in the control of pattern scales, but it is only recently that specific soil parameters have been correlated with particular pattern scales. This largely results from the difficulties of analysing soil properties in large numbers of replicates. In many situations soil properties correlating with pattern scales will represent a complex of individual parameters. Where pattern is related to obvious surface micro-topographical variation a similar complex of interacting factors will also operate. For example Greig-Smith (1961*b*) showed that the pattern of soil level

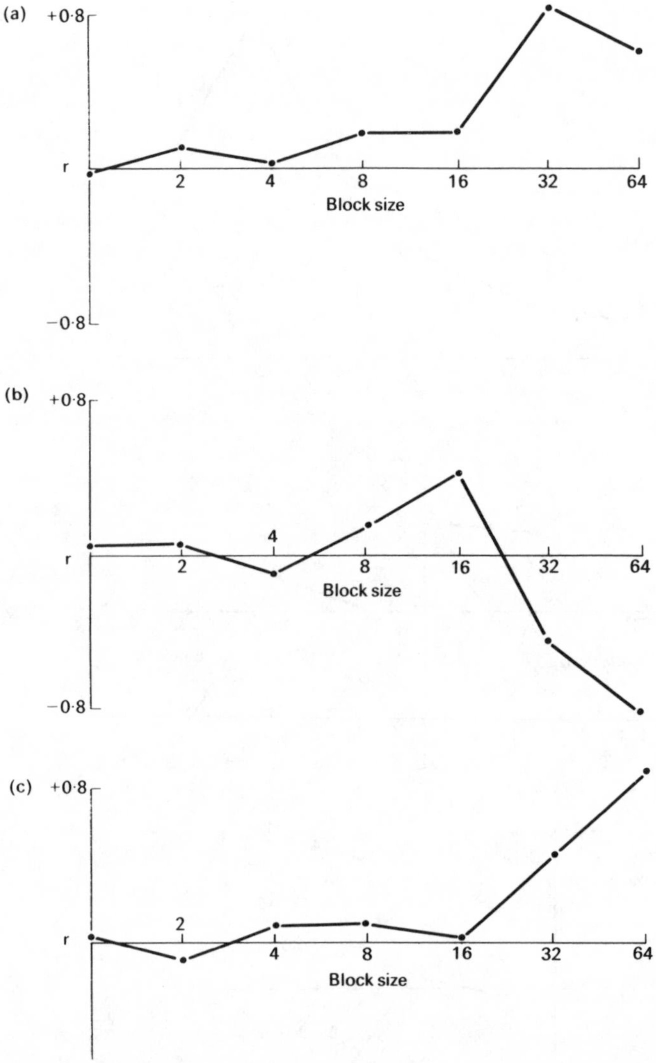

Fig. 8.10 The correlation of soil depth and pattern scales in (**a**) *Festuca* (**b**) *Bromus*, (**c**) *Brachypodium*. (From Hall 1967; Ph.D. Thesis, London.)

formed of alternating hollows and mounds was correlated with the distribution of several dune slack species. Similarly, Kershaw (1962*c*) obtained a relationship between the micro-topography of a *Rhacomitrium* heath in Central Iceland and the distribution of *Carex bigelowii* and *Festuca rubra* (see above). In some reported investigations there is a marked pattern at the largest block size caused by an almost imperceptible slope in the area sampled and such a large and intense scale of pattern may often obscure smaller patterns due to other causes (see Greig-Smith 1961*a*). Similar, but more obvious, patterns have been reported by Billings and Mooney (1959) and Warren Wilson (1952).

Thus it seems likely that drainage, availability of water, leaching, nutrient supply and pH would tend to vary together with changes in micro-topography and it would be difficult, if not impossible, to decide which factor or group of factors was directly controlling the distribution of a species. However, it is suggested that a slight fluctuation of pH or ion concentration would result in a detectable vegetational pattern, and it is interesting that Pigott has demonstrated considerable pH gradients over very short distances in the Burren Peninsula in Ireland. Snaydon (1962) has shown a similar steep pH gradient and also a gradient of calcium, phosphate and potassium concentration, varying by a factor of up to 3 in a distance of only 60 cm. It seems possible that similar or even steeper gradients will prove to be a general characteristic of undisturbed soils, producing the corresponding variations in the vegetation (see also Chapter 5). Thus the demonstration of a correlation between soil oxygen diffusion rate, and concentration of soil iron or boron (see above) are only three of the probable soil factors controlling pattern. Undoubtedly micro-gradients of other parameters will in the future be demonstrated as important in the control of pattern.

Other environmental parameters are of equal importance in controlling pattern scales. Kershaw and Rouse (1971) examined the pattern in the homogeneous and often pure stands of the ground lichen *Cladonia alpestris*, in spruce-lichen woodland in N. Ontario. It had been shown previously that the lichen mat acts as an extremely efficient mulch, the moisture content of the soil layers underlying a sample quadrat effectively denoting the relative amounts of integrated net radiation at that point, over the preceding period of weeks. Furthermore, since the metabolism of a lichen is entirely dependent on it being wet or at least moist, the relative drying rate of the lichen canopy will also control the growth pattern of the lichen. Pattern analysis of the depth of lichen mat, soil moisture (as a measure of incoming energy and hence drying rate of the lichen) and per cent cover of an associated woody shrub species *Ledum groenlandicum*, are given below (Fig. 8.11). The covariance data (Fig. 8.12)

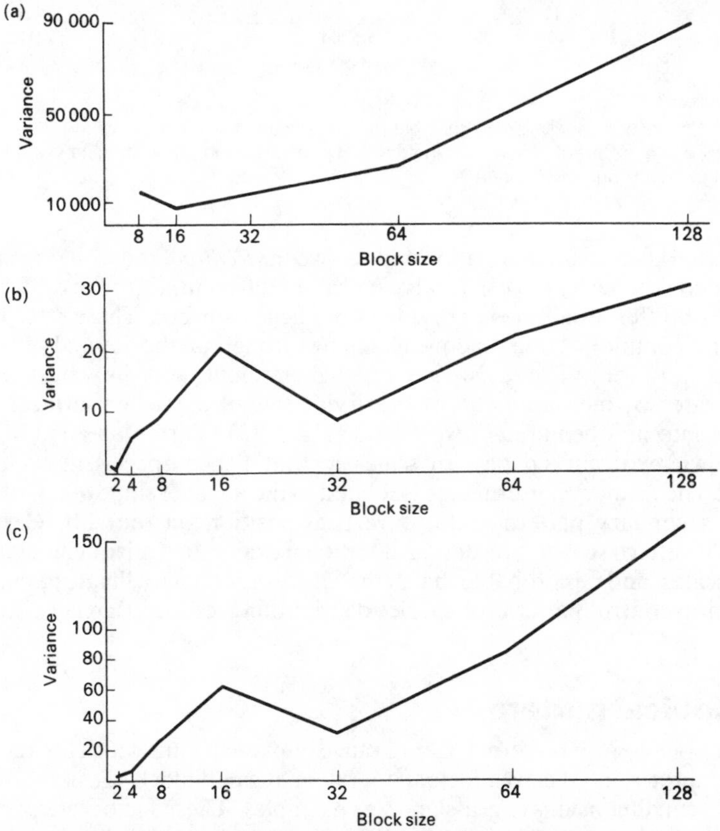

Fig. 8.11 The pattern analysis of the ground layer of uniform spruce-lichen woodland. (**a**) Soil moisture; (**b**) *Ledum groenlandicum*; (**c**) depth of lichen mat. (From Kershaw and Rouse 1971; reproduced by permission of the National Research Council of Canada from *Can. J. Bot.*, **49**, pp. 1389–1399).

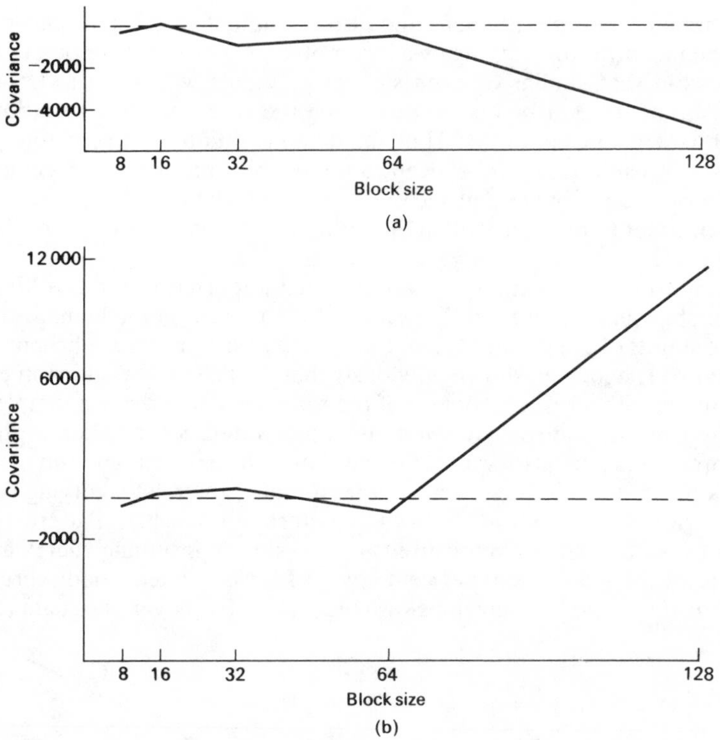

Fig. 8.12 Covariance analysis of the pattern interactions of soil moisture with (a) *Ledum groenlandicum*; (b) depth of lichen mat, in spruce-lichen woodland. (From Kershaw and Rouse 1971; courtesy of *Candian Journal of Botany.*)

show a marked positive relationship between *Ledum* and soil moisture. The homogeneous lichen surface in fact is a mosaic of differential growth rates induced by a mosaic of differential rates of drying of the lichen surface. These rates in turn are related to a complex of tree shadows changing throughout the day and year as the sun alters its position. The existence of this gradient and its effectiveness was demonstrated by measurements of the drying rate of the lichen surface, at varying distances into an open area of woodland (Fig. 8.13). Yarranton and Green (1966) have shown two marked pattern scales in four lichen species growing on rock surfaces. The primary pattern scale is related to the angle of slope of the rock surface and the secondary pattern scale to vertical position on the cliff. Certainly the secondary pattern scale represents an additional example of micro-climate control of pattern scales, and it is probable that in the future other microclimate parameters will be shown to control patterns of species distribution (see also Barclay-Estrup, 1971, p. 75).

Sociological pattern

The term 'sociological pattern' covers a range of pattern units which are the products of several inter-related causal factors operating at small block sizes (up to a pattern scale of 80 cm dimensions in grassland, for example). These factors are partly due to intrinsic properties of the plants themselves and partly due to the micro-environment. Thus sociological pattern is a product of the interaction of species on species and individual on individual which may or may not be directly modified by micro-environment.

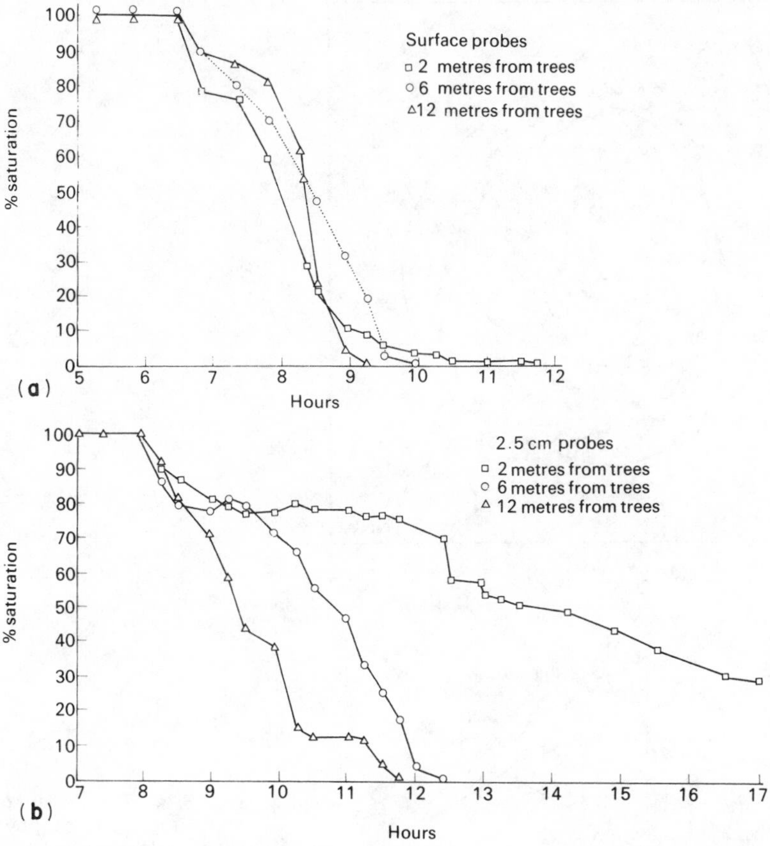

Fig. 8.13 Drying rate of the lichen canopy at (a) the surface, (b) 2.5 cm down on three sites, 12 m, 6 m, and 2 m into an open area in spruce-lichen woodland. (From Kershaw and Rouse, 1971; courtesy of *Can. J. Bot.*)

These small-scale sociological effects often take the form of a mosaic of patches of different density levels dynamically related to each other in time and space. The causal factors of sociological pattern are not only dependent on the competitive ability of an individual (which may also be, in part, dependent on the micro-environment), but also on the possible presence of toxins exuded by an individual, and its age. Several of these relationships can be examined by correlation methods and have been discussed previously (Chapter 5). However, they are usually evident in studies of pattern and are included here for completeness.

Almost every effect of one species on another (excluding parasitism) is in the form of a modification of the environment and it is possible that the environmental pattern which limits the distribution (in terms of density) of a species was in the first place imposed on the environment by another species. Zinke (1962) investigated the modification of the environment under *Sequoia gigantea, Pinus contorta* and other species. These modifications were expressed by pH, nitrogen content and exchangeable bases in the surface mineral soil. There was a marked radial relationship between these soil factors and tree cover (Fig. 8.14) and it is readily understood how such a pattern of environmental factors would impose a corresponding pattern of ground vegetation detectable for several years after the removal of the tree cover. The steady accumulation of leaf litter and corresponding increase in soil nitrogen is a well marked and expected result of tree establishment in a given site. It is very likely that similar less obvious gradients are developed in

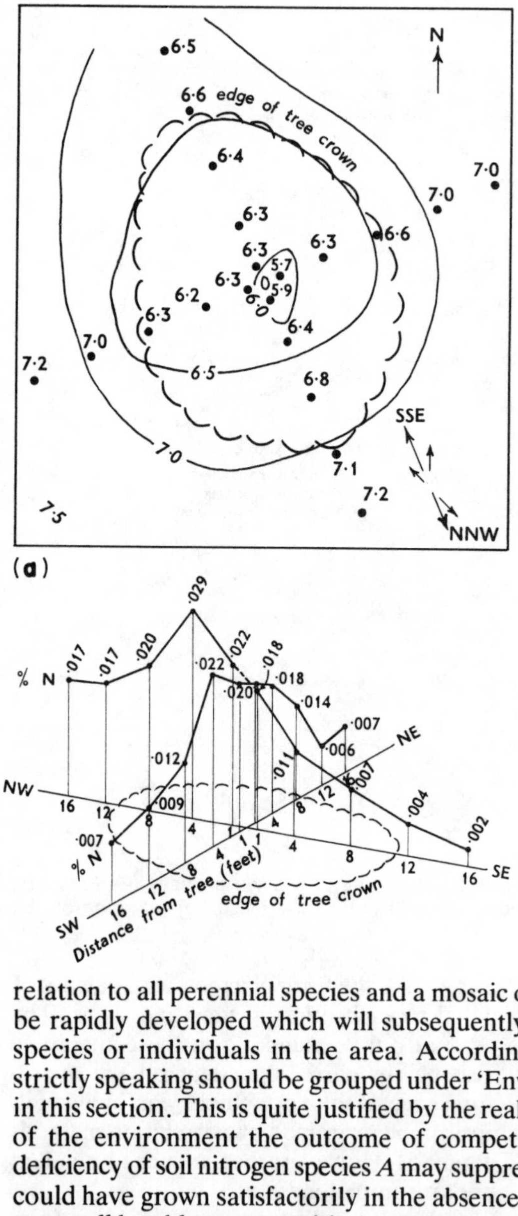

Fig. 8.14 The gradient of pH (A) and nitrogen (per cent by weight) (B) of surface mineral soil under a tree of *Pinus contorta*. The relative proportions of wind directions in the area are indicated. (From Zinke, 1962; courtesy of *Ecology*.)

relation to all perennial species and a mosaic of micro-environmental variations will be rapidly developed which will subsequently influence the establishment of other species or individuals in the area. Accordingly some small-scale patterns, which strictly speaking should be grouped under 'Environmental Pattern', may be included in this section. This is quite justified by the realization that under different conditions of the environment the outcome of competition may be different: thus under a deficiency of soil nitrogen species *A* may suppress species *B* in competition although *D* could have grown satisfactorily in the absence of *A*; at a high level of soil nitrogen *B* may well be able to grow with *A* or even suppress it.

Since Watt (1947*a*) outlined several systems in which species were related in cyclical phasic development (see Chapter 4), such cyclic phases have been suggested as one of the possible causes of pattern in vegetation. Relatively few further examples of cycles involving several species have been reported in the literature and it would appear that such cycles are rather exceptional and therefore one has to look elsewhere for a general explanation of patterns which are not causally related to environment and morphology. On the other hand the description of cyclical *density* phases of *Pteridium* in a pioneer, building, mature and degenerate series (Watt, 1947*a* and *b*) (pp. 63–5), where the performance fluctuates, has been confirmed by several workers since (Chapter 4). This is distinct from cycles in which the actual species change in a recognizable sequence. Watt (1955) shows that *Calluna* follows the same phasic development and furthermore that *Pteridium* as an associated species is inversely related in its distribution to *Calluna*. Thus where *Calluna* is at its building/mature phases, bracken is at its minimum development; conversely where *Calluna* is in its degenerate phase bracken is very much more in evidence. Concurrently, the competitive ability changes with the age series described as pioneer, building, mature and degenerate.

The phasic development of *Pteridium* and *Calluna* have been confirmed by Anderson (1961*a* and *b*), Nicholson and Robertson (1958) and the 'marginal effect' of 'advancing front' as described by Watt (1947) and representing the pioneer and building phases, has also been reported in several additional species. Thus Ovington (1953), Caldwell (1957), Pénzes (1958; 1960) and Kershaw (1962*a*) have described many examples of advancing fronts of even-aged rhizomes at an optimum level of performance (p. 67). Quantitative data showing the relationship between performance and age in *Alchemilla alpina* and *Carex bigelowii* (Kershaw 1960*b*, 1962*c*) (Figs. 4.11 and 4.12) together with the general observations on the occurrence of advancing fronts, suggest that the performance of most if not all perennial plants will vary with age. Correspondingly the variation of competitive ability with age (Watt, 1955) is likely to be a general phenomenon. Data on the variation of density of *Salix herbacea* as it is invaded by an 'advancing front' of *Eriophorum angustifolium* are given by Kershaw (1962*a*) and confirm the relationship between age and competitive ability. The variation in density of *Salix* taken from a line of contiguous quadrats running through a circular patch of *Eriophorum* actively invading a *Salix herbacea* sward has been discussed previously (p. 69) and the data show the density of *Salix* to be negatively related to the density of *Eriophorum* (Fig. 4.9). Furthermore the performance of *Eriophorum* (leaf length) is negatively related to the density of *Salix*. Thus when *Eriophorum* is at an optimum level of performance and density, the density of *Salix* is at a minimum. Austin (1968) in a similar situation has provided evidence of competition between *Zerna erecta* and *Carex flacca*. The internode lengths of the rhizomes of *C. flacca* were markedly shorter when growing within the clumps of *Z. erecta* than elsewhere ($\bar{x} = 1.32 \pm 0.11$ and $\bar{x} = 2.48 \pm 0.18$) respectively, $P < 0.1$ per cent). Since performance varies with age, and the ability to compete varies with the general level of performance of an individual, most if not all perennial plants will have gradually diminishing competitive ability with increase in age. Age as a causal factor of pattern would appear to be of considerable and widespread importance in perennial species, and a mosaic of *density phases* of a species imposed by the varying competitive ability of an associated species is of frequent occurrence and quite independent of any other environmental consideration. The findings of Barclay-Estrup and Gimingham (1969), and Barclay-Estrup (1971), showing strong microclimatic variability induced by the different cyclic phases in *Calluna*, strongly supports this conclusion (see Figs 4.13, 4.14, etc.). It also indicates the rather artificial boundary drawn in some instances between 'environmental patterns' and 'sociological patterns'.

A similar pattern effect has been described by Cooper (1960, 1961) who suggests that pattern in Ponderosa pine stands results from cyclical development governed by fire. As an old-aged stand degenerates young trees invade and establish, rapidly forming an even-aged group. Such groupings are maintained by the inability of young pines to establish underneath mature parents, and also by periodic fires which remove any young small scattered saplings that do occur. An even-aged stand once established remains as a single unit until the degeneration phase is reached. An area of pine thus consists of a mosaic of phases at different density levels, each group being of a different age.

Sociological pattern is also caused by the effect of an individual on the environment which can be quite independent of the age factor of the individual. The reaction produces a modification of the environment which reduces or enhances the chance of survival of progeny either of the same species or of other species merely, for example, by the annual deposition of leaf litter (see p. 150, Zinke, 1962).The interpretation of such instances is by no means as clear cut as would be imagined. The dependence on shrub cover for some herb species can be interpreted as a response to the accumulation of leaf litter or shelter, or protection from grazing, or protection from grass competition, etc., but considerable argument has developed around this phenomenon as to the likelihood of the direct inhibition of associated species by certain shrub species. The case of *Encelia farinosa* is capable of two apparently reasonable interpretations. The lack of ground vegetation under the shrub may be a reflection of toxic substances in its leaves but it may equally reflect the failure of this shrub to accumulate beneath it a fine soil tilth. Other species which contain equally powerful toxins in their leaves and yet can accumulate leaf litter and good soil tilth may also have a dense associated ground flora. The existence in the field of a chemical substance actively exuded by plant roots, in quantities sufficiently great to control the establishment or development of associated species, has yet to be demonstrated (see also p. 83), but the possibility of such control of one species by another seems sufficiently probable to necessitate its inclusion in any consideration of pattern.

A very puzzling feature of some rhizomatous species has been reported by Kershaw (1958; 1962*b*) where the aggregation of aerial shoots, arising from different and separate rhizome systems as small 'clumps' 5–10 cm in diameter, is responsible for what at first sight would be regarded as merely a very small-scale morphological pattern.

A typical arrangement of rhizomes and tillers is given in Figs 8.15 and 8.16 for an area of both recently established and mature *Calamagrostis neglecta* in central Iceland (Kershaw, 1962*b*). The pattern is quite marked, the small clumps being composed of at least three, and often four or five different rhizome systems. It has been suggested that the consistent orientation of rhizomes and the production of tillers in the form of a clump, may be related to micro-gradients of some environmental factor as yet undetermined, down which the rhizomes grow; finally, tillering occurs at a point where the factor responsible is at its maximum level. However, the possibility does exist that a more orderly and positive mechanism exists. It could be due to the chemical stimulation of a rhizome apex by an established aerial shoot and accordingly this type of pattern is included temporarily at least as a sociological pattern. Whatever the mechanism, it should be emphasized that small-scale pattern should not be immediately dismissed as morphological pattern.

Contagious distribution of individuals (clumping) in vegetation is a common feature, yet little evidence has appeared to suggest that regular distribution of individuals (underdispersion) is of more than rare occurrence in vegetation. The most likely situation which would produce regular distribution is one where there is a high density of individuals within a uniform area; the resulting active competition would limit both the total density over the area, and at the same time the spatial distribution of an individual relative to its neighbours. The lack of data demonstrating regular distribution suggests that either the theoretical conditions leading to competition between individuals for environmental factors are rarely fulfilled, *or are over-shadowed by extreme variations of environmental factors which are able to impose a marked pattern on the species as a whole*. Thus it should be emphasized that even though there is marked contagion at one particular block size, at a similar or larger block size there may be an indication of regular distribution of individuals or of pattern units. As Greig-Smith (1952*a*) states, 'Professor Bartlett (personal communication) has made a preliminary examination of the problem and his provisional conclusions may be briefly summarized as follows: (1) Variance may

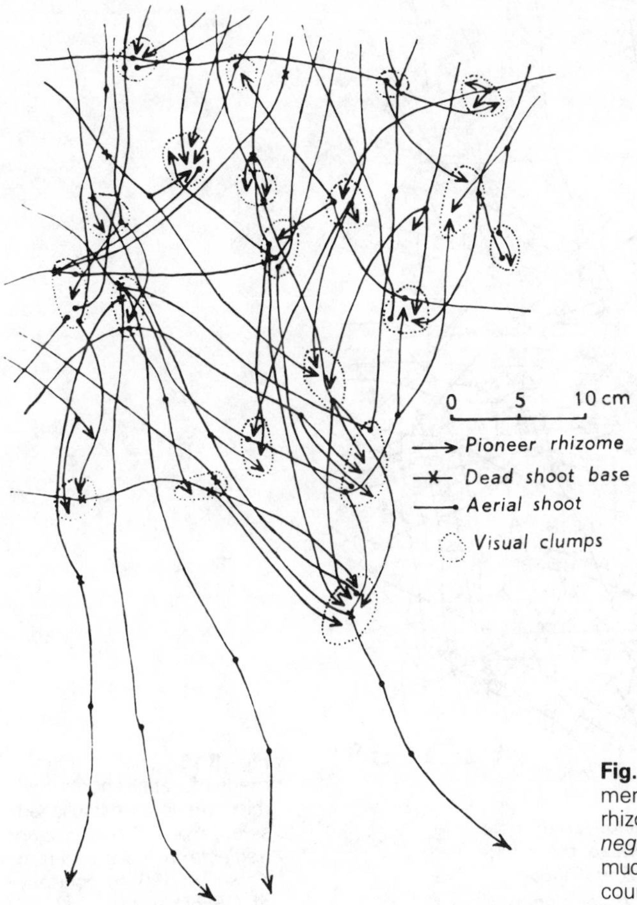

Fig. 8.15 The arrangement of aerial shoots and rhizomes of *Calamagrostis neglecta* invading bare mud. (From Kershaw, 1962*b*; courtesy of *J. Ecol.*)

be expected to rise to a maximum value as block size reaches the *mean area of clump*, diminishing thereafter if the clumps are underdispersed . . . .' Thus if there is a sharp fall of variance following a marked peak, it can be assumed that the units of pattern have a regular distribution in the area examined.

It is important, however, to bear in mind that such a regular distribution of pattern units is relative to the size of the area investigated, and it is almost certain that an increase in size of the sampling area would show the existence of higher scales of pattern. The evidence to date suggests that frequently there occurs a regular distribution of pattern units up to a limited scale beyond which a further scale of contagion appears. Thus regular distributions of individuals or groups of individuals, rather than being completely absent, are over-shadowed by the scales of contagion present.

In general it is difficult to categorize causalities which are included here under 'sociological' factors for it is virtually impossible to discriminate in the field between effects due to competition and effects due to micro-variations of the environment. The demonstration of positive or negative association between species is straightforward, but determination of competition between species very often necessitates experimentation (see Chapter 3). In certain instances some indication can be obtained from field data: thus Kershaw (1959) showed in *Dactylis* and *Lolium* a sociological scale of pattern at 80 cm. His decision was based on a significant negative association at high densities and nil association or even a slight positive association at

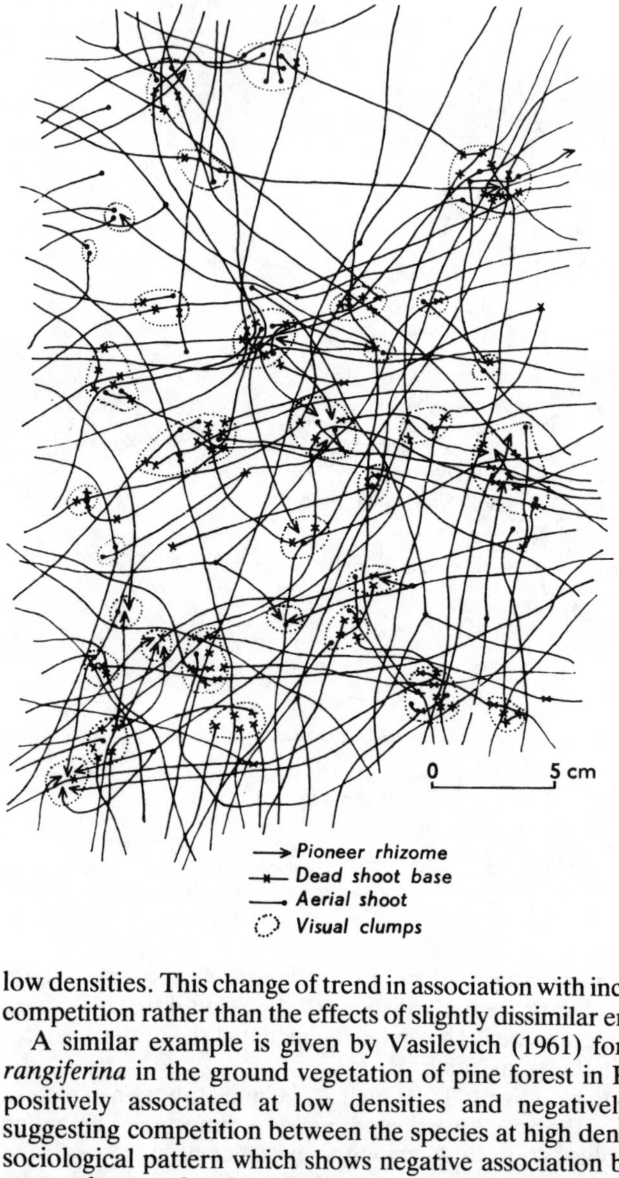

Fig. 8.16 The arrangement of aerial shoots and rhizomes in an established sward of *Calamagrostis neglecta* in Iceland. (From Kershaw, 1962*b*; courtesy of *J. Ecol.*)

low densities. This change of trend in association with increasing density does suggest competition rather than the effects of slightly dissimilar environmental requirements.

A similar example is given by Vasilevich (1961) for *Cladonia sylvatica* and *C. rangiferina* in the ground vegetation of pine forest in Russia. The two species are positively associated at low densities and negatively at high densities, again suggesting competition between the species at high densities. A further example of sociological pattern which shows negative association between species which is, in part at least, related to competition is taken from the relationship between *Carex bigelowii* and *Festuca rubra* (Kershaw, 1962*c*) (p. 134). Both the density and performance data of *Carex* are positively related to micro-topography and negatively related to the density of *Festuca*. However, the performance of *Carex* is more markedly affected by *Festuca* than is the density of *Carex*, which again suggests an effect of competition (see also Fig. 7.11).

General considerations

The widespread detection of non-randomness in vegetation, usually in the form of contagion or clustering of individuals, has led to the realization that all vegetation will contain one or many scales of pattern. The only possible exception to this is in the colonization of a homogenous bare area when the initial migrules may be distributed more or less at random. Morphological scales of pattern will always be present, and it is rare to find an area with such a completely uniform environment as not to produce one or more scales of environmental pattern. The presence of pattern in environmentally uniform areas is also to be expected since intrinsic properties of the plants themselves will interact to produce pattern. Thus it is difficult to envisage an area where all the individuals are of an even age and hence have equal levels of performance and competitive ability. Accordingly we are forced to the conclusion that pattern will always be present in vegetation except in a few rare and extreme cases.

9

The Detection of Natural Groupings of Species: Classification Methods

The two opposing schools of thought which on the one hand consider the climax association to be a complex organism or quasi-organism and on the other hand support the individualistic concept, also reflect the present-day ecological approaches to the delimitation and definition of community units. Thus the concept of the complex organism implies considerable interaction between the species which jointly modify the environment and form a distinguishable vegetational group. Conversely the individualistic concept regards no two communities as being identical, but considers that they show a continuous variation and accordingly can not be readily delimited as definable units. Numerous attempts have been made in recent years to extract statistically or objectively definable units from vegetation, by means which range from complex statistical procedures to rather simple subjective methods. On the other hand there has been a growing awareness that classification of units was in many instances an artificial step. Intermediates were being ignored and this variation, very often of continuous form, was important. Accordingly methods of *arranging* samples in order of similarity appeared in the literature, and recently the application of multi-variate techniques to this field has resulted in a bewildering array of coefficients, techniques and terminology. An attempt will be made to cover this area in the following chapters.

Of fundamental importance is the difference of purpose between classification of plant units or arrangement – *ordination* – of samples. The nature of the unit and its spectrum of species is the main aim of classification. Conversely, the continuous variation expressed in an ordination implies environmental control and the interest is centred around the environmental parameters concerned in the ordination.

The Braun-Blanquet school of plant sociology

The aim of the Braun-Blanquet school (Braun-Blanquet, 1932; 1951) is a world-wide classification of plant communities based on a number of concepts which have been critically reviewed by Poore (1955; 1956). The initial attitude of the British ecologists to these methods and to the results was one of extreme criticism, but more recently attempts have been made to understand more fully what the Continental workers are attempting to achieve and the vegetation of the Scottish Highlands has been comprehensively treated by McVean and Ratcliffe (1962) on this basis.

The foundation of the classificatory system is the *association* which is the basic vegetational unit and is an abstraction obtained from a number of lists of species from selected sites in the field.

Table 9.1 gives an association table for the *Rhacomitrium*/*Carex bigelowii* association made up of twelve lists based on the Domin scale with details of each stand incorporated at the head of the table. From such an association table it is possible to pick out those species which occur in a given number of stands, i.e. the *degree of presence* can be assessed for all species. Braun-Blanquet expresses the degree of presence of a species in a five-degree scale:

5 = constantly present (in 80–100% of the stands).
4 = mostly present (in 60–80% of the stands).
3 = often present (in 40–60% of the stands).
2 = seldom present (in 20–40% of the stands).
1 = rare (in 1–20% of the stands).

Thus in the *Rhacomitrium/Carex* association table (Table 9.1) *Carex bigelowii, Dicranum fuscescens, Polytrichum alpinum, Rhacomitrium lanuginosum, Cetraria islandica* and *Cladonia uncialis* occur with a degree of presence 5; similarly *Vaccinium myrtillus* and *Festuca vivipara* have a degree of presence 4, and so on down to such species as *Lycopodium selago, Nardia scalaris, Coriscium viride*, with a degree of presence 1. Where the data of species presence are derived from samples taken from plots of known and limited area the degree of presence is referred to as the *degree of constancy*.

Constancy

It is clear that if a very large sample area is used many more species will be 'constant'; this leads to a discussion of the minimal area concept which will be postponed to a later section (see below). The method of obtaining the lists and measures used have been described fully in Chapter 1, but several further concepts are of considerable and fundamental importance.

Choice of site

The association tables are derived from a large number of *carefully chosen sites* and it is clear that the whole basis of the system is established on a markedly non-random sampling procedure. The first important proviso in the choice of a sample plot is its homogeneity. Dahl and Hadac (1949) define homogeneity in the following way:

> 'A plant species is said to be homogeneously distributed in a certain area if the possibility to catch an individual of a plant species within the test area of given size is the same in all parts of the area. A plant community is said to be homogeneous if the individuals of the plant species which we use for the characterization of the community are homogeneously distributed.'

The accumulated evidence relating to the species distribution patterns present in all vegetation types so far examined (Chapter 8), throws considerable doubt on the existence of the homogeneous plot ideally suitable for formulation of the association table. Dahl and Hadac recognize the difficulty and state:

> 'In nature plant communities are never fully homogeneous, i.e. the density of the species of importance by the characterization of the community, especially the dominating species, have not exactly the same density in all parts of the area. We may thus talk of more or less homogeneous plant communities. Measurements on the homogeneity of the plant community are rarely carried out, but the human eye, badly adapted to measurement but well to comparison, rapidly gives the trained sociologist an impression whether a plant community he has before his eyes is highly homogeneous or not.'

Again this seems extremely doubtful since a large number of patterns have been detected in areas which appeared absolutely uniform even to the most careful inspection. This aspect of the method has also been severely criticized by Goodall (1954*a*) who states '. . . and it is pointed out that no area of vegetation has ever been

Table 9.1 *Rhacomitrium/Carex bigelowii* association table. (From Poore, 1955; courtesy of *J. Ecol.*)

No. of sample plot	520010	520011	520035	520064	520068	520076	520117	520162	520167	520144	520006	520193
Empetrum hermaphroditum	2	–	–	–	4	1	–	–	–	–	–	–
Salix herbacea	–	–	–	–	–	–	3	–	–	–	–	–
Vaccinium myrtillus	5	3	–	–	4	3	3	3	3	4	4	–
V. vitis-idaea	–	1	–	–	2	2	2	–	1	–	–	–
Lycopodium selago	3	–	–	–	–	2	–	–	–	–	–	–
Agrostis canina	–	3	3	–	–	–	–	–	–	–	–	–
A. tenuis	1	–	–	–	2	3	–	–	–	–	–	–
Deschampsia flexuosa	5	–	–	–	–	–	3	–	–	–	–	–
Festuca ovina agg.	3	–	–	–	–	–	–	–	–	–	–	–
F. vivipara	–	–	–	3	3	4	2	3	1	2	2	1
Nardus stricta	1	–	–	–	–	–	–	–	–	–	–	–
Carex bigelowii	4	4	4	4	3	3	3	5	4	4	6	9
Luzula multiflora	–	–	–	–	–	–	–	–	1	–	–	–
L. spicata	–	–	–	–	–	1	2	–	–	–	–	–
Alchemilla alpina	–	–	–	–	–	3	1	–	–	–	–	–
Galium hercynicum	1	–	2	–	–	2	2	–	–	–	2	–
Dicranum fuscescens	1	1	2	2	2	–	3	1	4	3	2	1
D. scoparium	–	1	–	–	–	–	1	2	–	–	3	3
Aligotrichum hercynicum	1	3	–	–	–	–	–	–	–	–	–	–
Pleurozium schreberi	–	–	–	–	–	1	1	–	–	–	4	–
Pohlia? annotina	–	1	–	–	–	–	–	–	–	–	–	–
Polytrichum alpinum	3	–	1	3	2	2	3	3	1	3	3	2
Rhacomitrium lanuginosum	9	7	10	10	9	9	10	10	10	10	6	9
Rhytidiadelphus loreus	–	–	1	–	–	–	–	–	–	–	1	–
?Aplozia sphaerocarpa	1	–	–	–	–	–	–	–	–	–	–	–
Diplophyllum albicans	1	–	–	–	1	1	–	–	–	–	–	–

Nardia scalaris	–	–	–	–	–	–	1	–	–	–	–	–
Lophozia? ventricosa	–	–	–	–	–	–	1	–	–	–	–	–
Ptilidium ciliare	–	–	–	–	–	–	–	–	–	–	–	–
Alectoria nigricans	–	–	–	–	1	–	–	–	2	–	–	–
Ceramia vermicularis	–	1	1	2	–	2	–	2	–	2	–	–
Cetraria aculeata	–	–	–	1	–	1	3	2	2	–	1	1
Cetraria islandica	2	1	2	3	2	1	2	2	3	3	2	3
Cladonia alpicola	–	–	2	–	1	–	–	–	–	–	–	–
C. bellidiflora	1	–	–	1	–	1	1	–	–	–	–	2
C. coccifera	–	–	–	1	1	–	–	–	–	–	–	–
C. crispata	–	–	–	–	1	–	–	–	–	–	–	–
C. furcata	–	–	–	1	–	–	–	–	–	–	–	–
C. gracilis	–	–	–	1	–	–	2	1	2	–	1	2
C. pyxidata	1	–	–	–	1	–	–	–	–	–	–	–
C. rangiferina	–	–	–	–	–	–	–	–	–	2	1	–
C. sylvatica	–	–	–	–	2	3	–	3	2	1	1	–
C. squamosa	–	1	1	1	–	–	–	–	1	–	–	1
C. uncialis	4	3	3	2	3	1	3	3	3	3	1	–
Coriscium viride	–	1	–	–	–	1	–	–	–	–	–	–
Icmadophila sp.	–	–	–	–	–	–	–	1	–	–	–	–
Ochrolechia frigida	–	–	–	1	–	1	–	2	–	1	–	–
Sphaerophorus globosus	–	–	–	–	–	2	–	1	–	–	–	–

shown to be fully homogeneous. It is even doubted whether areas with greater and smaller variance between replicate samples may be distinguished in natural vegetation.'

Minimal area and homogeneity

Assuming there is a reasonable degree of uniformity the second proviso is that the stand must be sufficiently large to enable a representative sample to be taken. This minimum size requirement of the sample is termed the *minimal area* and is related to the number of species which occur as the sample plot increases in size. With a small plot the number of species present is also small but as the size of the plot is increased the number of species increases very noticeably. Eventually the number of new species found in successively larger plots becomes progressively fewer. This general trend is apparent in a graph of the number of species against the area of the sample plot, or a species area curve as it is usually called. The Braun-Blanquet concept of minimal area is related to the area at which the species-area curve becomes approximately horizontal. A number of examples of species-area curves are given below (Fig. 9.1) all broadly similar and of characteristic

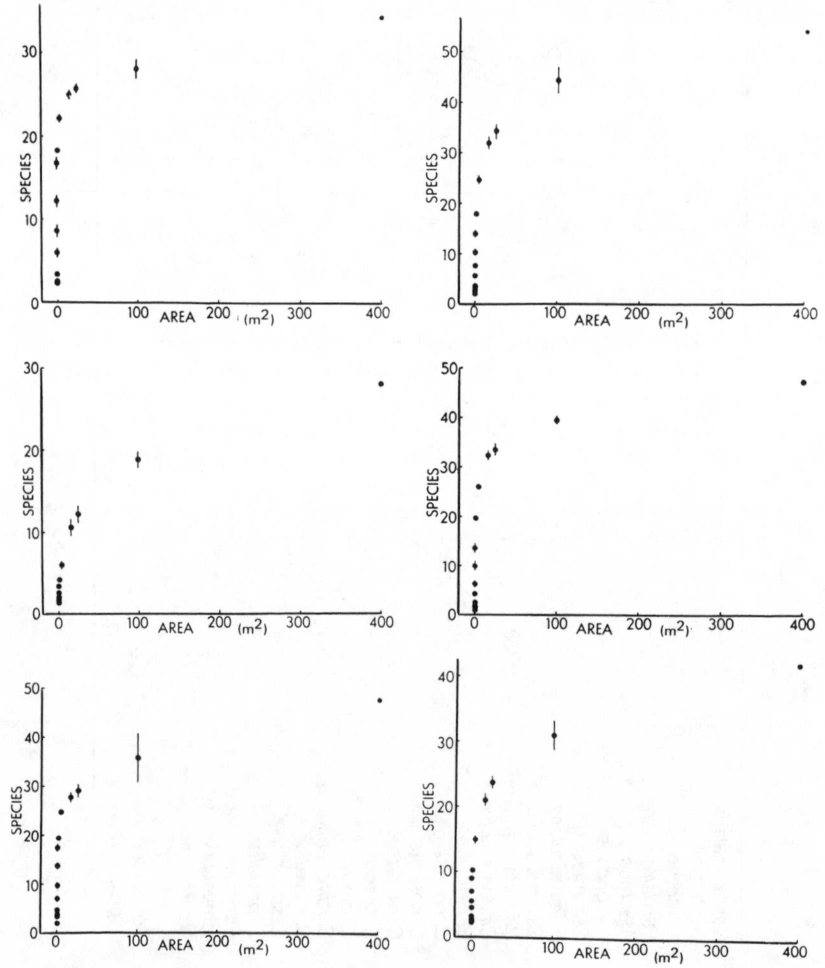

Fig. 9.1 Species area curves from six sample sites. (From Hopkins, 1957; courtesy of *J. Ecol.*)

form. It is apparent that though the slope of the curve changes fairly markedly there is a continued increase in the number of species as the area of the sample plot is further enlarged and this is particularly evident in rain forest (see for example Proctor *et al.* 1983). Hopkins (1957) presents a clear account of the use of species-area curves relative to the determination of minimal area (see also Goodall, 1952, Greig-Smith, 1957, 1983 and Dietvorst *et al.*, 1982) and draws attention to the extreme difficulty of establishing any certain definition of the minimal area. Thus the concept of minimal area depends on a definite 'break' in the curve. Cain (1932; 1934) claimed that such a break could be 'recognized' but later (1938) pointed out that its position depended on the ratio of the axes used. Following this a number of attempts to determine the minimal area based on the ratios of species and area have been made. A similar approach was made by DuRietz (1921) who related the number of constant species to area in constancy-area curves, 'constants' being defined as species which have a percentage frequency greater than 90% and the minimal area defined as 'the smallest area on which an association attains a definite number of constants'. Hopkins (1955) gives a number of constancy-area curves (Fig. 9.2) of a very similar form to the species-area curves and with the same attendant difficulty in determining the position of the minimal area. He concludes: 'Thus minimal area cannot be determined objectively from the species-area or constancy-area curve.'

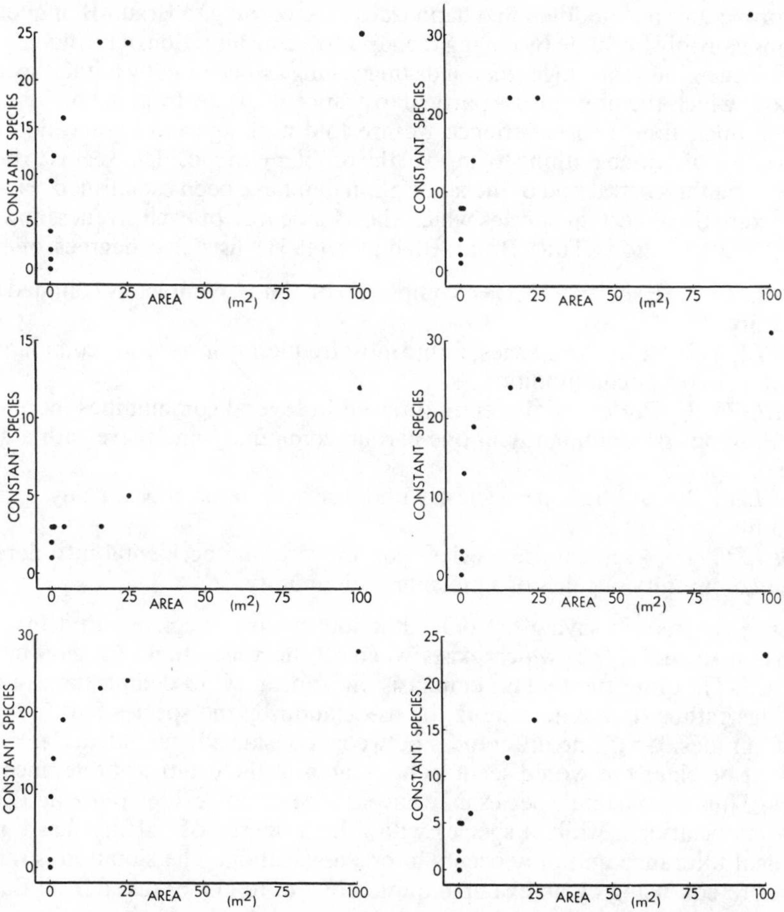

Fig. 9.2 Constancy-area curves from six sample sites (From Hopkins, 1957; courtesy of *J. Ecol.*)

The problems of homogeneity and minimal area outlined above are interrelated. If true homogeneity existed then minimal area would be a valid concept. However it is clear from the existence of pattern at numerous scales in apparently homogenous vegetation that minimal area can never be more than a gross approximation and accordingly some subjective judgement is necessary in assessing whether the area sampled is large enough to reflect the characteristics of that particular community.

A more recent review of the Braun-Blanquet system by Moore (1962) avoids the problems of homogeneity and minimal area by accepting the subjectivity of the approach and suggesting that a long 'tail' of isolated occurrences at the foot of an association table is a sign of a badly chosen stand. The fact that this does occur is clear evidence of the impossibility of visually assessing homogeneity and the minimal area required for sampling even by an experienced worker.

The association

The mechanical construction of the association table has been described above and the subsequent steps are a careful examination of the tables and separation of any lists which differ consistently and noticeably from the rest. These differences vary considerably from minor variations to more obvious differences which may necessitate the establishment of another association. Poore (1955) sums up this stage of the procedure as 'muddled and haphazard'. 'According to Braun-Blanquet these decisions can only be made by having recourse to a consideration of the fidelity of the various species, and this stage (that of distinguishing associations by faithful species) is one about which the literature is particularly reticent. Apart from the proviso that it requires much time and experience we are told nothing, and observation of the system in use has done nothing to unravel the problem for me.' This seems to be a fair comment on the abstraction of the association that have been established. Fidelity is a characteristic of certain species which shows a degree of exclusiveness towards a particular association. Thus Braun-Blanquet (1951) lists five degrees of fidelity

FIDELITY 5 Exclusive species, completely or almost completely confined to one community.

FIDELITY 4 Selective species, found most frequently in a certain community but also rarely, in other communities.

FIDELITY 3 Preferential species, present in several communities more or less abundantly but predominantly in one certain community and there with a greater degree of vigour.

FIDELITY 2 Indifferent species, without a definite affinity for any particular community.

FIDELITY 1 Accidentals, species that are rare and accidental intruders from another community or relics of a preceding community.

Poore is correct in saying the only situation in which degrees of fidelity can be properly assessed is that which exists when all the vegetation of region has been described. He quite rightly lays emphasis on 'constancy' to delimit the association groupings rather than determining the association by the species faithful to them (fidelity grades 3–5). The differences between 'constancy' and 'fidelity' may not at first sight be clear but would seem to be related to the ecological tolerance of the species. Thus a 'constant' species may have a wide ecological tolerance and occur in several associations, while a species with a high degree of 'fidelity' has a narrow ecological tolerance and only occurs in one association. The situation is rendered even more confusing by an example quoted by Poore (1956) taken from Barkman (1949). The example is an epiphytic association and the association table comprises fourteen lists with seven faithful species. Of the fourteen lists, one contains 5 of the faithful species, one 3, two 2, eight 1, and two 0. Poore expresses his difficulty in deciding which criteria were adopted in retaining the majority of these lists in their association. It would appear that either the associations are based in fact on constancy, in which case degree of fidelity seems unnecessary, or the sites are chosen because species of known narrow ecological amplitude are present, and these species subsequently, by circular argument, are defined as the faithful species. Thus faithful species are those which are confined to one association and the association is recognized by these faithful species; this is perfectly legitimate if the association is delimited by other characteristics in the first place. If in fact the association is recognized by its faithful species it is then not valid to define the faithful species from the associations. Moore (1962) denies the over importance of fidelity and quotes Braun-Blanquet (1959): 'It should not be concluded, as has sometimes been maintained, that the associations are based on fidelity. This is emphatically not the case. The association is an abstraction based on the totality of more or less homogeneous relevés (lists) which floristically correspond closely to one another. It is however, characterized, not merely floristically but also ecologically, dynamically and geographically. Nevertheless, in distinguishing from one another the associations which have been conceived on a floristic basis, far greater importance and more general significance is attributed to fidelity than to purely quantitative characteristics especially when fidelity is combined with high constancy.' This latter instance would seem to be of importance and considerable value, but as has been pointed out only possible when the complete vegetation of a region has been described. Accordingly Poore's approach based on constancy seems fully justified.

However, it is still a subjective decision to separate out a number of lists from an association table which have one or two species in common and set up a further association. There seems to be no objective reason why the splitting should stop there. Moore (1962) attempts to elucidate this point but states: 'In order to understand this (how the association under discussion is arrived at), it is essential to grasp the principle of the differential species.' It is this tendency for emphasis of fidelity and the suspicion of a circular argument which so reduces the value of the approach. Poore's insistence on the use of the 'constant' species to delimit the 'association' is a great improvement but subjective decisions still have to be taken as to the'level' of the association and whether a certain group of lists legitimately represent a fragment of a further association. Poore (1956) used a method of successive approximation; if no obvious constants were present in the lists they were taken together and divided on probable differential species (species consistently present in one group of lists). Provisional constants were named and sites were sought which possessed these combinations of constants. If the constants were real such stands were readily found but if not, they proved impossible to find.

Braun-Branquet, having established a number of associations then classifies them into a hierarchy, which is a convenient if unnatural system. The shortcomings of the continental system are rather numerous and have been discussed above under each step of the procedure. The modifications as proposed by Poore (1955; 1956) to a large extent remove some of the major criticisms and a workable system is available for the rapid description and characterization of a large area of vegetation. However, the sampling procedure, i.e. the choice of a sample site, is entirely subjective, and the apparent sharpness of the limits of associations are in fact a result of this non-random method of sampling. The sites which are clearly intermediate are avoided as 'not homogenous' and as large numbers of intermediates exist, even if they occur rarely, the sociological approach involving avoidance of such sites over-simplifies the picture to a considerable extent. *This by no means invalidates the real existence of the associations that have been erected and they are obviously of frequent and widespread occurrence.* A more real interpretation of the association should be based not only on its floristic composition but also on its frequency of occurrence in a region. The term 'noda' introduced by Poore is a satisfactory one and does imply the existence of intermediates of less frequent occurrence relative to the 'central type'. The difficulty of finding an area which is homogenous and of accurately defining the minimal area relative to it is unavoidable. It is worthwhile to accept these limitations within the system outlined by Poore, since it can readily contribute considerably to our knowledge of the vegetation especially in those regions where we have little or no floristic information at all.

Early objective methods

The subjective nature of the delimitation of vegetational units as proposed by the Continental and Scandinavian sociologists have led to numerous attempts at establishing similar units by objective methods. If we examine two noda which are positively related to some extent, in both their species composition and also their environmental relationships, we can consider them as an abstraction in terms of their species and also in terms of their relationships with *other* noda or plant associations.

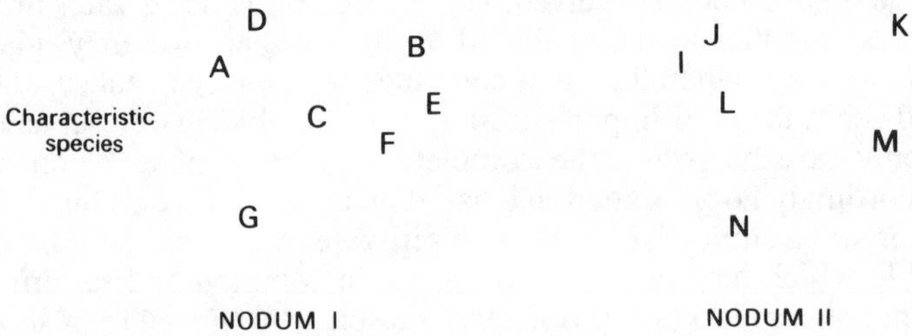

In fact this situation has several important mathematical properties, but of immediate relevance is that species A–G and species I–N are postively associated together, but group A–G is negatively associated with group I–N. One further important property of this simple model is that only 'characteristic species' are being utilized. Species H and species O–Z may be present or not but do not contribute any significant difference between the two groups. Diagrammatically then:

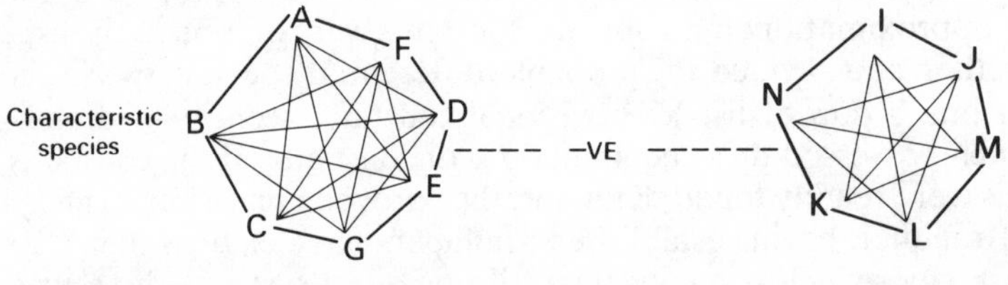

Objective methods of classification utilize and optimize this situation and several conveniently use χ^2 as a measure of positive and negative association. Two techniques will be considered illustrative of this approach.

Association analysis

The initial contribution to these methods was made by Goodall (1953). It is of considerable interest and will accordingly be described in some detail. 2 × 2 contingency tables were prepared from presence and absence data of all species in a *Eucalyptus oleosa/E. dumosa* association in Australia. The area was topographically uneven with sandy ridges running east and west separated by hollows with a heavier soil. An area 640 m^2 was delimited and divided into 64 square plots within each of which four 5 m quadrats were placed at random. Fifty-eight species were recorded from the area but 27 only occurred in less than 5 quadrats and were eliminated. χ^2 values were calculated for the occurrence of the remaining species in pairs. Only positive associations (in the sense that the species were positively correlated) were used in the classification of the quadrat data. Goodall points out that homogenous groups of quadrats can be obtained (which is the object of a system of classification) by eliminating correlation between species in either of four ways.

(*i*) Quadrats with one of two species which are correlated can be separated out as an initial potentially homogeneous group.

(*ii*) Quadrats not containing one of the species represent a similar potentially homogeneous group.

(*iii*) Quadrats containing both the species.

(*iv*) Quadrats containing neither of the species.

Goodall analysed the data by the four methods and found reasonable agreement in the results, and concluded that the first system, the presence of a pair of correlated species, was the most convenient method of approach.

The first step was to separate out from the total sample (256) those quadrats containing *Eucalyptus dumosa* which showed a number of highly significant correlations with other species, [This in fact is 'association' between species that is measured but the term correlation is employed here to avoid confusion with the use of 'association' in the Braun-Blanquet methodology.] and was the most frequently occurring species in the area (170 quadrats). Within this group of quadrats several highly significant positive correlations were still present and a second division was made, again on the most frequency species, *Triodia irritans*, which showed a highly positive correlation with some other species. One hundred and nineteen quadrats were separated out and again fresh contingency tables established and the heterogeneity of these 119 quadrats examined. The process was continued with successively smaller groups of quadrats until finally 53 quadrats selected on the presence of *Eucalyptus dumosa, Triodia irritans* and *Stipa variabilis* showed no significant correlation between species within the group and were finalized as a homogenous group (group *A*). All the remaining quadrats (203 in all) were pooled and a fresh series of contingency tables constructed. The same procedure was followed and the most frequent species *Eucalyptus oleosa*, in this residual group of quadrats was used as the initial division (121 quadrats). Subsequently 92 quadrats containing *Bassia uniflora* and finally 17 quadrats containing *Cassia eremophila* were separated out.

This final group of 17 quadrats showed no significant correlations and were established as group *B*. The remaining 186 quadrats were again examined by 2 × 2 contingency tables, divisions being based either on the most frequent species which showed significant correlation or on the next most frequent species with a significant correlation. After the final separation of a homogenous group the total remaining quadrats were subsequently examined each time. The data were divided by this means into five homogenous groups assessed by χ^2 tests. Each group was finally recombined with each of the other groups in turn and tested for heterogeneity. If the homogenous groups were satisfactory then recombination with other groups would immediately introduce correlation. This was found to be so, with the exception of groups *A* and *E* which on combination showed no significant correlations and were accordingly combined as a single unit (Fig. 9.3). The species components of the groupings are given in Table 9.2. Goodall characterizes the four groups *A, B, C* and *D* by

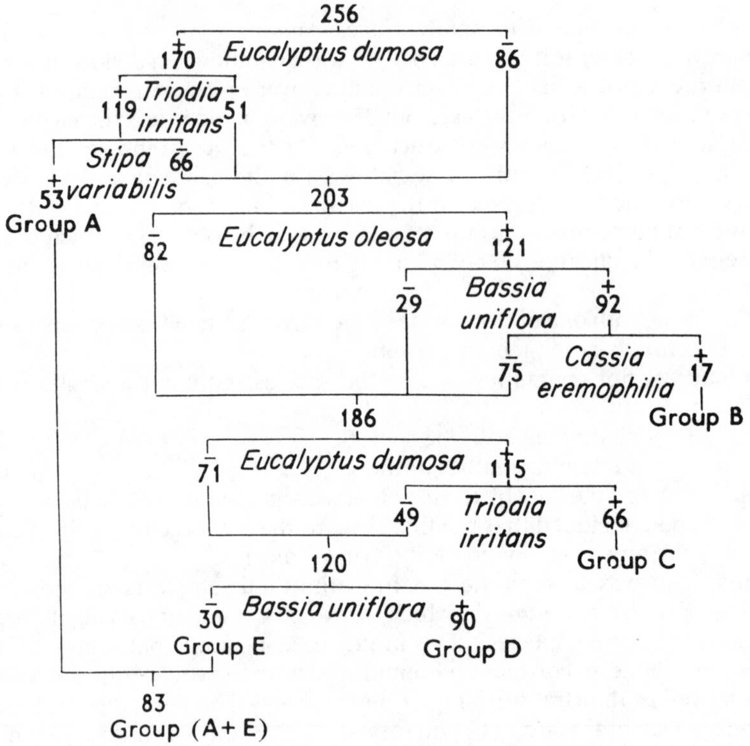

Fig. 9.3 Victoria Mallee vegetation. Stages in the classification of quadrats into groups. At each stage the number of quadrats in which the species in question is present (+) and absent (−) is indicated. (From Goodall, 1953; courtesy of *Austr. J. Bot.*)

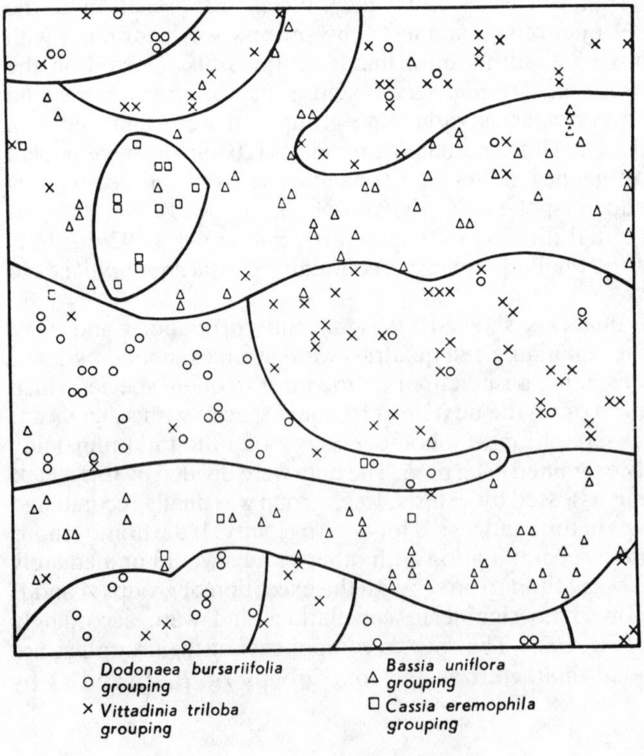

Fig. 9.4 Plan of the area of Mallee vegetation studies indicating the position of each quadrat and the grouping to which it was assigned. (From Goodall, 1953; courtesy of *Austr. J. Bot.*)

Table 9.2 Frequency (%) of species in four groups of quadrats based on presence of single species showing positive correlations. (From Goodall, 1953; courtesy of *Austr. J. Bot.*) Frequencies of 50% or more have been given bold type.

Species	Group			
	A+E	*B*	*C*	*D*
	Number of quadrats			
	83	17	66	90
Acacia rigens A. Cunn.	2	0	5	0
Bassia parviflora Anders	2	0	0	4
B. uniflora R. Br.	17	**100**	5	**100**
Beyeria opaca F. Muell.	1	6	5	0
Callitris verrucosa R. Br.	1	0	26	0
Cassia eremophila A. Cunn.	1	**100**	0	2
Chenopodium pseudomicrophyllum (1)	5	6	2	22
C. pseudomicrophyllum (2)	12	6	0	13
Danthonia semiannularis R. Br.	20	0	6	7
Dodonaea bursarifolia Behr.	**53**	0	**71**	17
D. stenozyga F. Muell.	1	0	0	6
Enchylaena tomentosa R. Br.	4	24	0	3
Eucalypus calycogona Turcz.	14	18	11	30
E. dumosa A. Cunn.	**81**	**12**	**100**	39
E. oleosa F. Muell.	49	**100**	24	**81**
Grevillea huegelii Meiss	5	35	8	4
Kochia pentatropis R. Tate	2	24	0	7
Lepidosperma viscidum R. Br.	1	0	9	0
Lomandra leucocelphala R. Br.	4	0	8	0
Melaleuca pubescens Schau.	4	0	8	0
M. ucinata R. Br.	**55**	0	**53**	3
Micromyrtus ciliatus J.M. Black	0	0	9	0
Myoporum platycarpum R. Br.	2	0	5	0
Olearia muelleri Bth.	5	**59**	0	6
Santalum murrayanum F. Muell.	1	12	5	1
Stipa elegantissima Lab.	1	12	0	3
S. molis R. Br.	7	0	5	10
S. variabilis Hughes	**87**	**88**	0	**78**
Triodia irritans R. Br.	**78**	6	**100**	3
Vittadinia triloba D.C.	46	24	11	37
Westringia rigida R. Br.	6	**71**	5	3
Zygophyllum apiculatum F. Muell.	4	**65**	2	47

Vittandinia triloba, Cassia eremophila, Dodonaea bursariifolia and *Bassia uniflora* based on those species of most frequent occurrence in those quadrats allotted to the same grouping by all four procedures. The distribution of the groupings in relation to the area studied (Fig. 9.4) shows a reasonable correlation with topography. Three ridges run across the area and are occupied predominantly by the *Dodonaea* and *Vittandinia* groupings while the hollows between contain the *Bassia* and *Cassia* groupings.

A similar approach leading to a hierarchical classification has been outlined by Williams and Lambert (1959; 1960). They have three main criticisms of Goodall's method: his use of positive correlations only, his use of the most frequent species as a basis of division of a potentially homogenous group, and the pooling of those quadrats which do not form the initial and successive homogenous groups. The use of negative correlations as well as positive ones greatly stengthens the analysis, as long as due regard is given to the size of quadrat used for sampling since negative correlation between species may frequently occur by virtue of the fact that no two individuals can occupy the same place at any one time. Their criticism of the 'pooling' of quadrats seems equally logical: 'Imagine a population divided on species *X* into (*X*) and (*x*). (*X*) is still found to be divisible on species *Y* into (*XY*) and (*Xy*). Goodall would "pool" (*Xy*) with (*x*) to form a new population. Our chief objective to this system is that information relating to the discontinuity (*X*)/(*x*) is being discarded, and on these grounds we would prefer a "hierarchical" system – a division, once

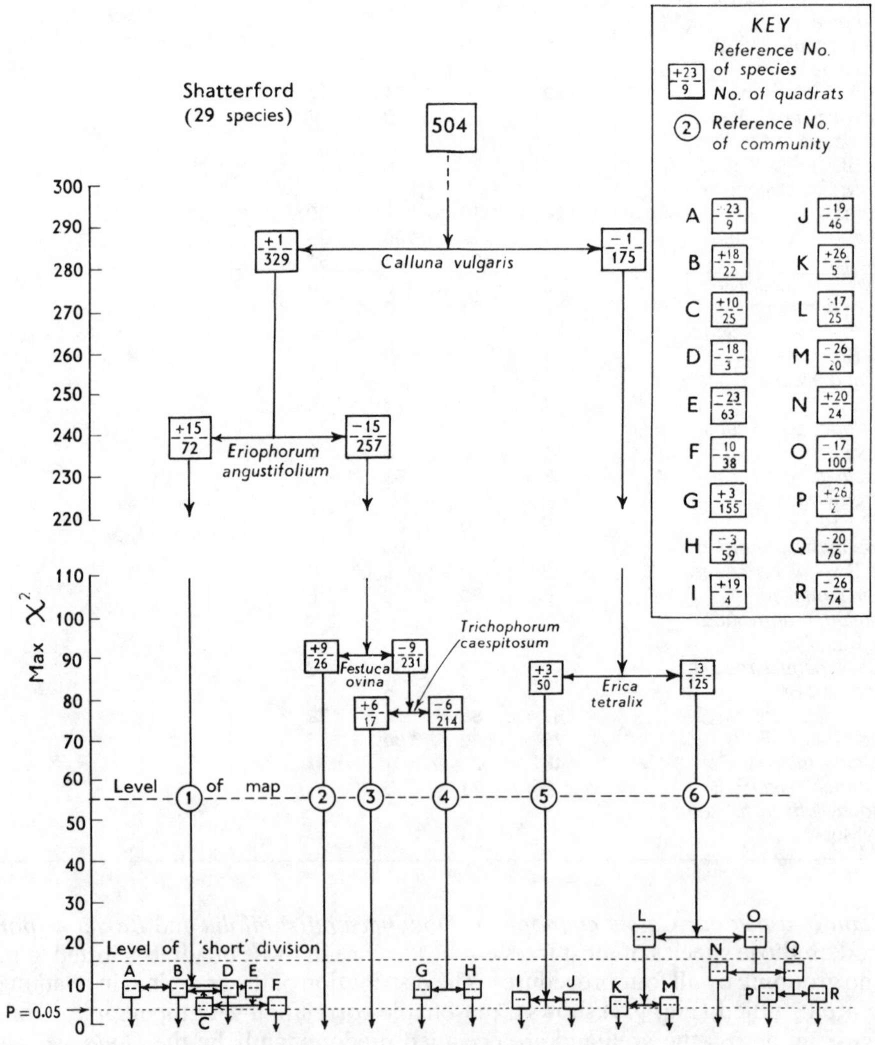

Fig. 9.5 The Shatterford community analysis. (From Williams and Lambert, 1960; courtesy of *J. Ecol.*)

made, would remain inviolate throughout the analysis.' It is also pointed out that Goodall's final groups are not necessarily capable of definition in terms of key species and also that the means by which the groups are obtained do not give any meaningful information.

2×2 contingency tables are used to establish a χ^2 matrix and the χ^2 values are summed irrespective of whether they reflect a positive or a negative correlation. The species with the largest $\Sigma\ (\chi^2)$ and not the most abundant species is used in the sub-division of the sample, the two quadrat groups thus formed then remaining separate throughout the remainder of the sorting procedure. Each of the two groups is re-examined and further divided on the species with the largest $\Sigma\ (\chi^2)$. The hierarchical division is continued until no significant correlations appear, at which point the group is designated as 'final'. Williams and Lambert give several convincing examples of the application of the method to actual field data. The Shatterford example involving 29 species is extremely clear-cut, the analysis (Fig. 9.5) separating out six major groups which are further subdivided into a larger number of smaller units. The spatial distribution of the individual quadrats in relation to the soil variations of the area is remarkably consistent (Fig. 9.6). The first division (community 4) represents the area of dry heath dominated by *Calluna vulgaris* with *Erica tetralix* and *Molinia caerulea*. Communities 5 and 6 are typical bog communities dominated by *Erica tetralix* and *Molinia caerulea* respectively with community 1 occupying an intermediate 'wet heath' position (Table 9.3). From the species composition of the groupings and the relative frequency of occurrence the 'constant' species of the grouping are readily apparent. Thus the analysis appears to accomplish all that was intended for it and has the great advantage (in common with Goodall's approach) that the position of each group of quadrats can be accurately plotted on a map and related to the detailed environmental variations of the area as a whole. Thus any group of quadrats can be picked out at a later date and the environmental interrelationships considered in detail.

In terms of our simple model:

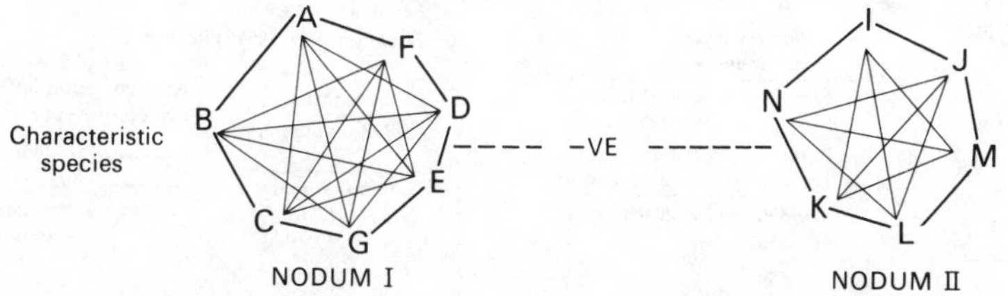

Goodall's approach simply uses the positive relationships existing within the two groups of species. Williams on the other hand utilizes in addition, the negative relationships existing *between* the two groups. Both methods worked satisfactorily and associated species together into groupings which in turn relate to the ecology of the area sampled. The methods, however, both follow a different pathway of sub-division and furthermore the divisions are based on the commonest species in one method and the species with the strongest level of association in the other. Additional methods of sorting out 'groupings' in objective ways are largely concerned with optimizing the efficiency of classification, and two of the evaluation criteria used are in fact the pathway followed and the method of subdivision. The major factor for assessment of the techniques, however, is the credibility of the result.

In an ecological context, variation between samples is continuous and there is no implicit evolutionary relationship between the noda which are present. Conversely in a taxonomic situation discontinuities *are* usually present and one objective of

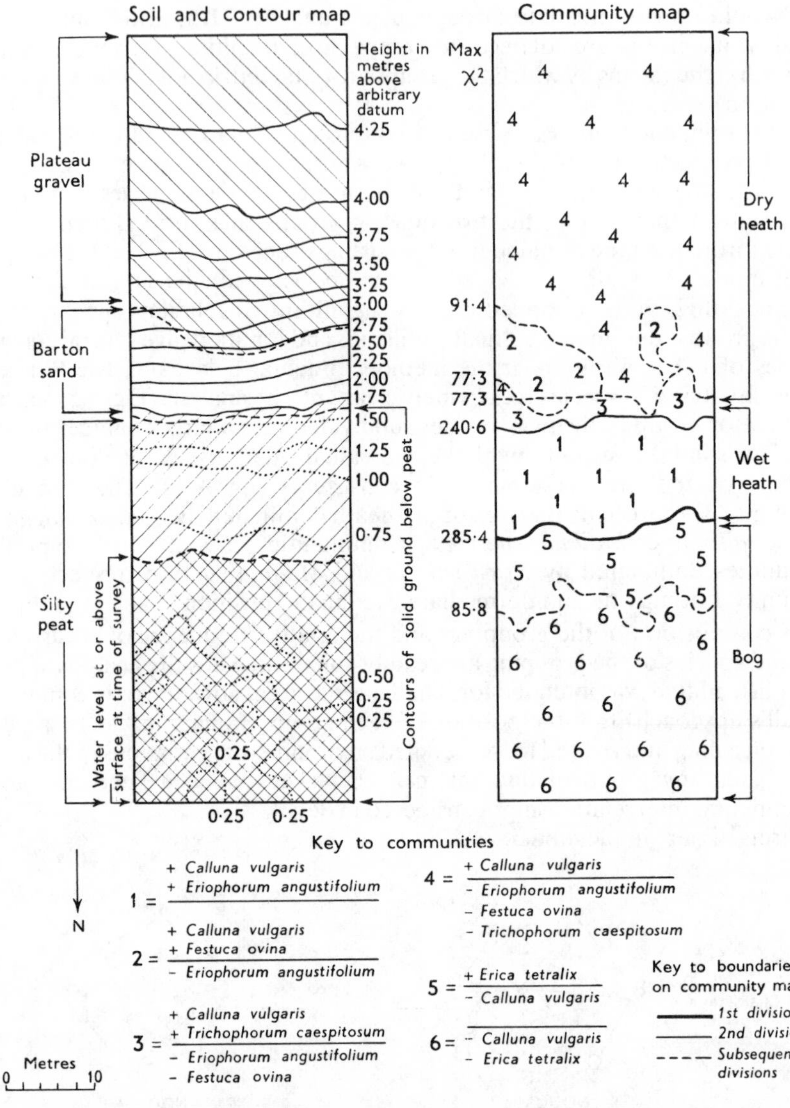

Fig. 9.6 The Shatterford community map, soil type, and contours. (From Williams and Lambert, 1960; courtesy of *J. Ecol.*)

classification is the uncovering of possible phylogenetic relationships. In the sense of continuous variation, classification of vegetational samples is thus an artificial concept which is achieved by either ignoring intermediates in the sampling or conversely by ignoring that determinate characteristic of a nodum, the relative abundance of its component species. Goodall and Williams both achieve groupings by utilizing presence-absence data and the resultant clarity is simply due to this. Thus in ecology, in a situation of continuous variation, there is much to be said for simplicity of approach if an arbitrary classification is required; the complexities of methodology should not be allowed to dominate. The use of χ^2 as a measure of association suffers the disadvantage that the trend of association can be affected by the size of quadrat used in the sampling. Kershaw (1961) has shown that the final quadrat groupings in association analysis depend largely on the size of the quadrat

Table 9.3 Species composition of Shatterford groupings at six-community level. (From Williams and Lambert, 1960; courtesy of *J. Ecol.*)

Ref. number of community	4	2	3	1	5	6
Number of quadrats	214	26	17	72	50	125
Erica cinerea L.	29	19	–	–	–	–
Ulex minor Roth	22	22	–	–	–	–
Festuca ovina L.	–	26	–	–	–	–
Pinus sylvestris L.	1	–	1	–	–	–
Calluna vulgaris (L.) Hull	214	26	17	72	–	–
Polygala serpyllifolia Hose	8	19	5	30	1	–
Erica tetralix L.	155	26	17	72	50	–
Molinia caerulea (L.) Moench	155	26	17	72	50	125
Trichophorum caespitosum (L.) Hartman	–	3	17	70	10	–
Drosera rotundifolia L.	–	–	7	58	23	4
Juncus squarrosus L.	–	–	–	5	–	–
Gentiana pneumonanthe L.	–	–	–	2	–	–
Epiophorum angustifolium Honck.	–	–	–	72	50	77
Narthecium ossifragum (L.) Huds.	–	–	–	57	49	25
Drosera intermedia Drev. and Heyne	–	–	–	9	15	–
Carex panicea L.	–	–	–	1	32	4
Potentilla erecta (L.) Rausch.	–	–	–	1	2	2
Myrica gale L.	–	–	–	4	47	76
Juncus bulbosus L.	–	–	–	–	1	2
Carex rostrata Stokes	–	–	–	–	1	8
Juncus acutiflorus Hoffm.	–	–	–	–	4	37
Potamogeton polygonifolius Pourr.	–	–	–	–	2	23
Agrostis? gigantea Roth	–	–	–	–	1	20
Carex echinata Murr.	–	–	–	–	1	28
Scutellaria minor L.	–	–	–	–	–	26
Cirsium dissectum (L.) Hill	–	–	–	–	–	15
Hypericum elodes L.	–	–	–	–	–	3
Potamogeton natans L.	–	–	–	–	–	2
Eleocharis pauciflora (Lightf.) Link	–	–	–	–	–	1

used. The primary divisions are not affected, but such overall differences are usually apparent on visual inspection of the area and do not necessarily need an elaborate technique to pick them out. On the other hand these methods of association analysis are objective, and for the primary survey of an area very effective. The initial problem of computing large numbers of χ^2 values has been largely removed by the subsequent development of computers with very large central stores and/or peripheral backing store. The original limitations on the size of sample and number of species which could be handled in any one analysis have now disappeared to a great extent.

The approach of Williams and Lambert (1959; 1960) is more efficient than that of Goodall (1953) in that negative associations are also included in the analysis, and despite the short comings of χ^2 as a measure, its relative simplicity recommends it. One further aspect is of importance. Williams and Lambert (1961) showed that in addition to the 'normal' analysis of plot data, a similar but 'inverse' analysis could be made of the species interrelationships.

For species A, B, C, D in sample plots 1, 2, 3, 4, calculation of the number of joint occurrences for entry into 2×2 contingency tables of the 'normal' or plot analysis are calculated as follows:

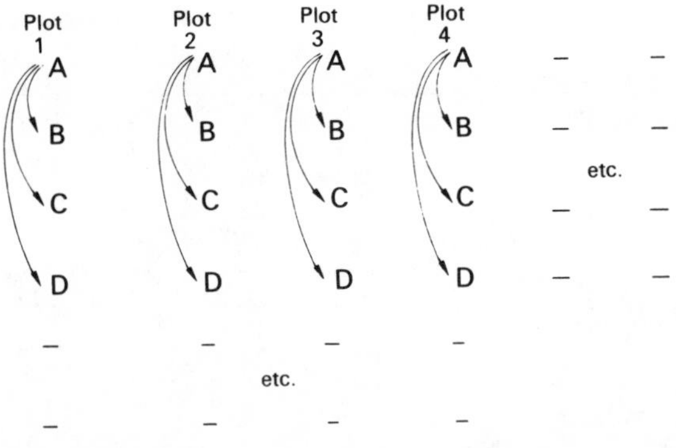

Species A + B, A + C, A + D, etc., in this case, each show four joint occurrences in the plots. From species comparisons, the species with the strongest level of association is selected and the *plots* subdivided on this species.

Conversely in an inverse analysis:

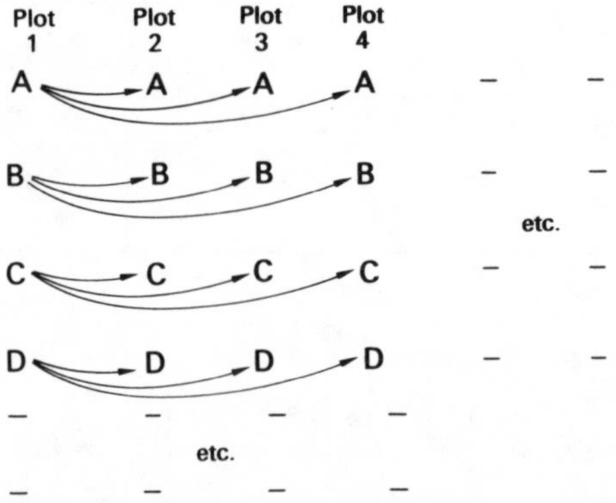

Plots 1 + 2, 1 + 3, 1 + 4, etc. in this case, each show four joint occurrences of species.

From plot comparisons the plot with the strongest level of association is selected, and *species* sub-divided in relation to this plot. Thus those species present in that plot form a group of species which are subsequently re-examined throughout the sample for further possible sub-division. The final outcome is species groupings whose ecological preferences are closely similar.

The results from an inverse analysis of the Shatterford data previously discussed are given below (Fig. 9.7) and show four species groupings. Group A comprised of *Calluna vulgaris*, *Molinia caerulea*, and *Erica tetralix*, a very characteristic combination of wet heathland species, Group B with *Trichophorum caespitosum*, *Eriophorum angustifolium* and *Narthecium ossifragum* typical of wet heath/bog, Group C composed of the dry heath species *Erica cinerea, Ulex minor, Festuca ovina*, and *Polygala serpyllifolia*. The final group D is a group of species of wet boggy habitats, but whose representation in the sample is poor. The 'inverse analysis' thus complements the 'normal' association analysis findings.

There has been considerable confusion of terminology following the application of the more recent multivariate techniques to the problem of vegetational groupings. The so-called R & Q techniques, despite an attempt by Williams and Dale (1965) to clarify the situation, are still utterly confusing to the non-mathematician. Accordingly it is advisable and much simpler, to use the terms *plot classification* and *species classification* which are readily understandable, together with sufficient detail

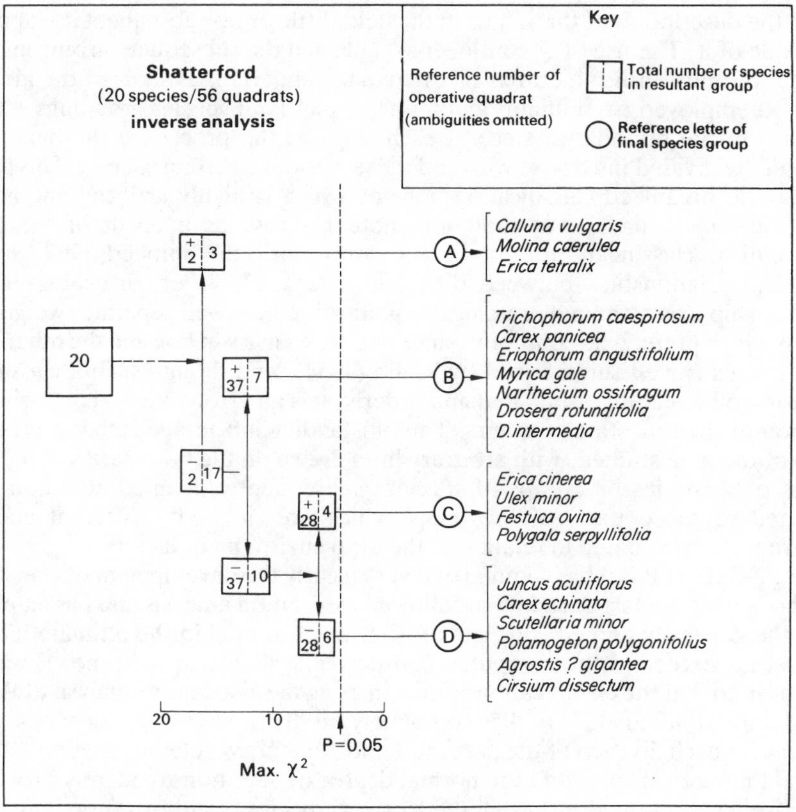

Fig. 9.7 The inverse association analysis of the Shatterford data. (From Williams and Lambert 1961; courtesy of *J. Ecol.*)

of the actual method used so as to avoid further confusion. This nomenclature will be followed here and also extended to plot and species ordinations subsequently (see below).

The simplicity and effectiveness of *plot* and *species* classifications probably accounts for their initial widespread use (for example, Gittins, 1965; Gimingham *et al.*, 1966; Harrison, 1970; Ward, 1970; Flenley, 1969; Hopkins, 1968; Poore, 1968; Greig-Smith *et al.*, 1967; and Kershaw 1968). If classification seems a useful approach then the oversimplification inherent in the χ^2 methods of analysis is probably beneficial in the first instance.

The use of a single species to divide data sets into appropriate subunits, is termed monothetic division and the association analysis of Williams and Lambert (1959) provides an excellent example of such an approach. However it clearly departs from ecological reality in that it is not merely the presence of a species which forms the basis for an association but the spectrum of species as well as their density, distribution, size, growth-form and their vigour. Thus, ideally any classificatory system should be based on the range of species present, but also on appropriate measures of their abundance, distribution, size and vigour. The use of a *number* of a species to classify vegetation is termed a polythetic method and certainly provides a more satisfactory although computationally more difficult approach.

On this basis the Braun-Blanquet approach also falls short of the ideal. It is based on non-random sampling, and it employs the measure of presence or absence only to characterize the association, and although some estimate of abundance is available from the description of the stands in the field, little or no subsequent use appears to be made of it. The use of χ^2 contingency table and the subsequent arrangement of a system of units, either hierarchical or spatial, similarly falls short of the ideal. The method employed by Williams and Lambert picks out species groupings which are given artificial boundaries since at each stage of the procedure the quadrats are repeatedly divided into those with and those without a certain species. This presents a tidy and organized classification but one which is highly artificial and has little ecological meaning.The method undoubtedly arose as a result of the parallel hierarchical classification used in taxonomy with the implied phylogeny and genetical relationships between different groups. However, in ecology no such relationship exists and a hierarchical classification may well separate two groups of plants which prefer acid conditions since one grows in a wet bog and the other in a dry heath. This is well summarized by Webb (1954), and though such a classification appears to be readily assimilated and orderly, it is an erroneous and over-simplified version of the real state of affairs. The outstanding advantage is that it produces a map of the area studied, with arbitrary lines drawn on the boundaries of the spatial limits of a species or groups of species. Subsequently detailed attention can be directed to those parts of the 'map' which are markedly different and these differences investigated in relation to the local environment factors.

Greig-Smith (1980) has summarized very clearly the development of classification methods over the last decade that followed association analysis and has emphasised that the generally agreed concepts of what was essential in the ultimate technique always far exceeded the computational power available at the time. He has also emphasised that the classification approach using the association analysis of Williams and Lambert (1960; 1961) relies completely on the presence or absence of a single species for each division of the data set. Unfortunately vegetation does not subscribe to rigid numerical rules and the normal degree of variation frequently provides for the case where a quadrat has all the 'correct' species of an association are present *except one*. Such quadrats can then be seriously mis-classified if the species in question happens to be one of the divisive criteria in the analysis.

As a result of these inadequacies, there has been a pronounced level of activity examining alternative objective methods for establishing and classifying vegetational units, resulting in a sharp focus on the desirable essentials of methods, certainly a plethora of techniques, but remarkably little gain in ecological understanding. The problem is quite staightforward and it has already been referred to several times in this and previous sections – it is the characteristic combination of *several* species occurring together at an appropriate level of *abundance* that visually substantiates the reality of the plant association. Unfortunately polythetic divisive methods, however appropriate ecologically rapidly become impractical as a result of the excessive computing levels involved when more than a few species are used for division.

The alternative approaches to *divisive* methods are the *agglomerative* methods where a single quadrat is 'matched' against all the others in the sample and identical or 'similar' quadrats fused into a group which finally is representative of a plant association. The difficulty of course lies in the definition of 'similar' since rarely, if ever, do identical vegetational samples exist. The nature of the definition not only determines the dimension of the 'cluster' but in part at least the distance from it, of other clusters or associations. Again, polythetic methods are more realistic than monothetic comparisons, as are the use of abundance values rather than simple presence or absence data.

There has been a remarkable amount of energy and ingenuity expended on a plethora of polythetic agglomerative techniques. This possibly reflects at least in part, simply the enthusiasm of the mathematical and computing fraternity. Certainly many of the fundamental concepts embodied in the apparent success of phytosociology appear to have been completely or largely overlooked, particularly the essential component of the initial non-random sampling, and the arbitrary rejection (based on experience) of non-homogenous sample plots. Thus it should be clearly recognized that the success of the Braun-Blanquet method of establishing clear-cut associations rests entirely on the careful choice of stand with the stipulations of minimal area and homogeneity. This is highly subjective, and initially at least, with only the central 'nodum' (see Poore, 1955–6) being isolated, characterized and described. This produces clear-cut associations, the numerous intermediates are ignored. At a later stage the more obvious intermediates could possibly be considered, and described as 'associations' or 'facies', but it is debatable whether this ultimate stage is possible in practice. Neither monothetic or polythetic agglomeratic techniques allow for this prior selection of the sample data to be used. As a result boundaries are arbitrary and are perhaps best viewed as a continuum in the sense of Curtis *et al.* where a random selection of sites necessarily includes 'noda' and intermediates alike (see below). For example, the universal presence of genetic variability in a population has never been specifically recognized as perhaps the ultimate reason why classification methods are at worst impossible, at best only partly successful. Thus population variability, either phenotypic or genotypic, is widely documented at the *sub*-specific level yet ecology uses only attributes at the *species* level to define plant associations. As a result even non-random sampling provides data that rarely will yield ideal 'clear-cut' plant associations. Fundamentally this is encouraging, although annoying – it does substantiate however the findings of an important branch of biology. Austin (1980) has also questioned the scarcely appropriate models that have been used by numerical methods particularly in terms of the inappropriate use of species distributions that perhaps are valid in theory but never realized in natural data sets (see also the comments of Greig-Smith, 1980; 1983).

Thus, the general ecological consensus in 1984 is to compromise between the positive combined features of both classification and ordination. But, without some level of human judgement, classification approaches however sophisticated, will always be prone to the limitations of field taxonomy as well as sampling error. As a result it is probably most conservative of energy either to follow the traditional Braun-Blanquet approach, most certainly with computer assistance (Maarel *et al.*, 1978) or to ordinate the data set with the association boundaries subsequently overlayed. These approaches are discussed in the subsequent chapters here.

10

Ordination Methods I

Early Techniques

In our simple nodum model:

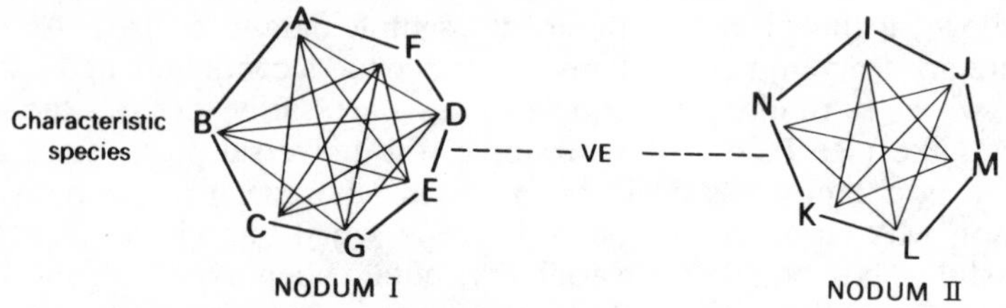

there was a deliberate simplification – implicit in the term nodum, is a 'blurring' of the distinctness of the grouping and:

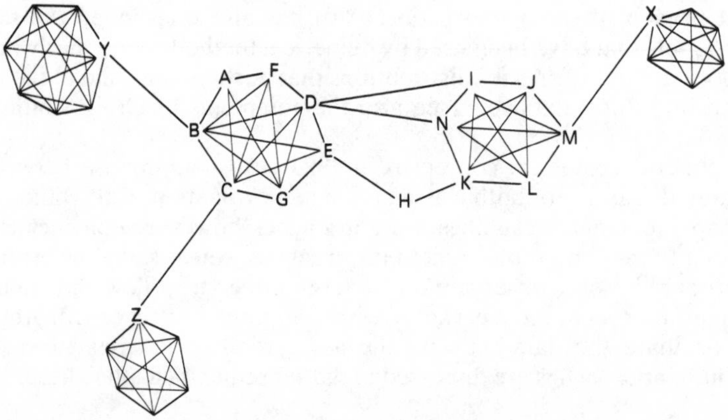

is a much more realistic conceptual model, where A, B, C, etc. are either species *or* plots. There is a spatial configuration implying dimensions as well as a sequence of noda which are not completely isolated but cross-linked. The dimensions can also relate to closeness of the positive associations in terms of whether the clusters are 'tight' or 'loose'. In addition, overall control by environmental factors is implied by the direction of the major axis. This is a simple ordination of species (or plots) and can again be readily obtained by use of χ^2 as a measure of association between species pairs.

The variation of floristic composition of communities dominated by *Juncus effusus* in North Wales was examined by Agnew (1961). He examined 99 stands chosen at random from a potentially large number of areas where *Juncus* occurred. The presence or absence of each species was recorded in each site. Species which occurred less than five times in the total sample were eliminated and 2 × 2 contingency tables constructed for the remaining 53 species. The correlations obtained (Fig. 10.1) were diagrammatically represented in the form of a two-dimensional spatial arrangement of species (Fig. 10.2).

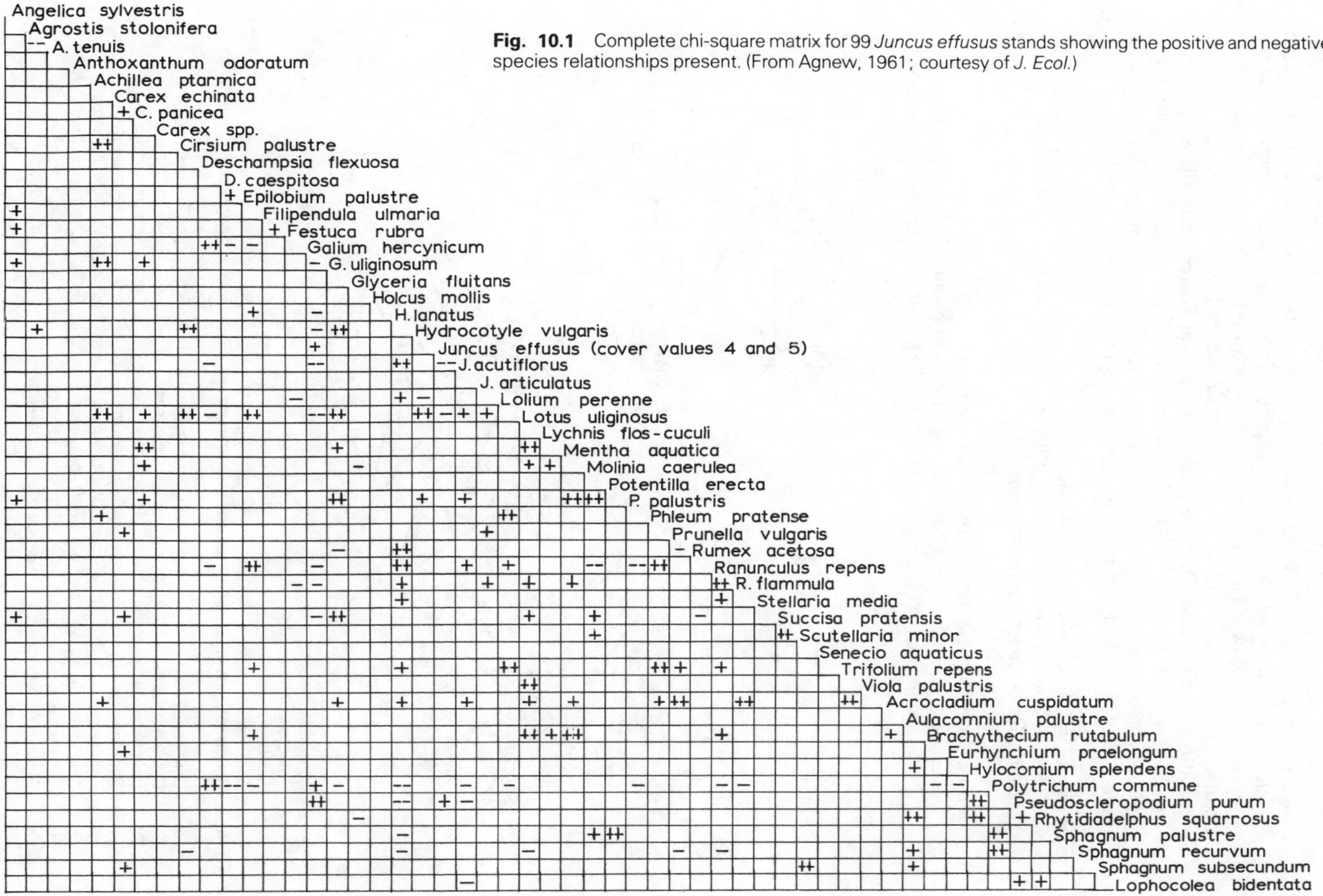

Fig. 10.1 Complete chi-square matrix for 99 *Juncus effusus* stands showing the positive and negative species relationships present. (From Agnew, 1961; courtesy of *J. Ecol.*)

The reciprocal of the calculated value of χ^2 between each species pair is used to construct the diagram. Thus two species highly positively associated and with a large χ^2-value are positioned close to one another. In practice some approximation is necessary (see below). The arrangement of species reflects positive correlations in the main, the general position of each group of species being related to the negative correlations also present. On pure chance alone some correlations will be significant (1 in 20 at a 5 per cent significance level) and accordingly those species with only a single correlation at a 5 per cent level are not included in the diagram. It is immediately obvious that three groupings of species occur (see Fig. 10.2).

Group 1 (left-hand side of the diagram)
Holcus lanatus, Lolium perenne, Phleum pratense, Ranunculus repens, Stellaria media and *Trifolium repens.*

Group 2 (lower right-hand side of the diagram)
Angelica sylvestris, Carex panicea, Galium uliginosum, Hydrocotyle vulgaris, Lotus uliginosus, Molinia caerulea, Potentilla palustris and *Succisa pratensis.*

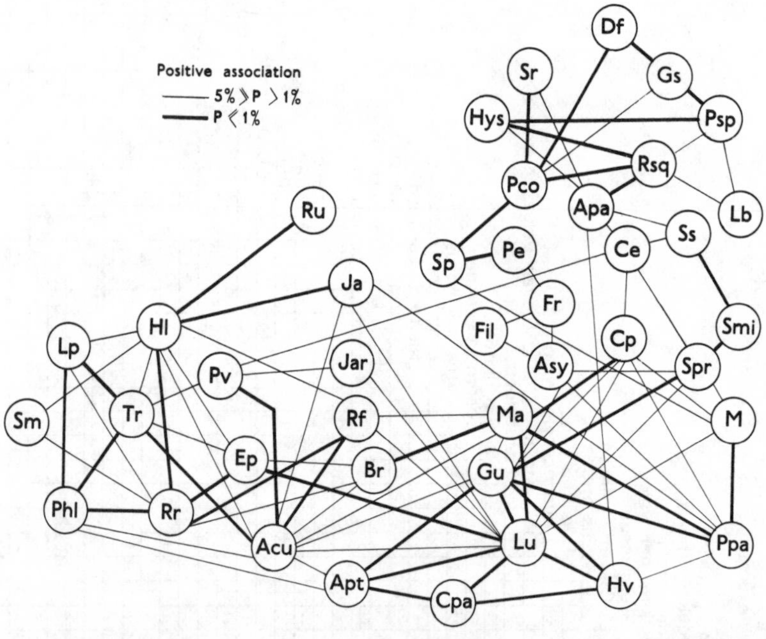

Acu	*Acrocladium cuspidatum*	Hl	*Holcus lanatus*	Psp	*Pseudoscleropodium purum*
Apa	*Aulacomnium palustre*	Hv	*Hydrocotyle vulgaris*	Pv	*Prunella vulgaris*
Apt	*Achillea ptarmica*	Hys	*Hylocomium splendens*	Pf	*Ranunculus flammula*
Asy	*Angelica sylvestris*	Ja	*Juncus acutiflorus*	Rr	*Ranunculus repens*
Br.	*Brachythecium rutabulum*	Jar	*Juncus articulatus*	Rsq	*Rhytidiadelphus squarrosus*
Ce	*Carex echinata*	Lb	*Lophocolea bidentata*	Ru	*Rumex acetosa*
Cp	*Carex panicea*	Lp	*Lolium perenne*	Sm	*Stellaria media*
Cpa	*Cirsium palustre*	Lu	*Lotus uliginosus*	Smi	*Scutellaria minor*
Df	*Deschampsia flexuosa*	M	*Molinia caerulea*	Sp	*Sphagnum palustre*
Ep	*Epilobium palustre*	Ma	*Mentha aquatica*	Spr	*Succisa pratense*
Fil	*Filipendula ulmaria*	Pco	*Polytrichum commune*	Sr	*Sphagnum recurvum*
Fr	*Festuca rubra*	Pe	*Potentilla erecta*	Ss	*Sphagnum subsecundum*
Gs	*Galium saxatile*	Phl	*Phleum pratense*	Tr	*Trifolium repens*
Gu	*Galium uliginosum*	Ppa	*Potentilla palustre*		

Fig. 10.2 The species 'constellation' based on χ^2 values and showing positive correlations from 99 *Juncus effusus* stands. (From Agnew, 1961; courtesy of *J. Ecol.*)

Group 3 (upper right-hand side of the diagram)
Aulacomnium palustre, Deschampsia flexuosa, Hylocomium splendens, Polytrichum commune, Pseudoscleropodium purum, Rhytidiadelphus squarrosus and *Sphagnum recurvum*.

It is equally clear that none of the three groupings are isolated and distinct entities; each is linked to the adjacent group by correlation between member species of the respective groups or through an intermediate species, e.g. *Juncus acutiflorus*. Agnew suggests it is possible that such intermediate species may prove on a more extensive survey to be part of additional groupings. The deduction that these correlated groupings of species represent significant 'communities' or plant associations (in the sense intended by Braun-Blanquet) is considerably strengthened by additional ecological information. Thus the general linear relationship of the three groups follows a pH gradient which is highly significant ($r = -0.509, p < 0.001$).

Welch (1960) present a similar series taken from woodland in Tanzania (Fig. 10.3).

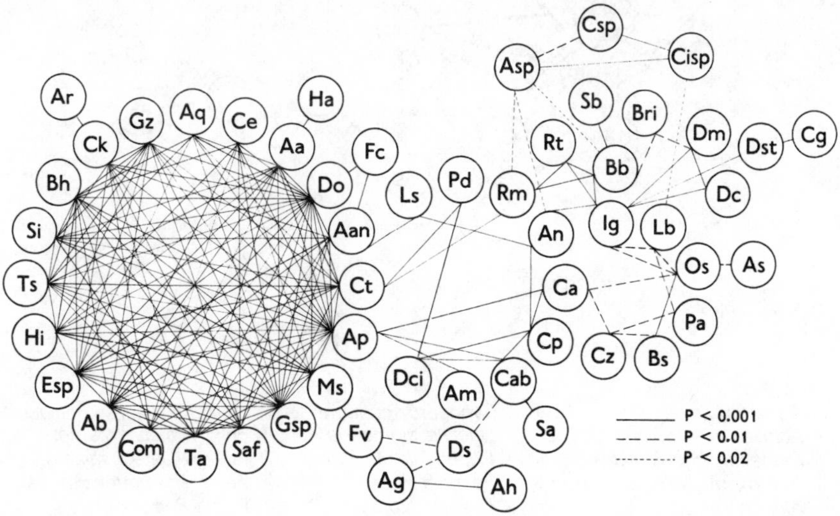

Aa	*Allophylus alnifolius*	Cg	*Combretum ghasalense*	Ha	*Harrisonia abyssinica*
Aan	*Albizia anthelmintica*	Cisp	*Cissus sp*	Hi	*Haplocoelum inopleum*
Ab	*Albizia brachycalyx*	Ck	*Capparis kirkii*	Ig	*Isoberlinia globiflora*
Ag	*Acacia gerrardii*	Com	*Commiphora spp*	Lb	*Lonchocarpus bussei*
Ah	*Albizia harueyi*	Cp	*Commiphora pilosa*	Ls	*Lannea stuhlmannii*
Am	*Acacia mellifera*	Csp	*Cordia sp*	Ms	*Manilkara sulcata*
An	*Acacia nigrescens*	Ct	*Canthium telidosma*	Os	*Ostryoderris stuhlmannii*
Ap	*Acacia pennata*	Cz	*Combretum zeyheri*	Pa	*Pterocarpus angolensis*
Aq	*Afzelia quanzensis*	Dc	*Dichrostachys cinerea*	Pd	*Phyllanthus discoideus*
Ar	*Asparagus racemosus*	Dci	*Danbeya cincinnata*	Rm	*Royena macrocalyx*
As	*Acacia senegal*	Dm	*Dalbergia melanoxylon*	Rt	*Randia taylorii*
Asp	*Acacia sp*	Do	*Dalbergia ochracea*	Sa	*Steganotaenia araliacea*
Bb	*Brachystegia boehmii*	Ds	*Dombeya shupangae*	Saf	*Spirostachys africana*
Bh	*Brachylaena hutchinsii*	Dst	*Dalbergia stuhlmannii*	Sb	*Sclerocarya birrea*
Bn	*Bridelia niedenzui*	Esp	*Euphorbia spp*	Si	*Strychnos innocua*
Bs	*Boscia salicifolia*	Fc	*Fagara chalybea*	Ta	*Thylachium africanum*
Ca	*Combretum apiculatum*	Fv	*Flueggea virosa*	Ts	*Teclea simplicifolia*
Cab	*Cassia abbreviata*	Gsp	*Grewia spp*		
Ce	*Carissa edulis*	Gz	*Gelonium zanzibarense*		

Fig. 10.3 Positive associations between plant species in woodland in Tanzania. (From Welch, 1960; courtesy of *J. Ecol.*)

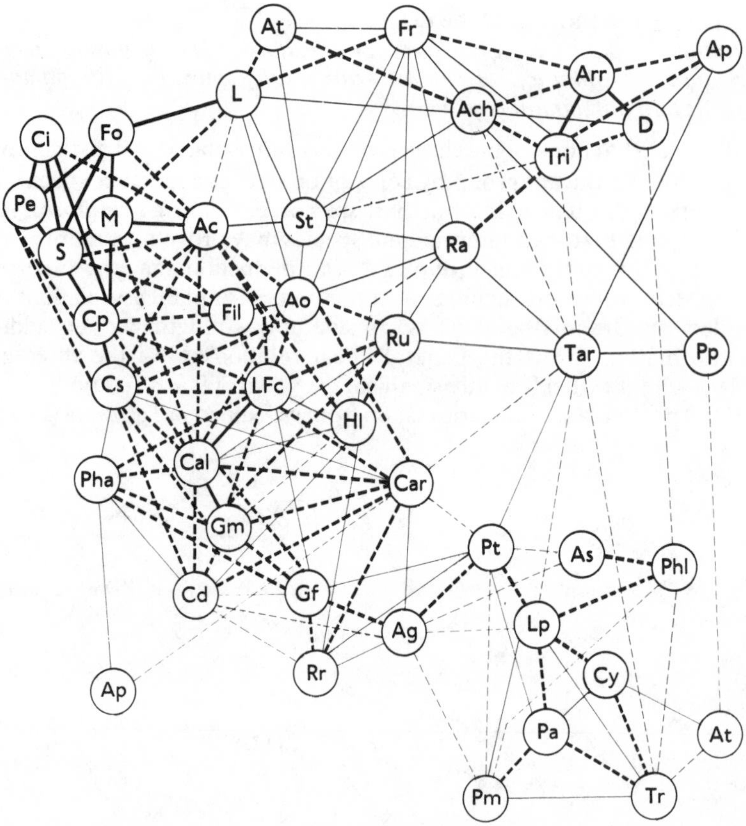

Ac	*Agrostis canina*	Cy	*Cynosurus cristatus*	Pha	*Phalaris arundinacea*
Ach	*Achillea millefolium*	D	*Dactylis glomerata*	Phl	*Phleum pratense*
Ag	*Alopecurus geniculatus*	Fil	*Filipendula ulmaria*	Pm	*Plantago major*
Ao	*Anthoxanthum odoratum*	Fo	*Festuca ovina*	Pp	*Poa pratensis*
Ap	*Alopecurus pratensis*	Fr	*Festuca rubra*	Pt	*Poa trivialis*
Arr	*Arrhenatherum elatius*	Gf	*Glyceria fluitans*	Ra	*Ranunculus acer*
As	*Agrostis stolonifera*	Gm	*Glyceria maxima*	Rr	*Ranunculus repens*
At	*Agrostis tenuis*	Hl	*Holcus lanatus*	Ru	*Rumex acetosa*
Cal	*Caltha palustris*	L	*Luzula campestris*	S	*Sieglingia decumbens*
Car	*Cardamine pratensis*	LFc	*Lychnis flos-cuculi*	St	*Stellaria graminea*
Cd	*Carex disticha*	Lp	*Lolium perenne*	Tar	*Taraxacum officinale*
Ci	*Cirsium dissectum*	M	*Molinia caerulea*	Tr	*Trifolium repens*
Cp	*Carex panicea*	Pa	*Poa annua*	Tri	*Trisetum flavescens*
Cs	*Carex stolonifera*	Pe	*Potentilla erecta*		

Fig. 10.4 Positive correlations between plant species in grassland, forming a constellation. (From DeVries 1953; courtesy of *Acta Bot. Neer.*)

The very obvious grouping on the left of the diagram represents a group of species which form dense thickets; the central group linked to the thicket species through *Albizia, Canthium* and *Acacia pennata* represents a group of pioneer species which invades abandoned cultivated land. Similarly the group of species on the right of the diagram represents a woodland association dominant in some adjacent areas. The three groups in fact reflect the initial and secondary stages of two lines of succession. DeVries (1953) examined a large number of grassland sites in Holland and estimated the percentage frequency of the species in each of the sample sites. Correlation coefficients were calculated and the data presented in the form of a two-dimensional diagram (Fig. 10.4) based on positive correlations similar to the *Juncus effusus* example above. Again a number of groupings are present, all of them interrelated yet forming fairly distinct ecological groupings.

Group 1 (top left of the diagram)
Molinia caerulea, *Sieglingia decumbens*, *Carex panicea*, *Potentilla erecta* and *Cirsium dissectum*, typical of acidic soils of low fertility.

Group 2 (bottom left of the diagram)
Glyceria maxima, *Caltha palustris*, *Phalaris arundinacea*, *Lychnis flos-cuculi*, *Carex disticha*, *Ranunculus repens*, *Glyceria fluitans* and *Alopecurus geniculatus*, typical of moist neutral-acidic grassland.

Group 3 (top right of the diagram)
Arrhentherum elatius, *Trisetum flavescens*, *Dactylis glomerata*, typical of dry hay meadows.

Group 4 (bottom right of the diagram)
Lolium perenne, *Cynosurus cristatus*, *Poa pratense*, *Agrostis stolonifera*, *Plantago major* and *Trifolium repens*, typical of fertile neutral pastures.

This method of approach has several advantages: the initial subjective choice of site is eliminated to a large extent and a partial-random choice can be made at least. The initial decision of the type of site (a grassland, woodland, etc.) is obviously unavoidable. Secondly the groupings of species are completely objective and present a more realistic picture with the interrelationships and almost continuous variation of the groupings readily appreciated. At the same time the element of constancy is present and Agnew (1961) suggests *Trifolium repens*, *Potentilla palustris* and *Galium saxatilis* represent constants for groups 1, 2 and 3 respectively. The major disadvantage of the approach lies in the lengthy computation necessary where the number of sites is large. Since the larger the number of samples the more complete and significant the interrelationships become, this is a serious disadvantage. The lack of over-all sensitivity of the method also detracts from its use as a standard approach. It is retained here however, as one of the original ordination methods that was used extensively, but particularly to illustrate dimensional distortion which is inherent to many ordination procedures (see p. 192).

Although environmental gradients are also implied, the species constellation diagram does not show them clearly. Corresponding plot constellation diagrams are usually too complicated and cumbersome to attempt, even though an environmental gradient could be evidenced as an equivalent abundance gradient of species components in the plot ordination. That species abundance can be equated to environmental parameters is an important concept with far reaching implications (see below).

A more effective way of ordinating plots in relation to an environmental gradient is that developed by Cottam (1949) and Curtis and McIntosh (1950; 1951). This development is usually called the *Continuum approach*. It implies a continuous variation of vegetation, and thus a viewpoint that vegetation is a *continuum* and cannot be classified into discrete entities (associations in the Braun-Blanquet sense). Furthermore rather than sampling the presence or absence of a species as a simple attribute of the vegetation, the method employs abundance measures. The definitive data were obtained from a randomized sample of woodlands in N. America. The choice of a sample site was based on criteria which avoided preconceived ideas and eliminated the conscious selection of sites inherent in the Braun-Blanquet approach. Thus any woodland stand was sampled which was 'natural' (not planted), of adequate size (6 hectares being the minimum) and not disturbed as far as could be determined by grazing, excessive cutting or fire. All the species of trees, herbs, shrubs were included and the frequency, density and dominance measured. Cottam (1949) defined frequency as the percentage occurrence of a species in the total sample,density as the total number of individuals of a species relative to the total area examined, and 'dominance' is expressed as basal area. From the frequency, density and dominance values a 'relative importance' value was derived by addition of the values, the DFD index (Curtis, 1947). subsequently these measures were modified by Curtis and McIntosh (1950) and used in the form of:

$$\text{Relative density} = \frac{\text{Number of individuals of species x}}{\text{Total number of individuals of all species}} \times 100$$

$$\text{Relative frequency} = \frac{\text{Frequency of species x}}{\text{Sum of frequency values of all species}} \times 100$$

and similarly relative dominance is used intead of dominance.

These measures are totalled and constitute the *importance value*, which can range from 0 to 300. The subsequent treatment of the data is most suitably described by reference to a specific set of data, given by Brown and Curtis (1952). The data relate to upland conifer-hardwood forests of northern Wisconsin, the sampling sites being chosen by the criteria listed above with the additional feature of the absence of standing run-off water. For each stand in turn, an importance value is calculated for each species present and by inspection of the importance values each stand is assigned a *leading dominant*, i.e. that species with the highest importance value. Having designated the leading dominant in each stand, the data are then sub-divided so that each group of stands has the same leading dominant. Obviously some of the subordinate species in any one stand will be the leading dominants of other stands and consequently a fresh importance value is then calculated for each species based on the group of stands with their common leading dominant. In the conifer-hardwood data the leading dominants (Table 10.1) are *Tsuga canadensis, Quercus rubra, Betula lutea, Acer rubrum, Betula papyrifera, Pinus strobus, Populus tremuloides, Pinus resinosa, Quercus ellipsoidalis* and *Pinus banksiana* in decreasing order of importance (importance values 25, 22, 21, 7, 6, 1, 1, 0, 0 and 0 respectively). *Acer rubrum* and *Betula lutea* do not achieve dominance in any stand and accordingly are not included by Brown and Curtis in the provisional arrangement of 'leading dominants'. These two species considered apart, the remaining mean importance value in those stands dominated by *Acer saccharum* gives a general basis for the arrangement of the leading dominants in a phytosociological order. Thus *Pinus banksiana* has an importance value of 0 and is generally ecologically accepted as a typical pioneer species in Wisconsin, the remaining species falling between this and the final 'climax' association dominated by *Acer saccharum*.

The next highest importance value to *Acer saccharum* is *Tsuga canadensis* and accordingly this is placed in an adjacent position to *Acer saccharum*. Confirmation of the validity of this is obtained from the relative size of the mean importance value of the species in the *Tsuga* dominated group. Thus (discounting *Betula lutea*) *Acer saccharum* has the highest importance value and should thus be close to the group which is in fact dominated by *Acer saccharum*. On this basis and with due

Table 10.1 Average importance value of trees in stands with given species as leading dominant – 104 stands. (From Brown and Curtis, 1952; courtesy of *Ecol. Monog.*)

Number of stand	Leading dominant	*Acer saccharum*	*Tsuga canadensis*	*Betula lutea*	*Acer rubrum*	*Quercus rubra*	*Betula papyrifera*	*Pinus strobus*	*Pinus resinosa*	*Populus tremuloides*	*Quercus ellipsoidalis*	*Pinus banksiana*
23	*Acer saccharum*	145	25	21	7	22	6	1	–	1	–	–
23	*Tsuga canadensis*	40	152	47	11	3	5	4	3	–	–	–
6	*Quercus rubra*	27	1	3	29	138	23	10	8	5	3	–
6	*Betula papyrifera*	48	8	7	27	16	108	19	1	29	1	–
19	*Pinus strobus*	12	6	2	24	12	12	150	39	9	5	–
9	*Pinus resinosa*	3	–	1	12	15	14	56	156	24	4	2
4	*Populus tremuloides*	11	–	–	10	29	34	14	19	140	–	–
4	*Quercus ellipsoidalis*	–	–	–	5	7	1	11	9	9	103	56
10	*Pinus banksiana*	–	–	–	3	3	3	13	12	14	36	213

consideration to the columns of numbers both horizontal and vertical, approaching at least a smooth curve, the remaining leading dominants are arranged. On the basis of the order of the leading dominants the remaining tree species are included, and an arbitrary climax adaptation number ranging from 1 to 10 is assigned to each species (Table 10.2). *Pinus banksiana* is given an adaptation number of 1 and similarly *Acer saccharum* 10 representing the two ends of the continuum. The remaining species receive adaptation numbers of intermediate value so that species which frequently occur together and have similar environmental requirements, have the same or closely similar adaptation numbers. Finally the importance value of each species in each sample is multiplied by the adaptation number, the total sum of these values for each stand being termed the *continuum index* of the stand. This index is regarded as a measure of the total environment of a stand of trees expressed in terms of species composition and their relative abundance. Brown and Curtis (1952) give a detailed table of data (Table 10.3) representative of the information extracted by the method from a mass of otherwise extremely variable data. The results are best appreciated as a series of graphs. A considerable scatter of points is usually present and the data are 'smoothed' by a standard procedure given by the formula

$$B = \frac{n_1 a + 2n_2 b + n_3 c}{n_1 + 2n_2 + n_3}$$

where B is the smoothed value of average b and n_1, n_2 and n_3 are the numbers of stands from which the averages a, b and c have been derived. The relationship between the importance value and continuum index is a Gaussian type of curve or part of such a curve (Figs 10.5 and 10.6), and clearly there is no grouping of curves in zones of the continuum index as should be the case if there were discrete and recognizable associations. On the contrary there is always a continuous overlap of

Table 10.2 Climax adaptation numbers of tree species found in stands studied. (From Brown and Curtis, 1952; courtesy of *Ecol. Monog.*)

Tree species	**Climax adaptation number**
Pinus banksiana	1
Quercus ellipsoidalis	2
Quercus macrocarpa	2
* *Populus balsamifera*	2
Populus tremuloides	2
Populus grandidentata	2
Pinus resinosa	3
Pinus pensylvanica	3
Quercus alba	4
Prunus serotina	4
Prunus virginiana	4
Pinus strobus	5
Betula papyrifera	5
* *Juglans cinerea*	5
Acer rubrum	6
* *Acer spicatum*	6
* *Fraxinus nigra*	6
* *Picea glauca*	6
Quercus rubra	6
Abies balsamea	7
* *Thuja occidentalis*	7
* *Carpinus caroliniana*	7
Tsuga canadensis	8
Betula lutea	8
* *Carya cordiformis*	8
Fraxinus americana	8
Tilia americana	8
Ulmus americana	8
Ostrya virginiana	9
Fagus grandifolia	10
Acer saccharum	10

* Climax adaptation number of these species is tentative only, because of their low presence in the stands studied.

several adjacent leading dominants. Similarly the species with intermediate and with lower importance values show the same overlap over a considerable range of the continuum index. It is of considerable interest that measures of environmental factors taken simultaneously with sampling of each stand show marked trends. The specific gravity of screened soil samples, the moisture holding capacity and the available calcium, in the A_1 horizon all follow a very definite trend in relation to the continuum index (Figs 10.7, 10.8 and 10.9). In addition to the inter-relationship of the species, some interactions between the vegetational continuum and the environment are thus suggested.

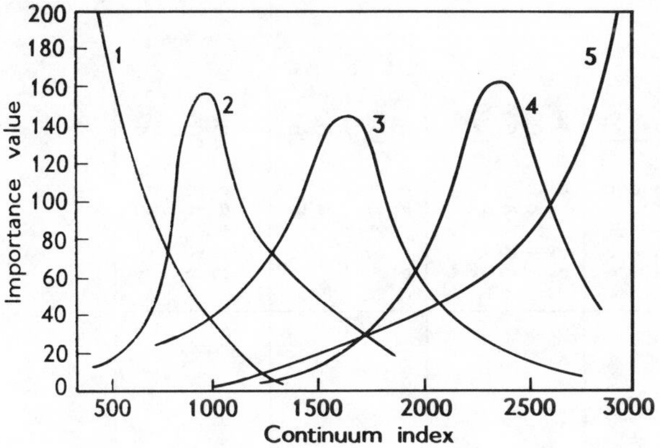

Fig. 10.5 Importance value curves for the five leading tree species (1) *Pinus banksiana*, (2) *P. resinosa*, (3) *P. strobus*, (4) *Tsuga canadensis*, (5) *Acer saccharum*. (From Brown and Curtis, 1952; courtesy and *Ecol. Monog.*)

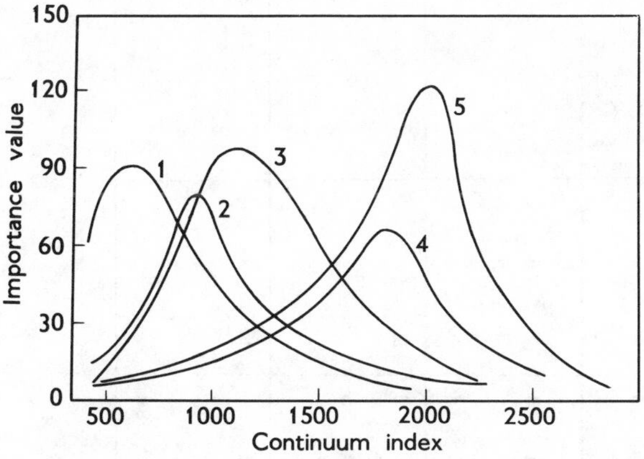

Fig. 10.6 Importance value curves for five intermediate tree species. (1) *Quercus ellipsoidalis*, (2) *Populus tremuloides*, (3) *Quercus alba*,(4) *Betula papyrifera*,(5) *Quercus rubra*. (From Brown and Curtis, 1952; courtesy of *Ecol. Monog.*)

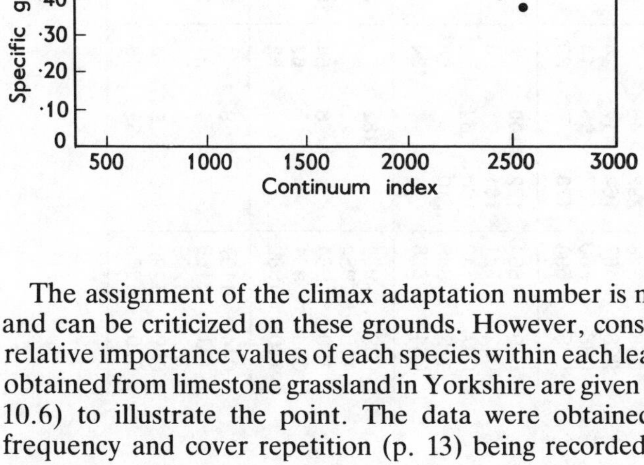

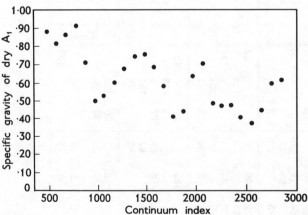

Fig. 10.7 Specific gravity of screened samples of the A_1 soil horizons of stands arranged in 100 unit classes. (From Brown and Curtis, 1952; courtesy of *Ecol. Monographs.*)

The assignment of the climax adaptation number is necessarily partly subjective and can be criticized on these grounds. However, considerable use is made of the relative importance values of each species within each leading dominant group. Data obtained from limestone grassland in Yorkshire are given below (Tables 10.4, 10.5 and 10.6) to illustrate the point. The data were obtained from a number of sites, frequency and cover repetition (p. 13) being recorded for each species and used

Table 10.3 Original data for importance values of seventeen species and for three soil characters. (From Brown and Curtis, 1952; courtesy of *Ecol. Monog.*)

Stand Number	Continuum Index	*Pinus banksiana*	*Quercus ellipsoidalis*	*Populus tremuloides*	*Populus grandidentata*	*Pinus resinosa*	*Quercus alba*	*Pinus strobus*	*Betula papyrifera*	*Acer rubrum*	*Quercus rubra*	*Abies balsamea*	*Betula lutea*	*Tsuga canadensis*	*Ulmus americana*	*Tilia americana*	*Ostrya virginiana*	*Acer saccharum*	Calcium in A_1 layer (100's lb. per acre)	ph of A_1 layer	WRC of A_1 layer
084	356	**272**	4	9		12				4									4.3	5.1	77
066	375	**237**	46	7	6		5												4.3	4.9	36
068	467	164	**117**	4			16												2.9	4.6	54
098	560	157	**95**	14				5	15	14									2.1	4.6	164
055	594	**178**		30	7	46		39											1.4	4.5	102
049	664	122	**100**				**62**	6		11									2.9	4.0	118
082	721	**181**		4		34		59	4	4		4							1.1	4.6	99
067	749	100	62	9	8	7	**101**				5								2.9	4.7	54
113	848	10		**68**		**204**		14	3										6.4	4.3	148
064	868			**194**	24	7		34	16	14	4								–	–	–
039	934		**132**		**27**		**107**		4	3	22								2.1	4.7	60
115	954				4	**268**		16	12										7.2	5.4	110
150	984		6	54	**162**	5			**30**	35	3								–	–	–
053	991			**63**		**159**		74	4										–	4.3	400
140	1082			**62**		116		121											3.6	5.1	281
044	1107		8	4	20	**153**	11	82	3		18								–	–	–
141	1168	7	4	31	**33**	65		145		3	7			3					3.6	5.2	400
132	1209				5	138		157											0.7	4.5	269
059	1290		5	4		96		**186**	7			3							–	–	–
070	1363		32	16			**109**	46		**84**									4.3	5.7	43

050	1403	3		3	41		**243**	10										7.1	4.9	49
089	1436				41		**238**	5	10	4								0.7	5.2	126
107	1508		15		38		**180**	3	15	3	**40**				3		3	6.4	5.2	106
106	1518		12	18	51		42	21	**54**	**84**	3						12	8.6	5.5	180
090	1573			12	15	11	166		39	13	3	3	16			3		3.2	5.5	65
105	1582		17		9		158	3	9	82				3	6		6	8.6	5.0	225
043	1640			5		39	121	7	**65**	22				20	3	3	3	5.4	4.9	89
147	1750		7	22			4	**52**	20	**143**		10	7		3	3	19	5.4	5.2	272
153	1807		10	10	3		56	**96**	6	47	**16**						57	2.1	5.3	288
086	1888				61		54	12	4			11	158					0.4	4.0	192
102	1950	4				10	13	20	26	**141**		3			19	7	35	2.7	4.9	91
095	1966			4			106		**58**			19	84				21	1.1	5.3	67
101	2060			12		11		3	9	**124**				3	17	6	89	12.5	5.5	80
119	2173		17						3		**29**	7	49		3		73	35.7	6.1	373
139	2215							11	41		6	12	**176**	4			11	3.2	4.6	270
138	2234						12		11		3	**81**	111				15	2.1	3.7	409
104	2287		10			11	8		19	59		15		17	20	12	102	16.0	4.5	289
116	2314	7		**28**				**29**	24	33	8					11	161	7.2	5.2	209
111	2366							11	32	5		52	157		4	3	33	8.6	5.0	223
063	2379							13	6			60	**206**				15	–	–	–
136	2404								12		**25**	**72**	145			3	33	5.4	4.6	400
085	2430						15	12				15	**198**			9	24	14.3	5.5	258
062	2464											**66**	**182**		16		31	7.1	5.6	210
134	2502								28		7	**84**	87		8		82	4.3	4.4	335
118	2520										7	29	160		19	11	66	12.9	5.0	497
125	2553											58	106	**25**	**34**	4	73	3.6	5.4	398
034	2591						10	35	5	8	3	8	27		13	**23**	115	45.7	5.6	365
121	2617								4			35	69		**49**	8	111	28.6	5.4	131
103	2636											8		**80**	**30**	6	124	42.8	5.5	305
135	2711										4	22	86	3	19	**18**	149	42.9	6.9	119
080	2747											40	34	11	22	**21**	162	6.4	4.8	158
094	2780											17	13	8	**44**	13	**183**	23.6	5.3	129
076	2817											3	3	**23**	8	**30**	**184**	51.0	7.6	137
133	2852											12		**32**	16	10	**220**	10.7	5.7	526
114	2938									3		6			8	7	**268**	10.0	5.0	157

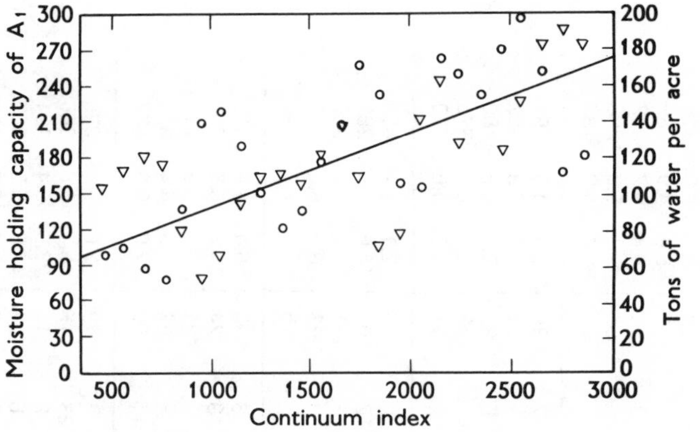

Fig. 10.8 Moisture holding capacity of the A_1 horizon in percentage, and total water held in the entire A_1 horizon at field capacity in tons per acre. Points are average values in 100 unit classes. (From Brown and Curtis, 1952; courtesy of *Ecol. Monographs.*)

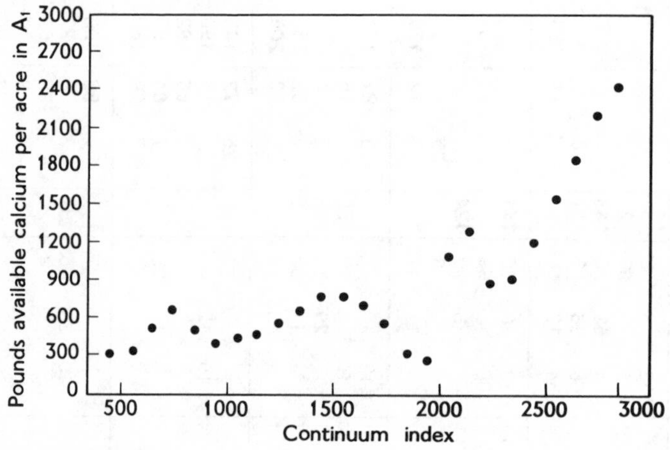

Fig. 10.9 Available calcium in the A_1 horizon in pounds per acre. Points are average values in 100 unit classes. (From Brown and Curtis, 1952; courtesy of *Ecol. Monographs.*)

jointly as a measure of the importance value. The limestone area is very variable, the soils ranging from pockets of acidic boulder clay to dark soils immediately overlying limestone rock with a pH above 7.0. The grass cover reflects this wide range of pH, varying from *Nardus stricta* dominated areas to almost pure stands of *Sesleria caerulea*. The importance value calculated from the relative frequency and relative cover of each species, shows *Festuca rubra, Agrostis tenuis, Nardus stricta* and *Sesleria caerulea* to be the leading dominants in the area. The data are accordingly split into these four groups and importance values calculated for each species (Table 10.4).

The leading dominants ecologically form a series with *Nardus* and *Sesleria* at the two extremes and *Agrostis/Festuca* falling in an intermediate position. The importance values in each group confirm this relationship *Agrostis* and *Festuca* having a similar relationship at the acidic end of the scale, but *Festuca* being clearly more closely related to *Sesleria* at the basic end. The remaining species are included in the comparison and importance values calculated for each species in each group

Table 10.4 Importance values of the four most abundant species in relation to the leading dominant in an area of grassland in Yorkshire.

		Nardus	Agrostis	Festuca	Sesleria
Leading dominant	*Nardus*	58.6	44.0	49.5	0
	Agrostis	36.0	58.1	50.2	4.5
	Festuca	15.5	37.5	53.9	12.7
	Sesleria	0	13.6	45.1	61.6

Table 10.5 The importance values of each species present in an area of grassland overlying limestone in Yorkshire in relation to four leading dominants.

Species	Festuca group	Agrostis group	Sesleria group	Nardus group
Festuca rubra	53.9	50.2	45.1	49.5
Agrostis tenuis	37.5	58.1	13.6	44.0
Sesleria caerulea	12.7	4.5	61.6	0
Nardus stricta	15.5	36.0	0	58.6
Koeleria gracilis	9.3	0	9.7	0
Thymus drucei	3.8	3.6	11.7	0
Linum catharticum	2.5	0	11.2	0
Viola riviniana / lutea	5.8	3.6	11.4	0
Carex flacca / panicea	11.9	7.9	18.7	0
Deschampsia caespitosa	3.9	7.9	0	0
Trifolium repens	6.2	0	5.5	0
Potentilla erecta	8.4	0	0	0
Anthoxanthum odoratum	5.1	0	1.0	0
Galium hercynicum	10.6	22.8	0	22.0

(Table 10.5). From Table 10.5 it is now possible to arrange all the species in their relative order and assign to each of them an adaptation number. *Nardus* is arbitrarily designated 1 and is most closely related to *Agrostis, Galium hercynicum* is similarly closely related to both *Agrostis* and *Nardus* and in Table 10.5 shows a slightly closer relation to *Agrostis* than to *Nardus*. Accordingly *Galium* is allocated an adaptation number of 3 and *Agrostis* 4 (Table 10.6). The species of maximum importance left in the *Agrostis* group are *Carex* and *Deschampsia* and reference to the *Festuca* group shows *Deschampsia* to be much more closely related to *Agrostis* and accordingly is allocated 5 as its adaptation number. At the other end of the series *Sesleria* is arbitrarily designated as 10, and clearly *Carex* is closely related to it and allocated an adaptation number of 9. Similarly *Thymus, Linum, Viola* and *Koeleria* are more or less equally related to *Sesleria* and are thus all give an adaptation number of 8. *Festuca* is allocated an adaptation value of 7 and the remaining three species referred to a value of 6. These three species, *Anthoxanthum, Trifolium* and *Potentilla*, all have low importance values and are unfortunately of relatively infrequent occurrence in the samples available. It is possible that a larger sample might indicate a reversal of the order of *Festuca* and these three species, but it is unlikely that the present arrangement will unduly influence the continuum indices of the samples. This final arrangement of species is given in Table 10.6.

Table 10.6

Species	Adaptation number
Nardus stricta	1
Galium hercynicum	3
Agrostis tenuis	4
Deschampsia caespitosa	5
Anthoxanthum odoratum	6
Trifolium repens	6
Potentilla erecta	7
Festuca rubra	7
Koeleria gracilis	8
Viola rivinina / lutea	8
Linum catharticum	8
Thymus drucei	8
Carex flacca / panicea	9
Sesleria caerulea	10

Thus the allocation of an adaptation number is based largely on the relative orders of the importance values of each of the species and the element of subjective assessment necessary is slight and restricted to the species with a low percentage occurrence. This can be circumvented by taking a more adequate sample but the labour of such an increase in sample size could be excessive in some instances. The general relationship of the four most abundant species again shows a considerable overlap of their range (Fig. 10.10) relative to the continuum index and is broadly similar to the results given by Brown and Curtis (1952). There is no discrete zone solely dominated by one species, or a group of species, but instead there is a continuous variation of composition along the continuum index.

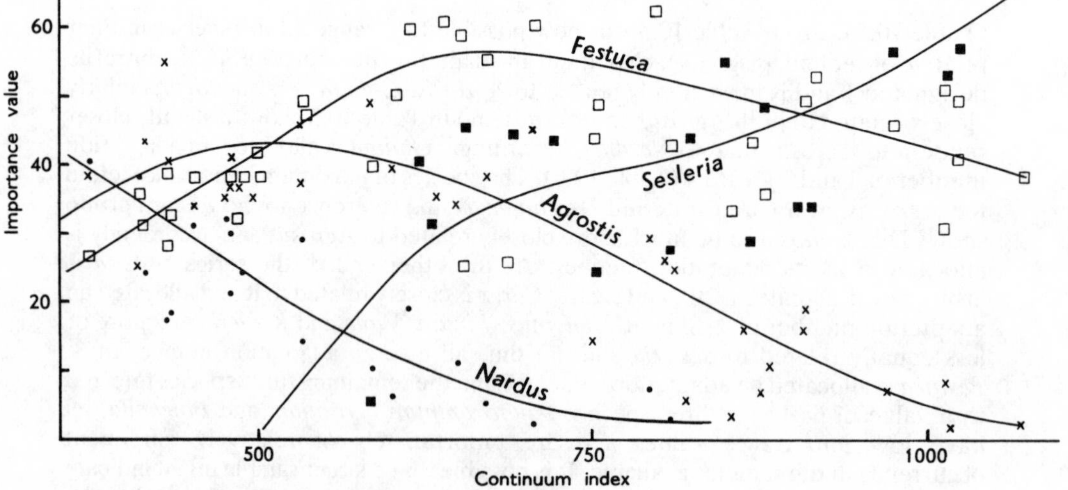

Fig. 10.10 The importance value/continuum index curves for the four leading dominants of sample plots in limestone grassland.

In essence the continuum approach is using a single axis which is proportional in length to the abundance of the commonest (dominant) species in the sample. Each plot sample is then positioned on that axis. The so-called climax adaptation numbers are simply *weighting factors* to 'magnify' the effect of inter-plot abundances and thus distances. All the species are used to assess inter-plot distance and where there is a *single overriding environmental parameter* controlling the vegetation, only then will the continuum approach produce a sensible result. However, when two or more separate parameters are controlling the plant associations sampled, a single axis will not form an effective framwork on which to ordinate the samples. Two or more axes will be required.

Using a simple vector representation, two environmental parameters A and B, that operate on vegetation independently can be represented as:

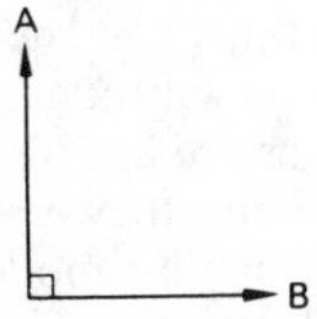

and similarly:

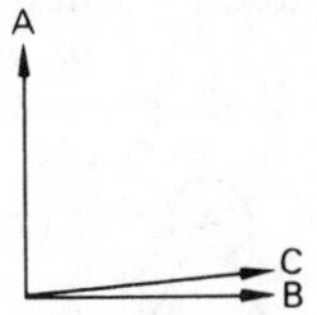

represents three environmental parameters A, B and C where B and C are closely correlated together but both unrelated to A. For a series of plant associations 'A' could be pH, 'B' aspect and 'C' temperature, and correlated with each axis would be a range of species abundances which in turn were controlled by the parameters pH and temperature. For example:

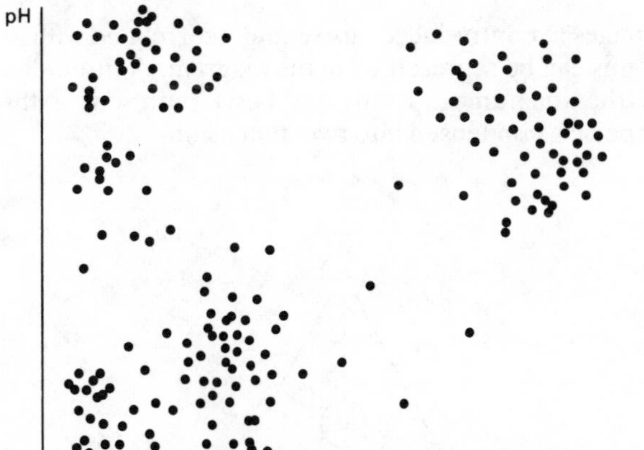

In this simple ordination model we have a number of plots on warm (south facing) slopes with high pH, a group with cooler temperatures and with high pH, and a large group of low pH, low temperature plots. A simple ordination using the continuum approach would give a very distorted image of the real situation. Since many environmental factors operate simultaneously on vegetation and only in a gross climatic sense are there usually overriding parameters, the continuum as a method will have a very restrictive usage; effectively only in geographical surveys where a single dominant factor is operating.

In the simple ordination model used above, instead of the axes being pH and temperature respectively they could be two of the commoner *species* whose distribution is controlled by pH and temperature. In general, ordination axes are usually related to the abundance of 'indicator' species, where hopefully, information about the ecology of each species is available and can be used to identify the environmental variables correlated with each axis. Often the correlation of an axis with an environmental variable is intuitive and necessitates a considerable amount of experimental and environmental work subsequently.

Our model is still simple. The axes are arbitrarily at right angles to each other by convention, there are only two of them, they lie on a single plane surface, and they are straight lines. This is a closely similar model to that used previously (p. 176 above). In addition, as was pointed out above, in the construction of a constellation diagram from the scaled reciprocals of the χ^2 values between species pairs, some approximation is necessary. This arises during the construction of the diagram in the following way.

If A, B and C are equally correlated together then this relationship can be simply represented as:

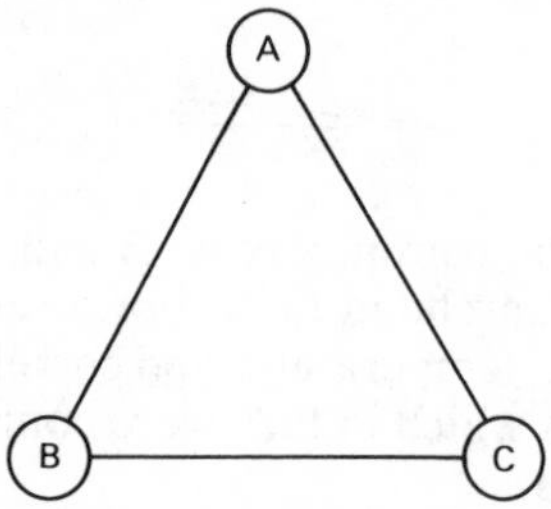

When a fourth species D is introduced, also equally correlated with A, B and C, there is no simple way this can be represented in the diagram. A three-dimensional model is required and the diagram can only (at best) represent a three-dimensional arrangement of species, condensed into two dimensions:

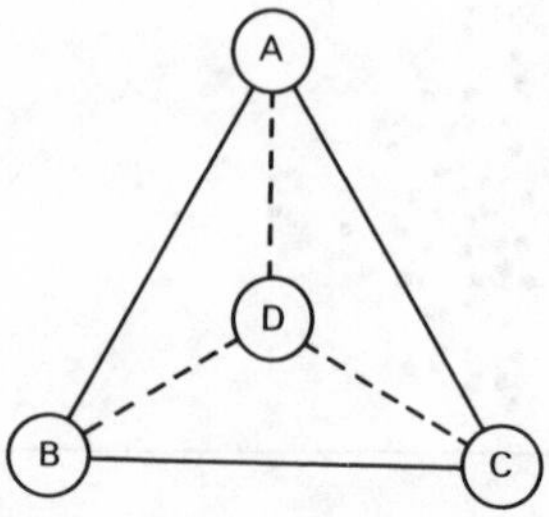

When further species are added to the diagram, numerous other distortions are introduced resulting in the necessary approximations of interspecies distances. This is a general situation. To represent accurately intercorrelation distances between either species or plots, necessitates many dimensions. Three species or plots can be represented in two dimensions, four requires three dimensions and so on. The difficulty of representation in more than three dimensions at once (other than in a mathematical sense) needs some simplification and usually only three axes are considered at any one time.

Thus our ordination model can now become:

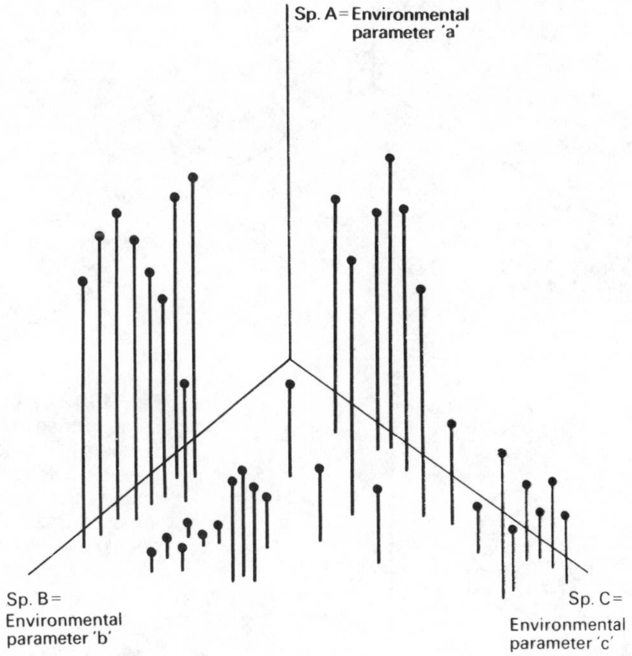

An example of such an ordination is given below (Fig. 10.11) of 130 plots sampled from lichen heath in Northern Ontario. The three species chosen as axes were three of the most abundant in the area sampled. *Cetraria nivalis* is representative of slightly moist conditions, *Cladonia rangiferina* characteristic of the general associations on the intermediate and dry slopes, and *Dryas integrifolia* dominant on the exposed and dry summits of mounds and low ridges. As a dimensional ordination approach its chief merit is its extreme simplicity. It is a gross approximation since only *three* of the N-species are being used. Having spent considerable time and effort sampling the total species complement, throwing most of it away subsequently is totally unacceptable. Even more serious, information relating to the structure of the ordination, carried by the other common species, is also being thrown away. The extent of this information loss can be seen by comparing Figs 10.11 and 10.12 which is an analysis of the same lichen heath data using a component analysis (see below).

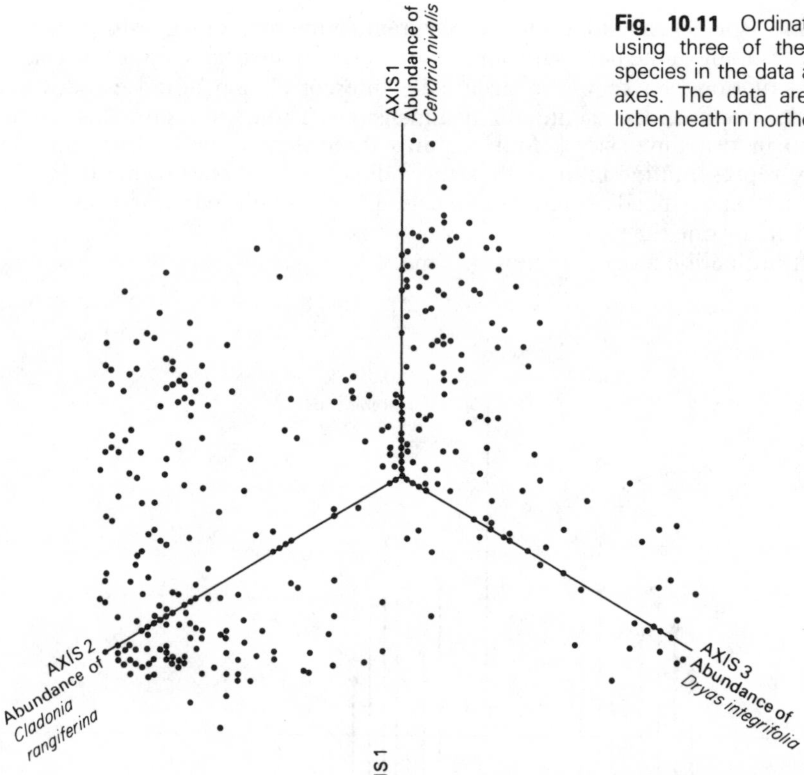

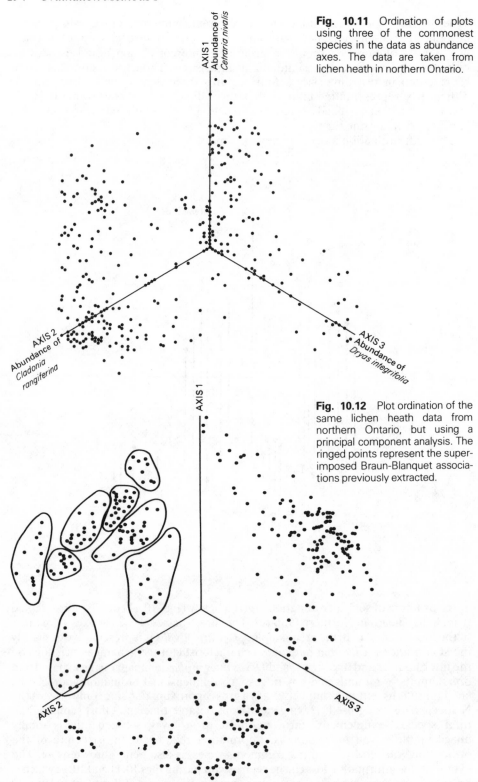

Fig. 10.11 Ordination of plots using three of the commonest species in the data as abundance axes. The data are taken from lichen heath in northern Ontario.

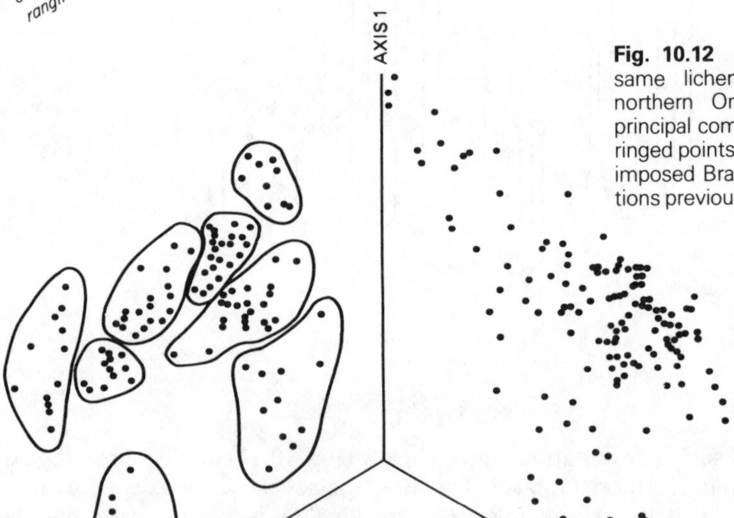

Fig. 10.12 Plot ordination of the same lichen heath data from northern Ontario, but using a principal component analysis. The ringed points represent the superimposed Braun-Blanquet associations previously extracted.

We arrive then, at our final model situation for a plot ordination:

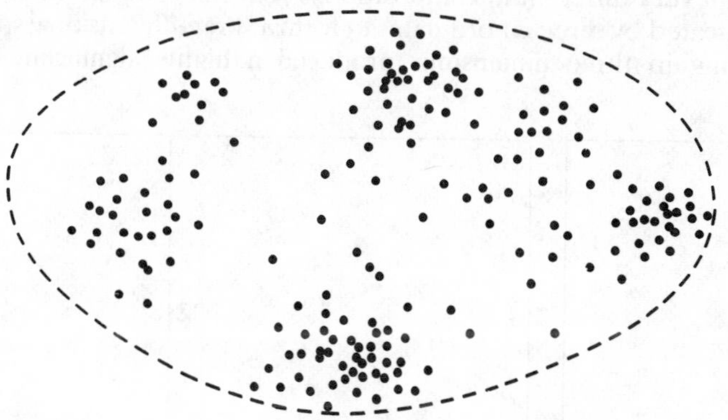

We know the sample plots lie in some type of spatial cluster of ellipsoidal form. We do not want to use simple species axes, because this is too wasteful. But certainly axes *are* required to which the data can be related. Three possible configurations could be:

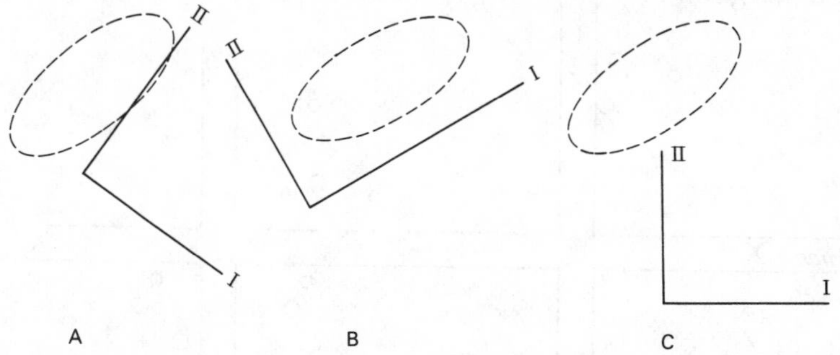

These are an infinite number of such possible axes to choose from. Intuitively 'B' would seem to be better than 'A' and 'C', but the objective reasons for this decision involve some complex mathematical concepts and techniques.

One simple approach is that given by Bray and Curtis (1957), using a series of coefficients of similarity to set up a primary axis and subsequent continuum axes. The sampling sites were chosen at random as before and quantitative measurements taken of the tree, shrub and herb population in each stand (density and basal area for the trees, frequency for the herbs and shrubs). Fifty-nine stands were sampled and each of these stands was compared with the other 58 and a coefficient of similarity calculated. The coefficient used was that proposed by Gleason (1920), $C = 2w/(a+b)$, where a and b are the *quantities* of all the plants found in the two stands to be compared and w is the sum of the lesser value for those species common to the two stands. Thus if the two stands were absolutely identical a and b and w would be equal and C then equals 1. Conversely with complete dissimilarity $C = 0$. From the 1711 values thus obtained the two most dissimilar stands are chosen as the end points of an axis and the remainder of the stands located on the axis by using the inverse value of the other coefficients as ordinates. From this arrangement several stands were found to be more or less equidistant from the two ends of the first axis but were still very dissimilar. These were examined and the most dissimilar pair again used to establish a second axis. The stands were repositioned by the two axes, again using the inverse coefficients as co-ordinates. A third axis was finally established on the most dissimilar pair of stands which occurred equidistant from the ends of the axes but which were still very different in composition. From the resultant three axes, each stand is thus located by three co-ordinates to form a three-dimensional spacing. This final positioning in three dimensions produced a highly significant correlation

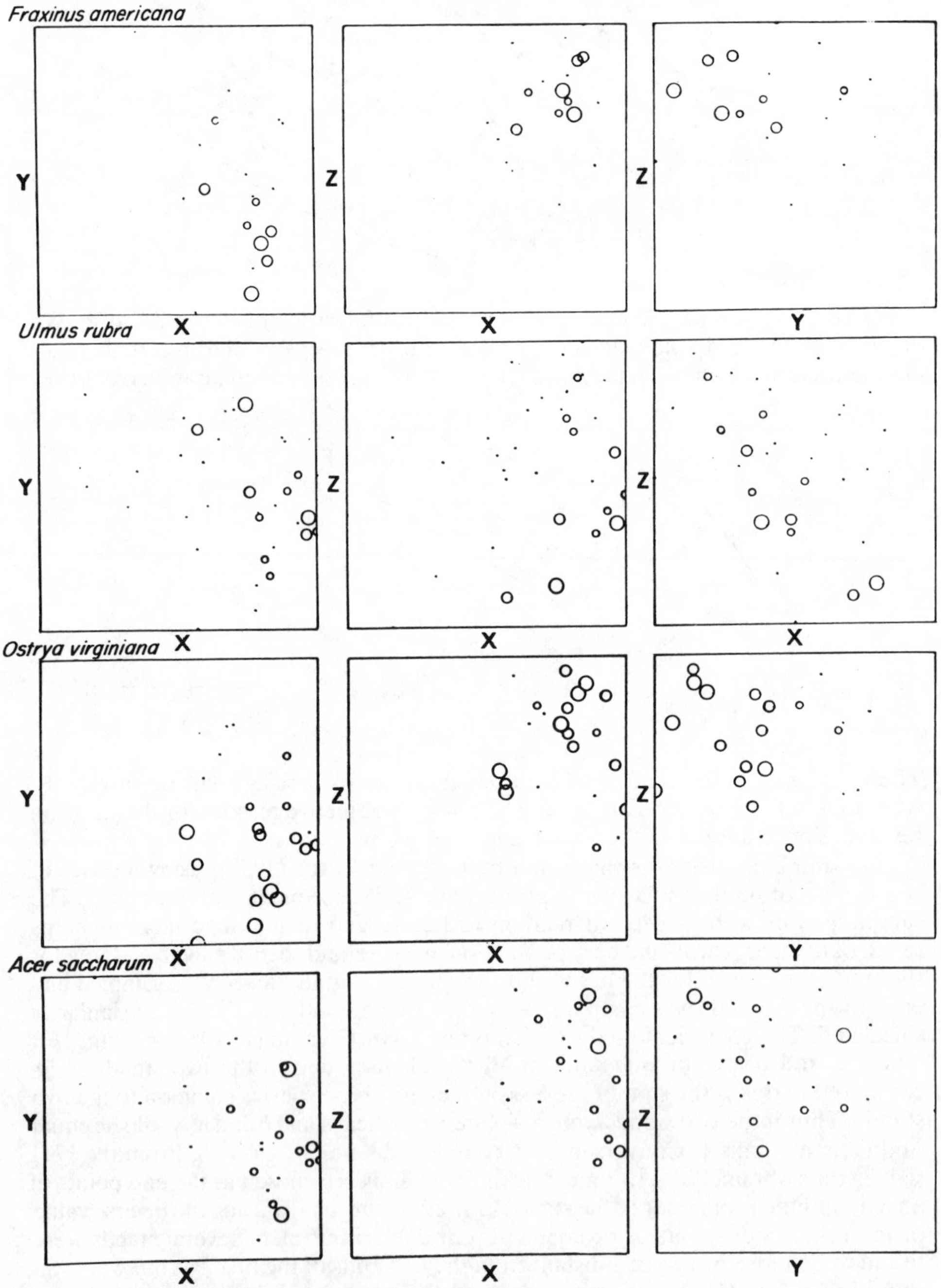

Fig. 10.13 and 10.14 (*opposite*) The dominance behaviour of the eight most important tree species at three levels of the ordination. (From Bray and Curtis, 1957; courtesy of *Ecol. Monographs.*)

between the spacing of the stands and the original coefficient calculated from the data. From the spatial arrangement of the stands the arrangement of the species can be readily determined although there is some difficulty in representing the three-dimensional spacing on paper. Bray and Curtis give data of the dominant tree species (Figs 10.13 and 10.14). Each species shows a definite spatial 'centre of importance' (related to the number and size of the balls). Away from the centre the concentration and importance of a species diminishes, but not equally in different directions. The

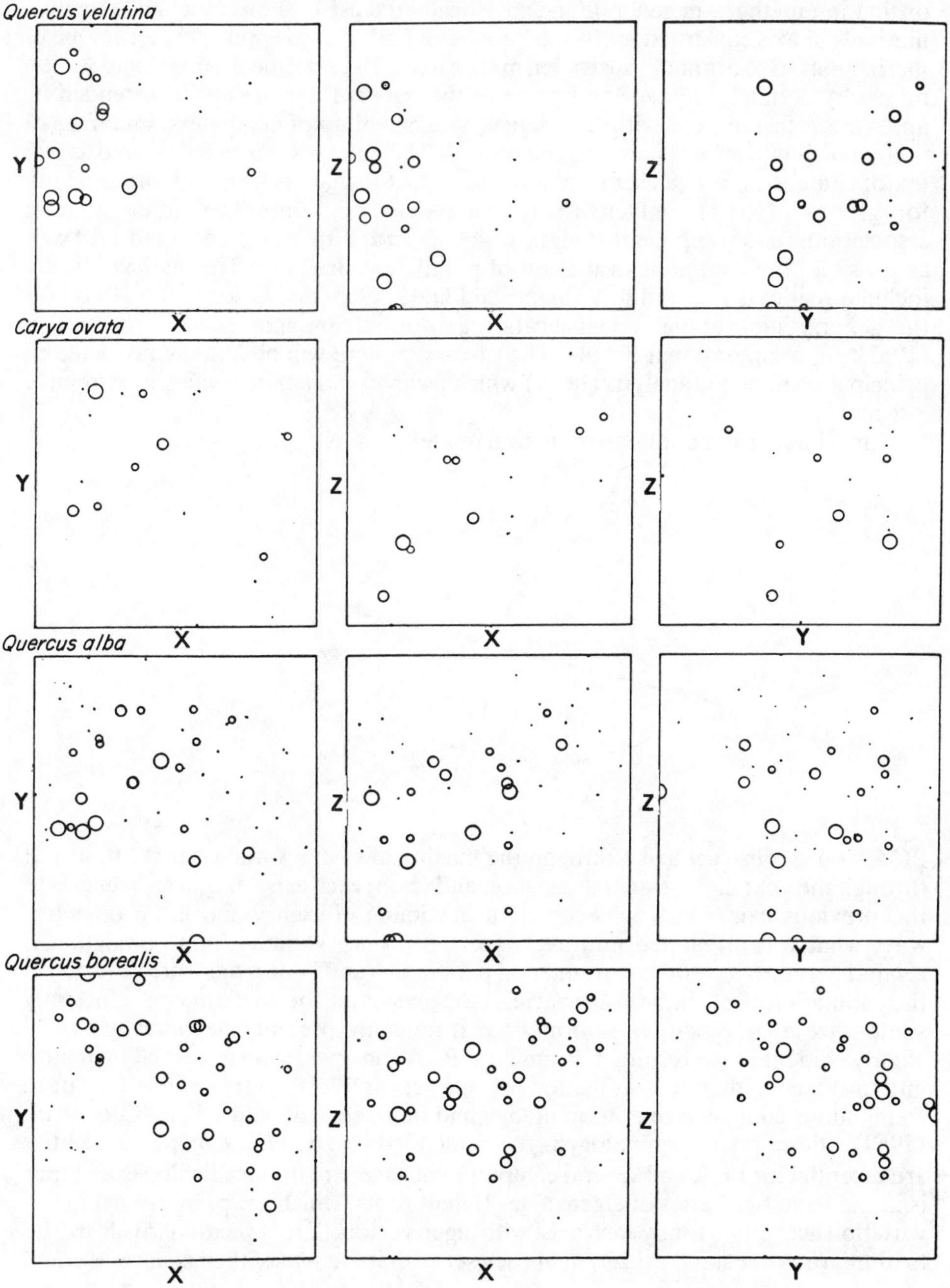

three axes are interpreted as: (1) Reflecting recovery from major past disturbance and showing a marked correlation with features of an increasingly mesic environment and with several soil factors – organic matter, pH, Ca, P, etc. (2) Related to drainage and soil aeration and showing correlation with soil water retaining capacity and ammonium ion concentration. (3) Related to recent disturbance and the influence of gap phase replacement.

The ordination procedure of Bray and Curtis (1957) is fairly crude and very unwieldy with a large sample. It did, however, stimulate a considerable interest in ordination methods in general, and specifically the use of more exact and effective methods of axis construction. In 1954, Goodall (1954*b*) had applied the technique of factor analysis to ordinate Australian mallee data. The extremely heavy computation necessary certainly limited the impact of the paper. Dagnelie (1960) extended its application further and with the increasing availability of computers a number of related techniques have now been employed. The methods can be considered as two major categories from a user's point of view. Factor analysis which attempts to look for groups of correlated environmental parameters controlling different plant associations, and component analysis which seeks to display, in as economical a way as possible, the continuous variation of population structure. The mathematically inclined reader is referred to Williams and Dale (1965) and Grieg-Smith (1983) for the general implications in classification and ordination approaches, and Kendall (1957) for computational details. The discussion here will be initially restricted to principal component analysis (PCA) which as an *ordination technique* is extremely efficient.

More simply from our final ordination model:

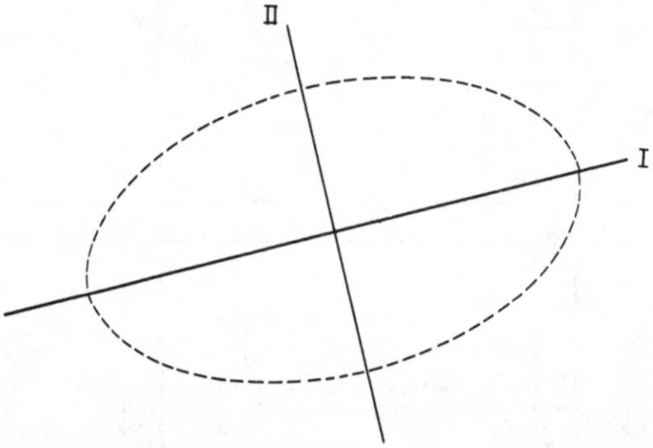

PCA would construct axis I through the maximum spatial variation present, axis II through the next highest spatial variation and so on, each axis being at right angles to the previous axis. PCA achieves, with maximum efficiency and in an objective way, what was attempted intuitively (p. 195). Since variation in a population is related to one or several environmental parameters and in turn these are reflected in the abundance of indicator species, the resultant ordination can efficiently summarize a large body of information. If noda are present, they equally will be clearly evident in the resultant ordination. PCA can operate on a correlation matrix and produce either an ordination of species or plots. Again there has been tremendous confusion over terminology and in agreement with Ivimey-Cook *et al.* (1969) a descriptive terminology is recommended (see p. 172). Computer facilities are essential for PCA and several computer packages are now available. The output is in the form of a series of eigenvalues (latent roots) which are proportional to the variation accounted for by each axis, with eigen vectors (latent vectors) which are the loadings or, after standardization in each set so that the sum of their squares is equal to the respective eigenvalue, simply the spatial co-ordinates of each plot (or species in a species ordination). Usually only the first three axes are examined, the fourth and subsequent axes only accounting for small and decreasing amounts of residual percentage variation. Most packages will operate on either correlation, covariance or a defined 'similarity' matrix, of which several have been proposed. The choice as to which is most effective has very largely been decided by the outcome of the analysis – from experience a simple correlation coefficient or Orloci's coefficient

$$\sum_{i=1}^{N} (x_{ij} - \bar{x}_i)(x_{ih} - \bar{x}_i)$$

where N = number of species x_i, x_{ij} and x_{ih} are the abundance of species i in samples j and h, and x_i is the mean abundance value for species i (Orloci, 1966; Austin and Orloci, 1966).

The presentation of the data output can take the form of a standard ordination on each of the three pairs of axes, or more effectively as Grieg-Smith *et al.* (1967) have done, as projections of the spatial cluster of plots on each plane surface in turn (Fig. 10.15).

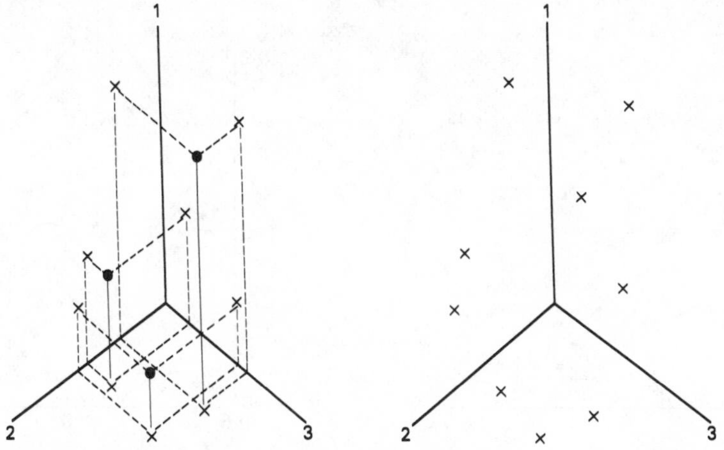

Fig. 10.15 Method of projecting the spatial plot locations on to the three plane surfaces composed of Axis 1–2, Axis, 1–3, and Axis 2–3.

The identification of the trends in the ordination is achieved by overlaying species abundance values or environmental measure on the plot ordination. Austin and Orloci (1966) ordinated dune slack vegetation from Newborough Warren, Anglesey, sampled using percentage frequency as a measure of abundance.The ordination is given below (Fig. 10.16) together with the corresponding plot frequencies of *Ammophila arenaria, Potentilla anserina*, and *Carex serotina*. These three species are characteristic of dunes and dry slacks, moist slacks, and wet alkaline slacks respectively. Accordingly axis I is related to a vegetational gradient from transitional dunes or dry slacks, to slack communities, whilst axis II follows a vegetational gradient from wet alkaline slacks to the more leached meadow type of dune slack. Similarly Grieg-Smith *et al.* (1967) ordinating tropical forest from density data sampled from Kolombangara Island in the Solomon archipelago in the S.W. Pacific showed a marked distribution of altitude and slope classes with the plot ordination (Fig. 10.17).

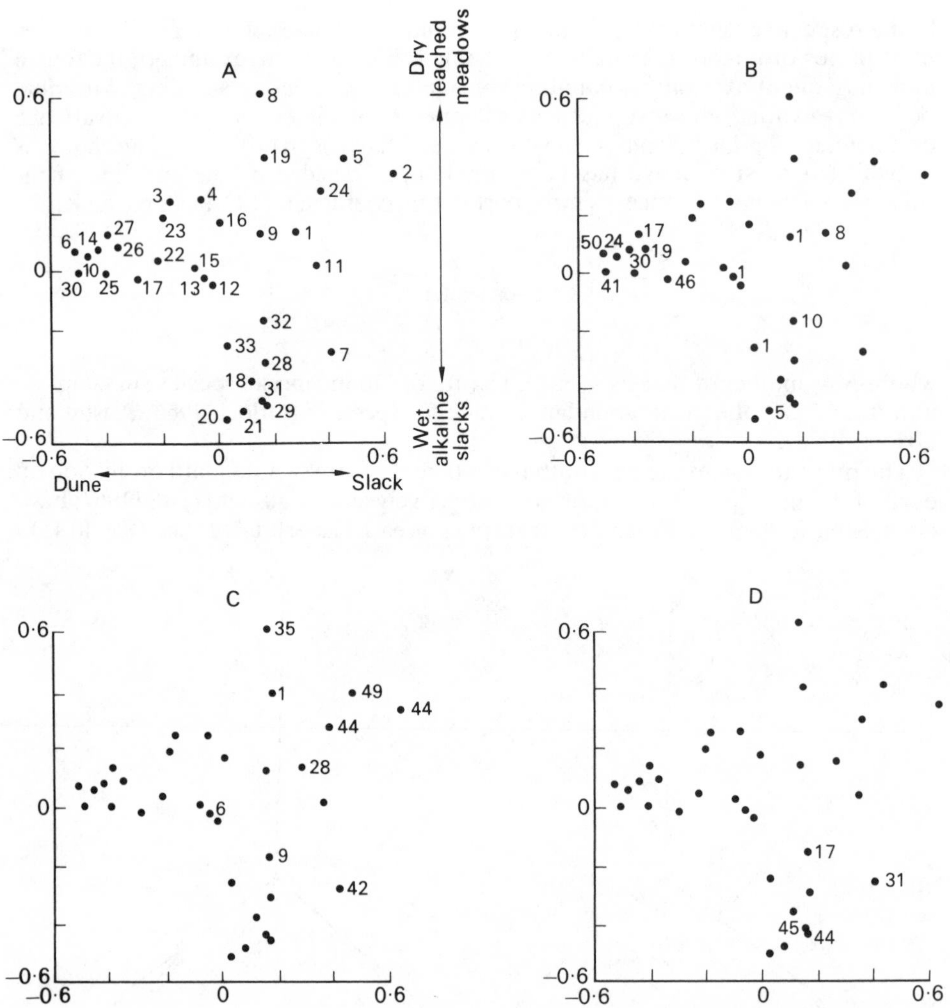

Fig. 10.16 The ordination of dune slack data on the first two PCA axes (A) with the abundance (per cent frequency) of *Ammophila arenaria* (B), *Potentilla anserina* (C) and *Carex serotina* (D) in each plot. (From Austin and Orloci, 1966; courtesy of *J.Ecol.*)

Kershaw (1968) ordinating Domin values sampled from savanna vegetation in Northern Nigeria showed the first three PCA axes were identified with low/high water availability, presence/absence of fossil ironstone cuirass material, and a northern/southern trend respectively. Traditional Braun-Blanquet associations extracted previously showed a close level of agreement with the ordination (Fig. 10.18).

The successful outcome of a PCA analysis is very dependent on the initial degree of heterogeneity in the data. If the data are markedly discontinuous, for example, with one or two associations which are very clear cut from the remaining central mass of data, these groupings will certainly be distinct in the ordination but the main bulk of the data will also remain as an unstructured 'central mass'. Use of PCA should be thus restricted to relatively homogeneous data. If the analysis is unsatisfactory then the variation should be reduced. This can be achieved in several ways: extreme associations can be simply deleted from the data and the residual bulk of the data re-analysed. Thus Austin and Grieg-Smith (1968) used an association analysis on

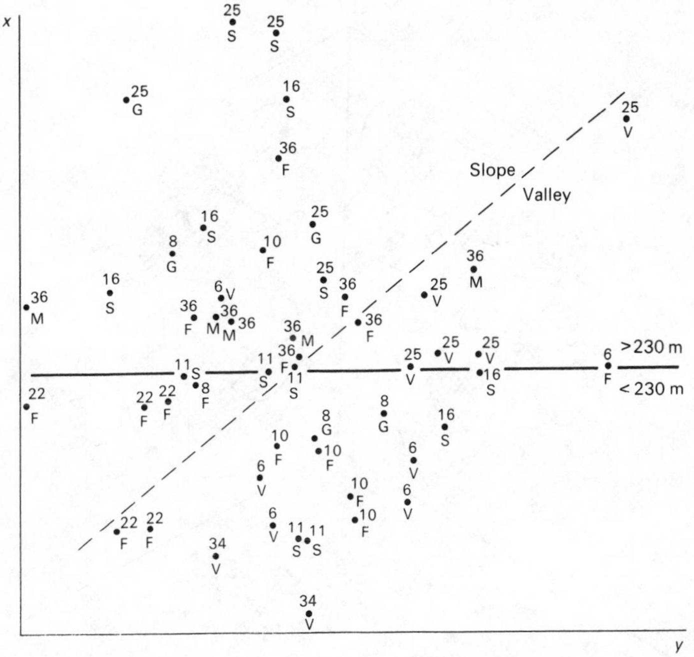

Fig. 10.17 Correlation of altitude and topography with the plot ordination on the first two PCA axes from tropical forest data in the Solomon Islands. S, Steep slope; G, gentle slope; F, flat/plateau; V, valley; M, varied topography. Figures indicate actual plot altitude in 10-m classes. (From Greig-Smith *et al.*, 1967; courtesy of *J. Ecol.*)

their Kolombangara Island data to divide it into more homogeneous units prior to PCA. Kershaw (1968) used square-root transformation of the data and achieved an improvement in the ordination. Thus it is decidedly more effective to analyse presence-absence data, or Domin values, than density or percentage cover data where the variance is inevitably much greater. Implied in this is also the additional advantage of the simplicity of sampling in the field.

Reduction of the total variation in the data is particularly effective in improving a species ordination and especially the removal of the less common species from the analysis.

Thus most of the information relating to the structure of the ordination is carried by the most abundant species. This fact is of considerable importance and immediately simplifies the orginal collection of the data enormously – the rare species can simply be ignored. Austin and Grieg-Smith (1968) simultaneously reached the same conclusion and showed that an efficient ordination could be achieved with 25% of the total flora. Using standardization procedures, species ordination was markedly improved, especially where the abundance gradient was removed, which usually is extracted by the first axis of a species ordination. Thus PCA extracts the maximum variance with each axis. In data which is markedly heterogeneous the maximum variance lies in the relative species abundances which will range from high levels of abundance to isolated occurrences. The first axis accordingly has the common species loaded at one end and rare species at the other and is little information value. However, the ecological interpretation of a plot ordination is quite insensitive to either data standardisation or to centering and this will be discussed further in Chapter 11.

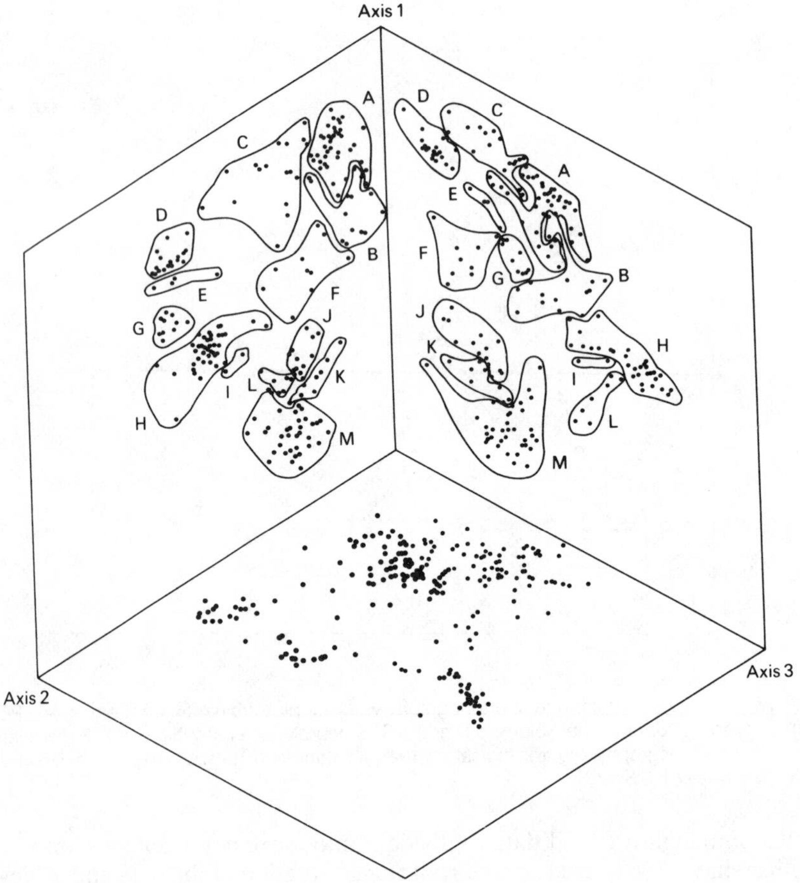

Fig. 10.18 Braun-Blanquet associations superimposed on the principal component plot ordination of savanna data from Northern Nigeria. (A), *Monotes* association (southern); (B), *Monotes* association (northern); (C), *Monotes–Parinari* association; (D), *Parinari–Gardenia–Detarium* association; (E), *Daniellia–Gardenia–Detarium* association; (F), *Monotes–Isoberlinia* association; (G), *Detarium–Gardenia* ironstone association; (H), *Combretum glutinosum–Dichrostachys–Entada* ironstone pavement association, *Anogeissus–Feretia* river bank association, *Bombax–Diospyros–Ficus* inselberg association, and the *Dichrostachys* erosion complex association; (I), flood plain association (J), *Isoberlinia–Parinari* association; (L), *Isoberlinia tomentosa–I. doka* association; (M), *Isoberlinia–Annona–Terminalia–Ximenia* association. (From Kershaw, 1968; courtesy of *J. Ecol.*)

Newnham (1968) applied PCA to climatic data consisting of 19 variables from 70 weather stations in British Columbia. Two components extracted from a correlation matrix accounted for 87% of the total variation. They were correlated with a general index of winter–fall climate/length of growing season, and spring–summer temperatures/precipitation (Fig. 10.19).

Overlaying the known distribution of Sitka spruce, Engelmann spruce or White spruce, Black spruce, Ponderosa pine, Garry oak, etc., reveals a very marked correlation with the ordination of climatic factors (Fig. 10.20). This is an unusual and interesting use of component analysis – in a normal ecological context rarely are there more than 3 or 4 environmental parameters available to correlate with a plot ordination.

The presentation of PCA output has followed the standard approaches which have been adopted by various workers (Figs 10.16–10.18). Both these procedures, in fact,

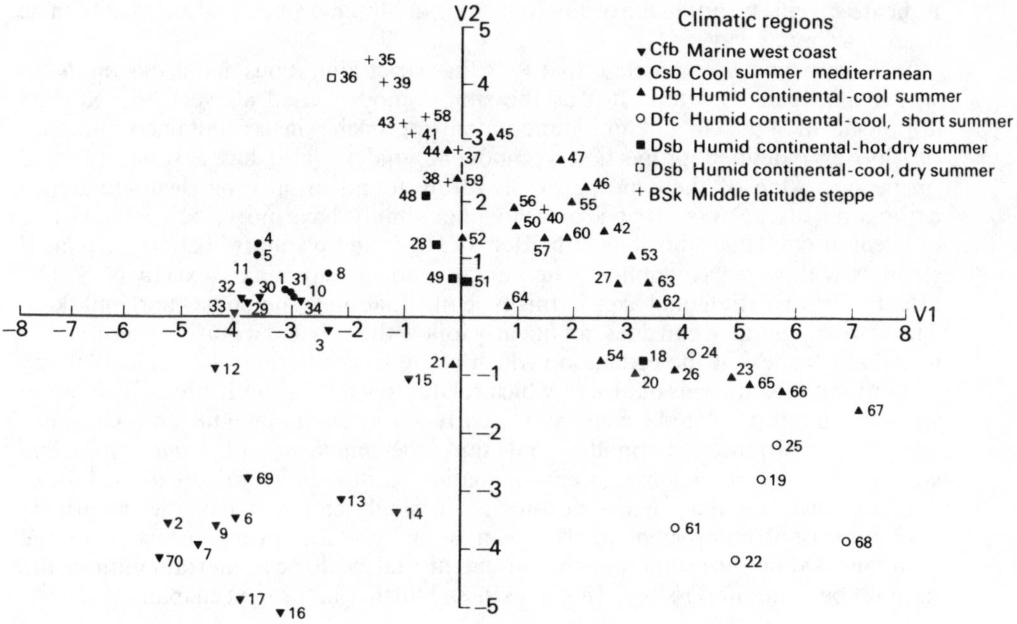

Fig. 10.19 Ordination of meteorological data from 70 stations in British Columbia on the first two principal component axes. (From Newnham, 1968; courtesy of *Forest Sci.*)

obscure the multi-dimensional structure derived from the PCA and certainly a three-dimensional physical model is considerably more informative. However, the construction of such models is extremely time consuming and only possible with relatively small samples. The approach of Kershaw and Shepherd (1972), utilizing the off-line graphics facilities of a computer to output perspective drawings of the three-dimensional spatial configuration as viewed from any desired observer's location, is an improvement. Considerable structural detail emerges which is completely obscured by projection of the spatial clusters on to the plane surfaces 1–2, 1–3, and 2–3 (Fig. 10.21).

Summary and conclusions

Although constellation diagrams offer a simple method of approach to an objective ordination, with large numbers of species or plots they become very impractical. The technique is insensitive and only utilizes one parameter of the population structure – presence and absence. The continuum approach, although utilizing species abundance values, is only valid for single, major environmental gradients. Since there are usually many environmental parameters involved, it is not effective in many situations. Use of the common species themselves as both environmental 'indicators' and axes, although introducing a multi-dimensional concept in ordination, ignores the information carried by the other common but unused species in the population. PCA is a very flexible technique which objectively and with maximized efficiency utilizes all the correlations present within a population of plots or species, to construct ordination axes. These axes related sequentially to variation extracted. Plot ordinations are little affected by either data standardization or the spectrum of species abundance. Species ordinations conversely are very sensitive to these aspects of data structure. Species abundances overlayed on the plot ordination indicate specific trends in the ordination, and equally environmental measures can be used in a similar way.

It is important to appreciate that PCA has strict limitations and is *one model* to which data can be fitted. Just as the simple models used above to develop an appreciation of the current multidimensional approaches in use contained a number of (loosely) defined properties, component analysis also has a series of rigid properties. One of these properties is rarely found in an ecological situation – orthogonal axes. Axes that are orthogonal simply have no correlation existing between them. Thus a north–south effect in Nigerian savanna *is* in effect correlated strongly with water availability. The rainfall gradient runs almost exactly N–S. The effect of using orthogonal axes in this type of situation is sometimes fairly marked. The environmental trend does not linearly follow the axis and it is usually more exact to delimit 'zones' of the ordination which relate to northern/low water availability rather than talk in terms of an *axis* which relates to a north–south effect. In addition species are rarely related exactly and linearly to *one* environmental parameter, nor are they distributed normally, and thus the *mathematical model, principal component analysis*, is only an approximation to the real world situation. Indeed over the last few years it has become increasingly apparent that the underlying assumptions of component analysis can often produce severe distortion of the resultant ordination, sometimes to an extent that ecological interpretation of the data set becomes impossible. This is discussed further in the next chapter.

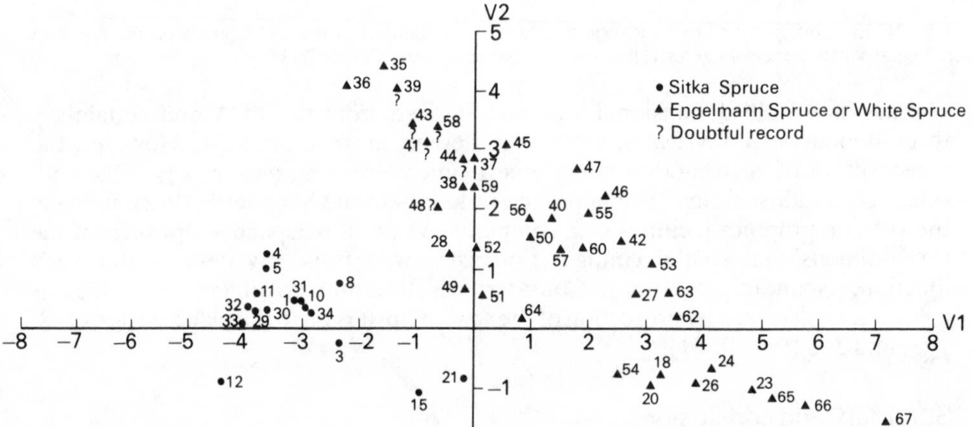

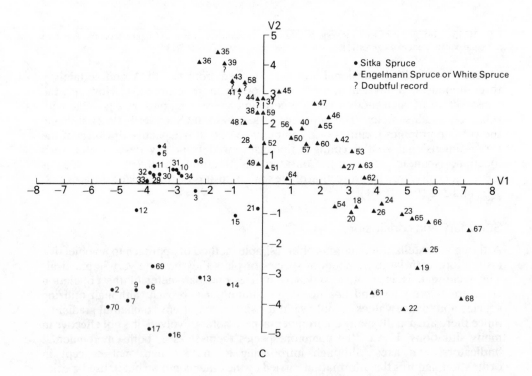

Fig. 10.20 (A) the correlation of Garry oak, Ponderosa pine, and Black spruce distribution with the meteorological ordination; (B) the correlation of Western hemlock and (C) the correlation of Sitka spruce; Engelmann spruce or White spruce distribution with the principal component ordination. (From Newnham, 1968; courtesy of *Forest Sci.*)

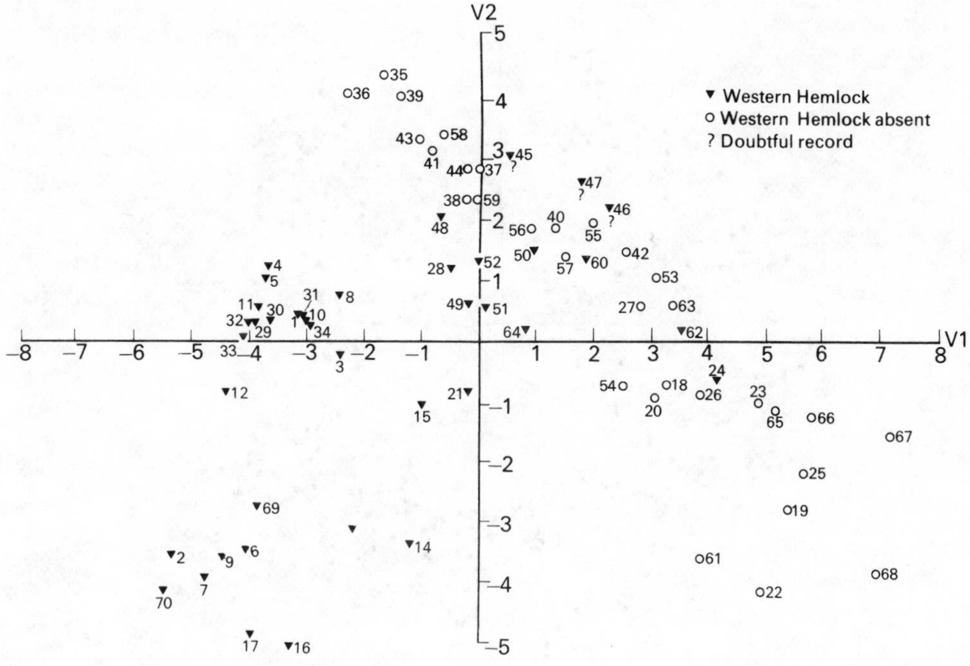
V2
V1
▼ Western Hemlock
○ Western Hemlock absent
? Doubtful record

B

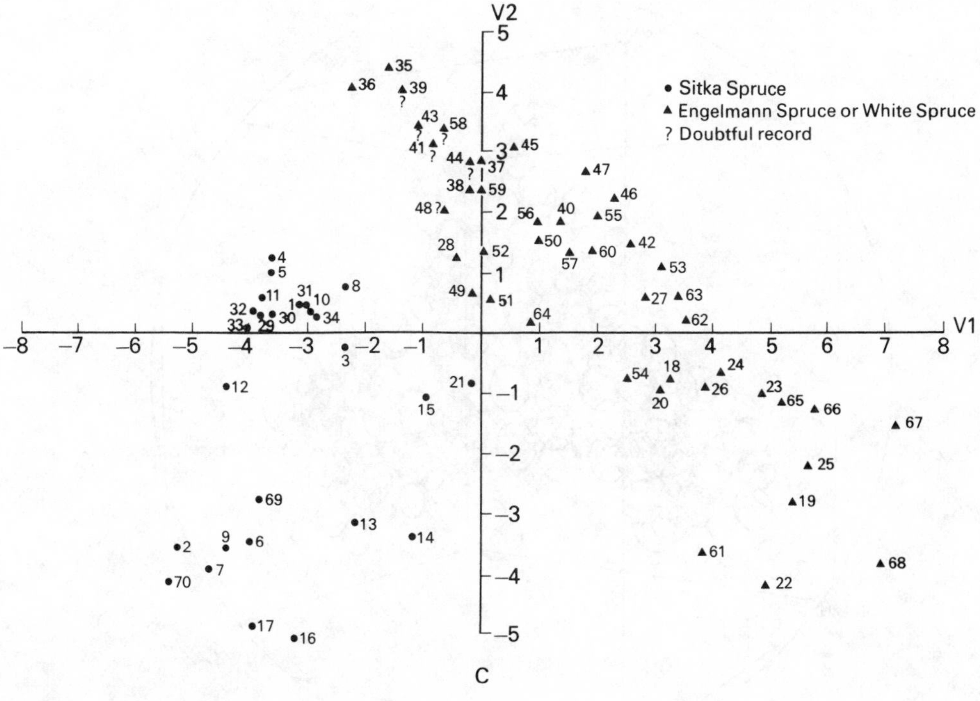
V2
V1
● Sitka Spruce
▲ Engelmann Spruce or White Spruce
? Doubtful record

C

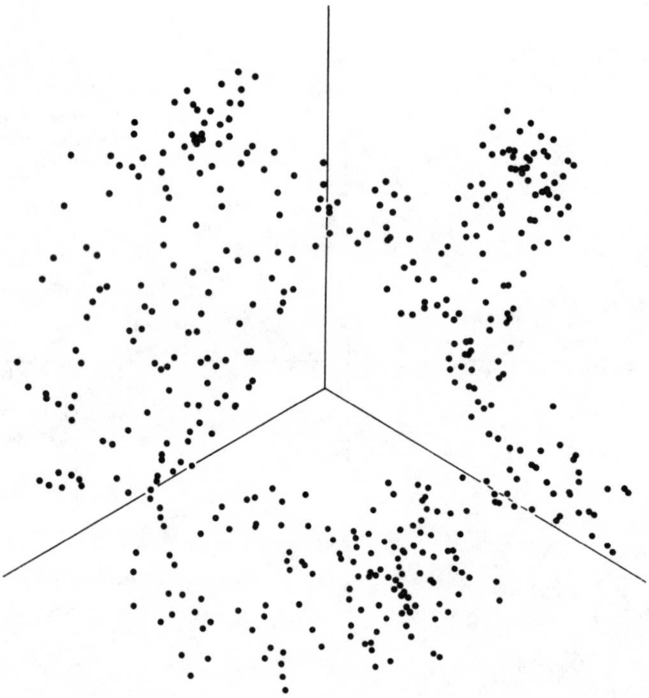

Fig. 10.21 (A) Ordination of lichen heath data from N.W. Ontario as a projection on to the plane surfaces bounded by the first three component axes; (B), (C) and (D) persepective drawings of the spatial configuration of the same data viewed from three different directions.

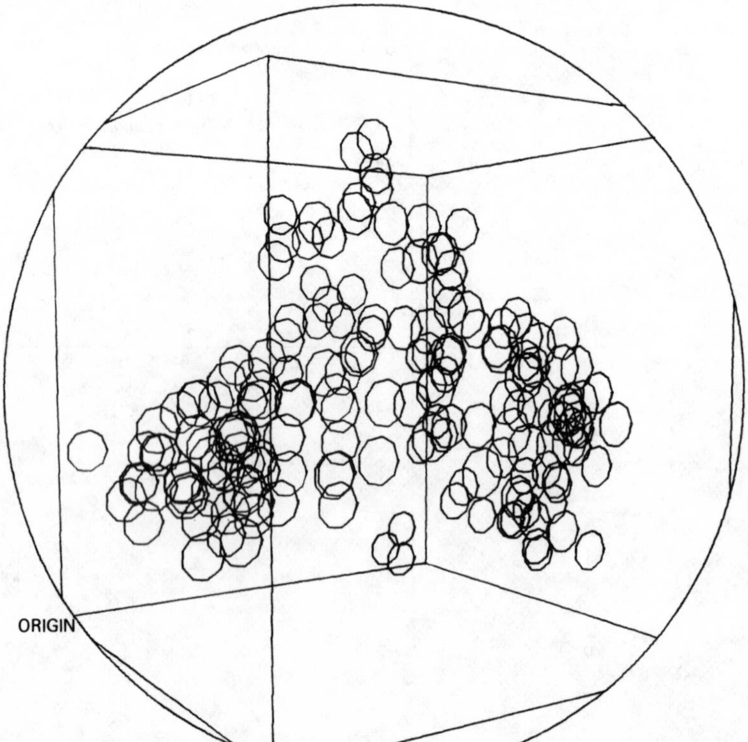

Fig. 10.21 (B) The origin of the ordination lies at the left centre (ORIGIN) of the model.

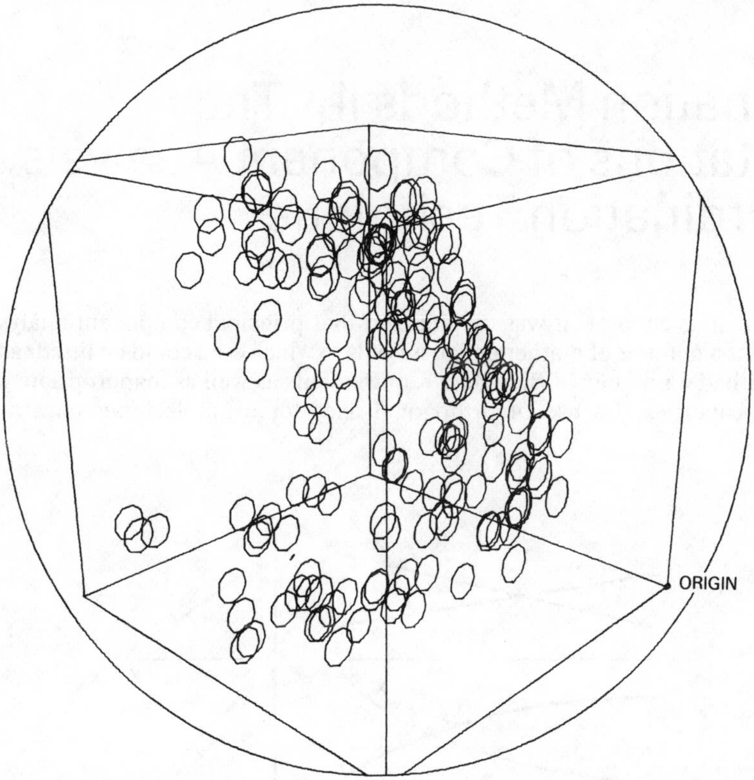

Fig. 10.21 (C) Perspective view of the same model viewed from 'behind'.

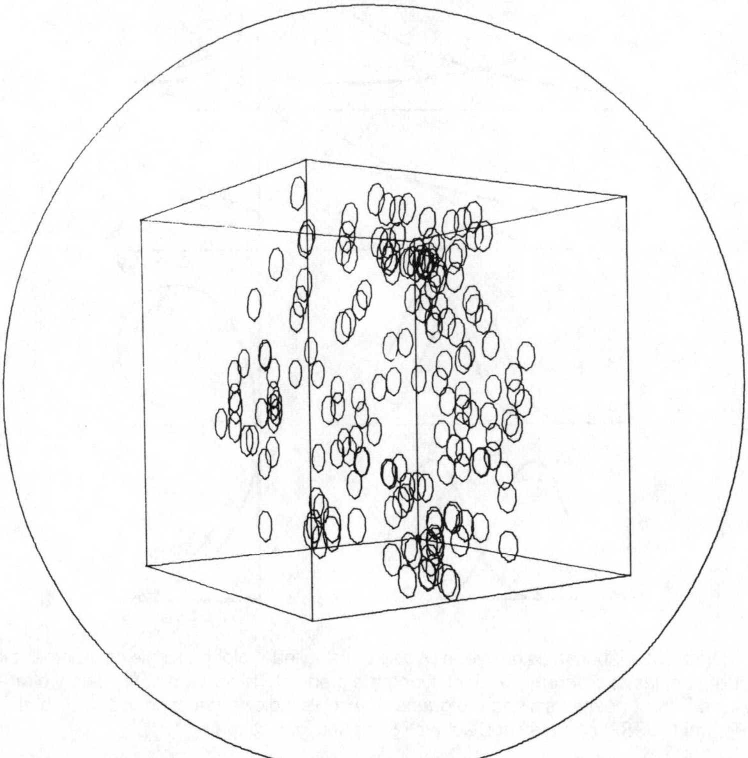

Fig. 10.21 (D) Perspective view of the model after a further rotation, the origin now lying at the right rear of the ordination.

11

Ordination Methods II The Limitations of Component Analysis as an Ordination Technique

In the previous chapter, it was emphasized that principal component analysis was only one of a number of mathematical models to which data could be fitted, and that equally it had a number of rigid data requirements as well as inappropriate mathematical properties. Its use of orthogonal axes for example, does constrain the

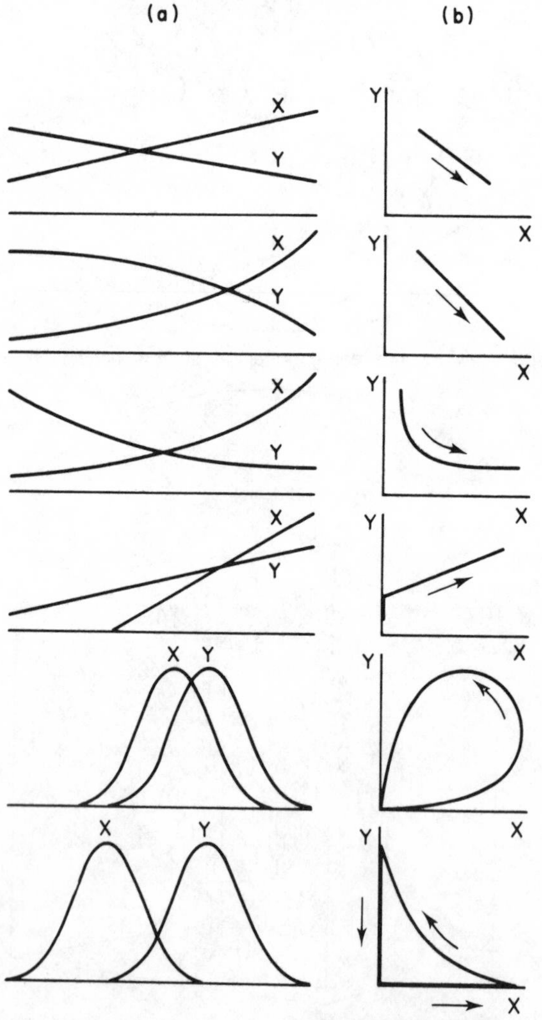

Fig. 11.1 Different relationships between two species, X and Y, along a single environmental gradient. (**a**) Amount of species at different positions along the gradient, (**b**) amount of species Y plotted against species X for different positions along the gradient. Arrows indicate direction moving 'up' the gradient. (From Greig-Smith, 1983; courtesy of Blackwell Scientific Publication.)

interpretation of the resultant ordination and the axes cannot always be directly or simply related to ecological parameters. PCA, strictly speaking, also demands normally distributed data as well as complete linearity between species relationships. Normally distributed data are rarely found in biological systems and linear relationships are the exception rather than the rule. Greig-Smith (1983) for example has shown very elegantly how bell-shaped curves, or even their distributional tails are usually related in a markedly non-linear manner (Fig. 11.1). In some isolated circumstances though and especially along short gradients, two non-linear species distributions can be related linearly (Fig. 11.1b). Accordingly an ordination derived using PCA will contain an unknown amount of distortion, arising from a number of sources (see below) but initially from forcing the non-linear data set into a model which is based entirely on linearity. As a result it is essential to consider the likely extent of non-linearity in a data set in terms of using PCA or an alternative analysis as an ordination method.

The interaction of non-linearity and ordination distortion is shown in Figs 11.2 and 11.3. The distribution of three species along an environmental gradient is

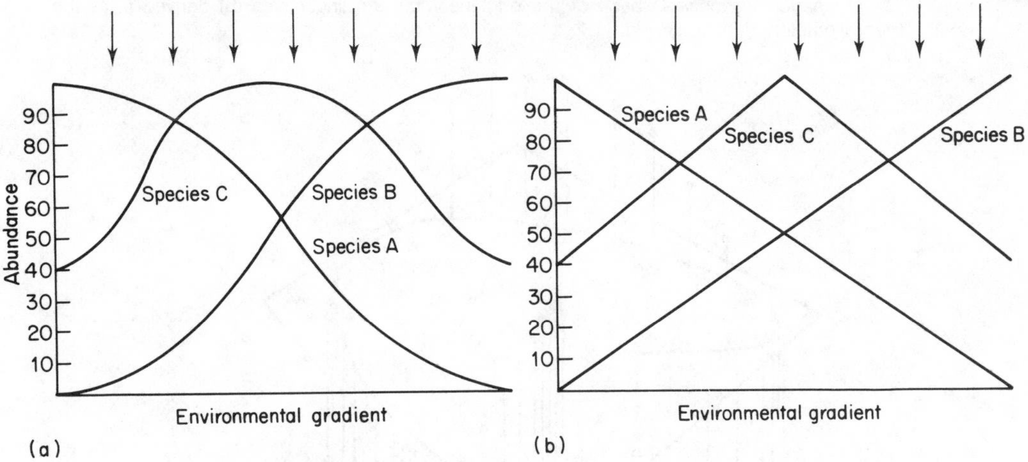

Fig. 11.2 Hypothetical changes in the abundance of three species A, B and C growing along an environmental gradient. The relationships between the species are largely non-linear in (**a**) and linear in (**b**). Sample points along the gradient are marked (↓).

represented both as bell-shaped curves and correspondingly, as linear and regular increases and decreases in a species abundance (Fig. 11.2). The resultant three-dimensional ordination using species abundance as axes (Fig. 11.3) shows clearly how the bell-shaped distribution of species A, B and C is reflected strongly as an arch-shaped configuration in the ordination. Conversely the linear species distribution is projected as a linear sequence (Fig. 11.3b). However it is evident from the discussion in the previous chapter than variation along an environmental gradient when examined by the continuum approach consistently occurs as a sequence of bell-shaped curves (compare for example Figs 10.5. 10.6 and 10.10) rather than as the linear relationships expressed in Fig. 11.2b. This widespread nature of bell-shaped species distributions with the concurrent distortion when PCA is used as an ordination technique, was first recognized by Swan (1970) and Noy-Meir and Austin (1970). Swan used simulated vegetational data along a supposed environmental gradient and showed that an ordination of samples from this simple sequence did not recover the linear data sequence, instead an arched distribution of sample points resulted. Noy-Meir and Austin (1970) using the same data set showed a considerable

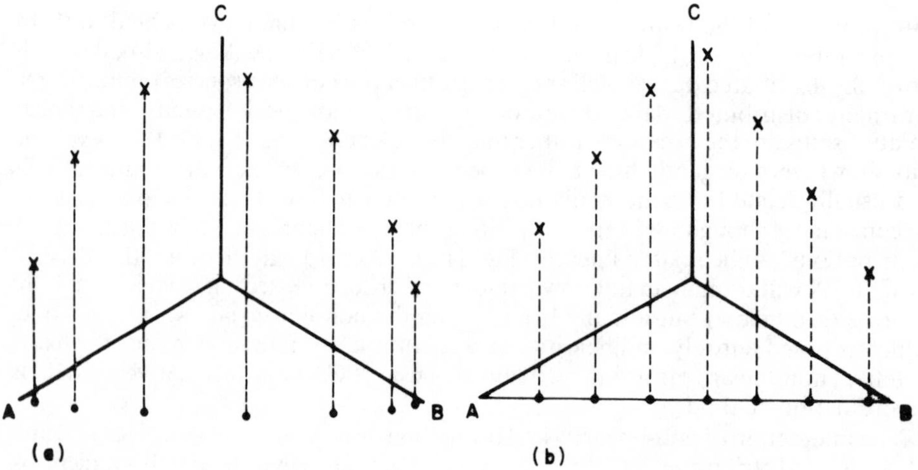

Fig. 11.3 Ordination in species space of the non-linear (**a**) and linear data (**b**) derived from the environmental gradient depicted in Fig. 11.2.

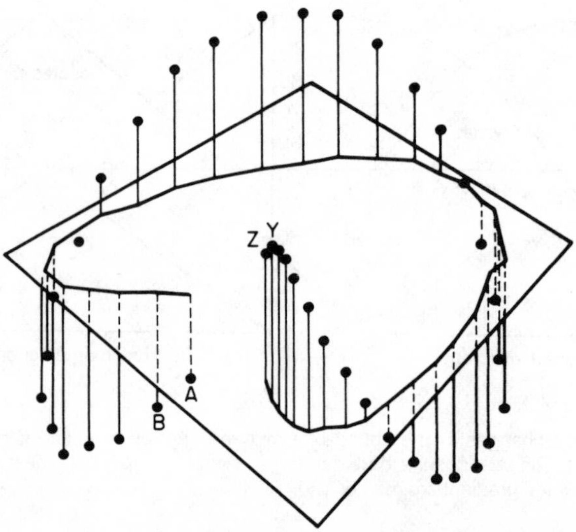

Fig. 11.4 Principal component ordination of Swan's model V data showing the first three axes; the original linear sequence A, B, C, . . . Y, Z becomes a complex three-dimensional curve. (From Noy-Meir and Austin, 1970; courtesy of *Ecology*.)

degree of distortion was present when PCA was used as the ordination procedure (Fig. 11.4). Thus the sequence of samples A, B, . . . Y, Z taken from a simulated regulated environmental gradient becomes a three-dimensional curve when ordinated on the first three axes.

Subsequently there has been extensive use of simulated data, particularly sequences of species whose species content was controlled in a defined way. Gauch and Whittaker (1972*a*) for example established a number of simulated coenoclines which were then examined by a range of ordination techniques providing information as to which particular set of properties affected the degree of distortion in subsequent ordinations. Some characteristic coenoclines together with their principal component ordinations are given in Fig. 11.5 and Gauch (1982) (see also

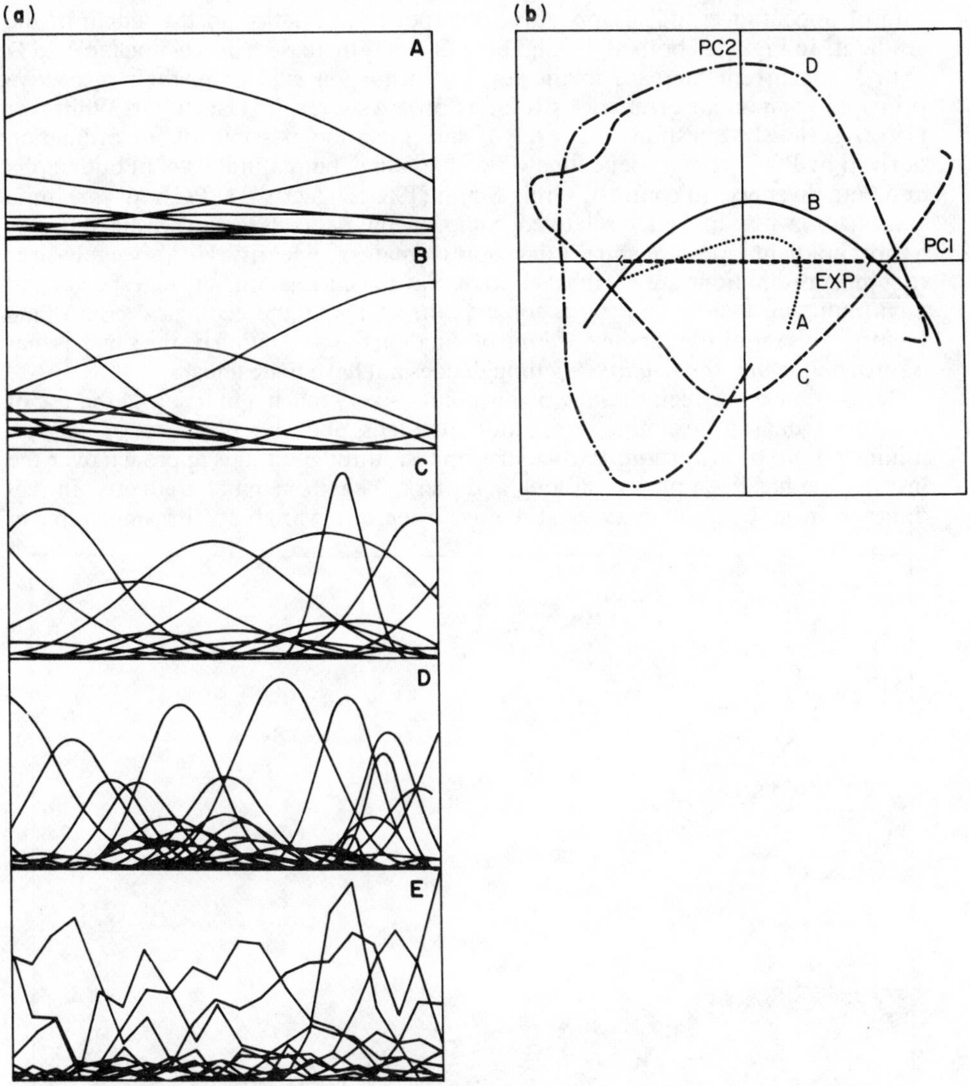

Fig. 11.5 (**a**) Five simulated coenoclines produced by the program Coenocline Simulation. C may serve as a standard which others are compared (and was produced using input parameters I = 52, J = 100, MTN = 260 JMX = 1609, Z = 6.0, STD = 0.3, AM = 1500, CL = 10 000, CO = 0.01, and ANL = 0). Alpha and beta diversities are doubled from A to B, B to C, and C to D; E differs from C in the introduction of sampling error (input parameter F = 0.227, producing a similarity of replicate samples ("internal association") of 77%, as measured in units of percentage similarity). (**b**) Principal components ordination of the samples from the simulated coenoclines. PC1 and PC2 are the first two axes of principal components analysis. The heavy broken line EXP gives the result to be expected if there were no distortion; A, B, C, and D show the increasingly complex distortion of the ordination as alpha and beta diversities increase. (From Gauch and Whittaker, 1972*a*, courtesy of *Ecology*.)

Whittaker, 1967) summarizes their form particularly in terms of the amplitude and position of a peak. The major contribution to distortion however is in terms of the number of species occurring in a sample which is termed alpha diversity as well as the rate of *species turnover* (beta diversity) in addition to any non-linear relationships between species pairs. The rate of species turnover in coenocline data represents the rate of appearance and disappearance of species in relation to the length of the gradient. In Fig. 11.5 both alpha and beta diversity increase from coenocline A to D with a concurrent increase in the level of distortion evident in their respective principal component ordinations (Fig. 11.5b). As a result, Gauch and Whittaker (1972) seriously question the level of interpretation possible of an ordination derived by PCA from a coenocline which has a high but natural level of both alpha and beta diversity. In contrast, Greig-Smith (1983) suggests that PCA may be quite satisfactory when applied to relatively homogeneous data sets where presumably only a short gradient is present and the data approximates to linearity. He also emphasizes that most ordinations are established to display the correlation between species and environmental factors as a basis for hypotheses about the ecological controlling factors. This requires that the *ordering* of stands in the ordination be the same as that of the coenocline, the *relationship* though does not have to be linear.

The variance between these two viewpoints is very much a function of the use of simulated data to examine ordination methods and the presence of, or the minimization, of distortion. Particular emphasis throughout this approach over the last decade has been placed on long and marked environmental gradients. In real data sets from such situations with their usual degree of variability, it is impossible to

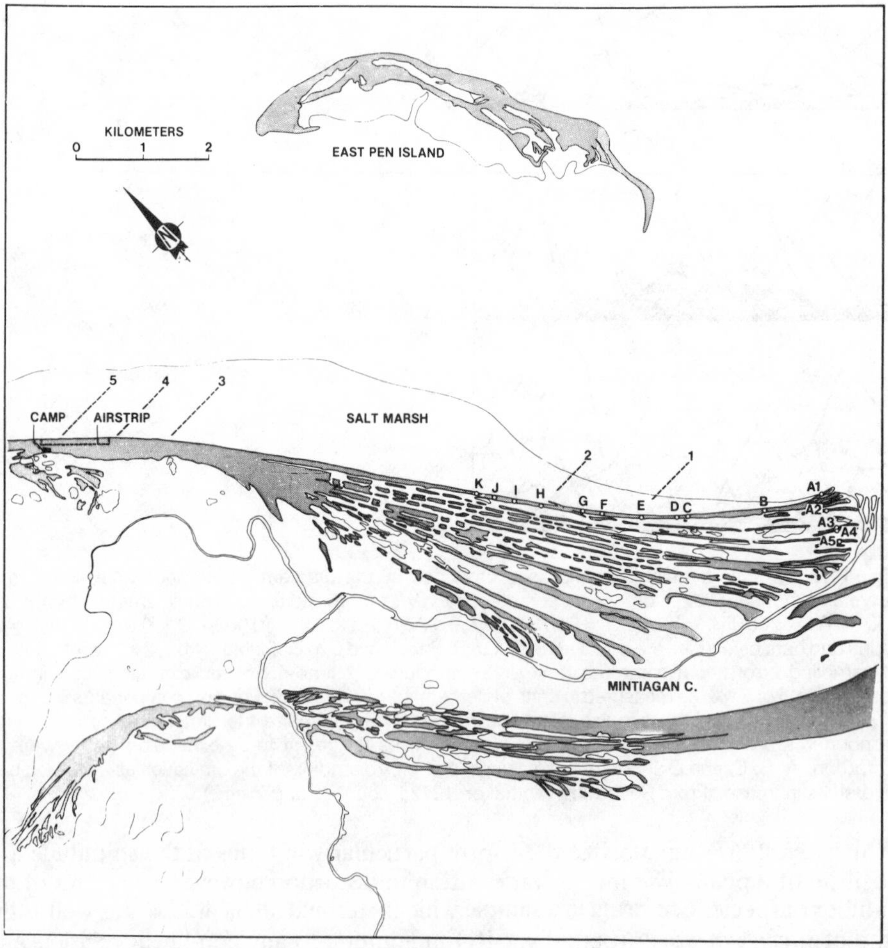

Fig. 11.6 Raised-beach and salt marsh formation adjacent to East Pen Island in Hudson Bay, in response to coastal isostatic uplift, showing the locations of the sample sites. (From Pierce and Kershaw, 1976; courtesy of *Can. J. Bot.*)

allocate a specific reason to difficulties of ordination interpretation. The difficulties may indeed reflect analagous problems that have been identified using simulated data sets, but they may not. However Looney and Kershaw (1985*a* and *b*) have provided two field data sets which do allow interpretation of distortion to be made and we use them here to explore more fully and at the same time, to illustrate the extent of distortion when PCA is used as an ordination technique. In the next chapter this same data set is again used to show how more recent techniques can resolve the problem of distortion and thus clarify to a considerable extent the ecological interpretation of an ordination.

The salt marsh data set

Kershaw (1976) has described an extensive salt marsh system on the coast of Hudson Bay. The marsh is developing in the lee of east Pen Island (Fig. 11.6) in response to isostatic uplife of the region following the disappearance of a massive ice-cap centred over Hudson Bay during the last ice-age. Thus the weight of the ice-cap depressed the earth's crust, which is slightly elastic, and this deformation has slowly been corrected by a continuous but gradual uplift of the land surface following the end of the glacial period. This has resulted in an extensive sequence of beaches, laid down by onshore gales and currents, and left as the land surface returned towards its original undeformed position. These raised beaches are found over 200 km inland forming a natural age sequence with progressive distance from the coast. Their initiation is often as a discrete sand bar or island and East Pen Island represents one such formation currently only separated by a narrow channel of open water from the mainland. Similarly West Pen Island is now completely joined to the mainland whereas 250 years ago it was a completely separate entity (Fig. 11.6). The protection from off-shore currents afforded by the incipient beach systems leads to extensive onshore silt deposition with the resultant widespread development of salt marshes along the whole of the southern coastline of Hudson Bay as well as on the western side of James Bay. All of these marshes are quite young and lack the secondary drainage developments and dissection so characteristic of many European salt marshes. The classical salt marsh linear successional sequence is thus very evident and has not been obscured at all by the development of drainage features, although there is a limited development of salt-pans in the lower marsh.

The zonation of the marsh is given in Fig. 11.7 as a simple histogram of species abundance against elevation on the marsh from one of the 5 transects originally sampled (Kershaw, 1976). The linear sequence of the marsh profile with the complete absence of any drainage channels is clearly evident in the regular occurrence of species zonations. The lower marsh is dominated by *Puccinellia phyryganodes* and *Carex subspathacea* which are only inundated by onshore gales, indicated by strand lines in the mid-marsh. The general marsh is dominated by *Carex glareosa* and *Calamagrostis neglecta* which largely occur above any recent strand lines, and which form visually a very uniform zone. The fresh-water marsh at the top of the salt marsh sequence is extremely well-developed as a result of the extensive sub-surface drainage from the landward side above the surface of the permafrost. Thus there is a continuous layer of frozen ground at this latitude in Canada and accordingly in summer there is an exceptionally large input of lateral land drainage which emerges at the top of the salt marsh inducing an extensive fresh-water zone dominated by *Eriophorum russeolum* and *Carex aquatilis*. Immediately below the first raised-beach ridge there is conspicuous evidence of a shallow permafrost zone resulting from the late-lying snow pack, with an abundance of ice-cored moss hummocks. This represents the final well-defined zone of the salt marsh, dominated by a number of moss species but particularly by *Aulacomnium palustre* associated with a number of

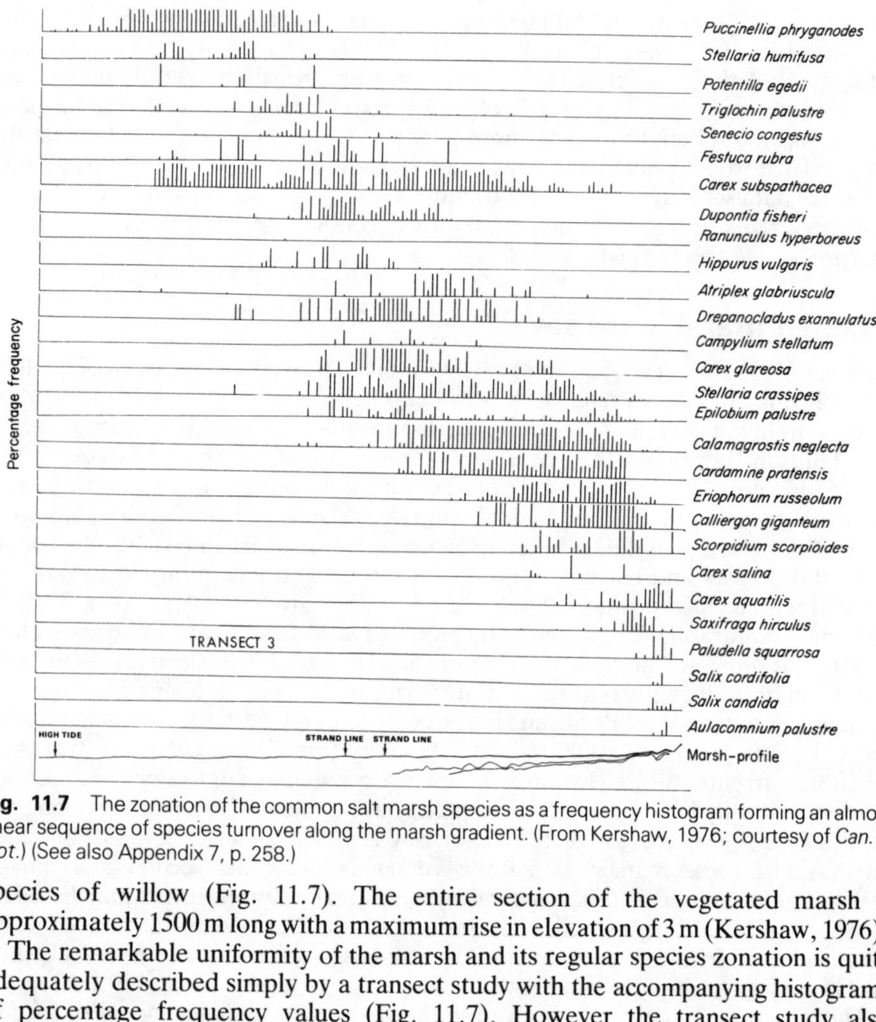

Fig. 11.7 The zonation of the common salt marsh species as a frequency histogram forming an almost linear sequence of species turnover along the marsh gradient. (From Kershaw, 1976; courtesy of *Can. J. Bot.*) (See also Appendix 7, p. 258.)

species of willow (Fig. 11.7). The entire section of the vegetated marsh is approximately 1500 m long with a maximum rise in elevation of 3 m (Kershaw, 1976).

The remarkable uniformity of the marsh and its regular species zonation is quite adequately described simply by a transect study with the accompanying histograms of percentage frequency values (Fig. 11.7). However the transect study also represents an ideal but completely natural data set, analagous to the simulated coenoclines that have been previously extensively used to examine distortion induced by different ordination methods. It has a completely regular sequence of species running from quadrat 145 at the top of the marsh to quadrat 1 at the bottom, and although in ordinary circumstances an ordination method for displaying the data would be rather contrived and unnecessarily complex, certainly any distortion in the ordination would be immediately apparent. For our purposes here then, it represents a *natural* regular data sequence ideal for displaying distortion induced by different ordination methods, but quite representative of field conditions. The full data matrix is included as Appendix 7, p. 258.

Prior to analysis the local frequency data, consisting of the number of sub-divisions of each quadrat occupied by a particular species (0–25), was edited to remove the 'rare' species (see discussion above, p. 201, and useful summary accounts in Gauch, 1982; Greig-Smith, 1983). The cut-off level used in the analysis here was < 1%. The first two axes extracted by PCA, operating on a similarity matrix (Orloci, 1966; Austin and Orloci, 1966) are given in Fig. 11.8 overlayed with the approximate plot sequence in the ordination. The linear sequence on the marsh is distorted into an almost closed loop on axes 1 and 2. Examination of the salt marsh sequence (Fig.

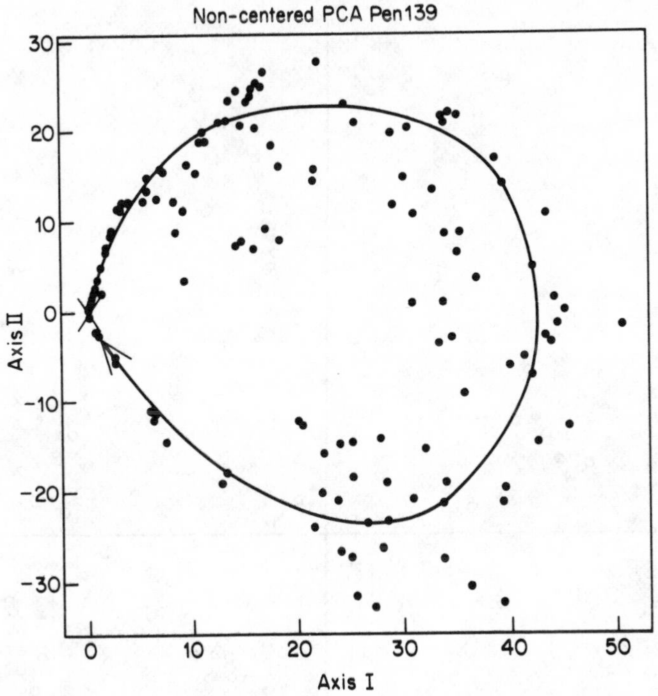

Fig. 11.8 Ordination of the salt marsh data on the first two axes extracted by non-centered and non-standardized principal component analysis. The direction of the gradient from quadrat 1 at the bottom of the marsh to quadrat 145 at the top is shown.

11.7) as well as the overlays of the abundance values of *Puccinellia phryganodes*, *Carex subspathacea*, *Calamagrostis neglecta*, and *Carex aquatilis* (Fig. 11.9), shows that the top and bottom of the salt marsh zonation are completely distinct from the middle in terms of species composition. Effectively PCA transposes these two strong negative correlations, to a close juxtaposition (similarity) in the ordination. Although the sequence of plots ordinated between axes 1 and 2 is still readily interpretable any attempt to interpret a less simple but analagous field situation would be difficult. To interpret higher order-axes would be out of the question since the ordered sequence of quadrats is completely lost when a third axis is used in the ordination (see additional discussion below, p. 221).

There has been considerable discussion over the last decade as to the relative merits of transformed or non-transformed data for use with PCA. Noy-Meir *et al.* (1975) summarize the range of possible standardizations and show indeed that the resultant ordination is to a greater or lesser extent dependent on the final method selected. However it is also emphasized that some of the species-standardizations result in a marked variance between the ordination actually achieved and a more normally-accepted ecological interpretation of a data set where 'rare' species are usually considered to carry little information. As a result Noy-Meir *et al.* (1975) recommend that site standardized abundance data, or non-standardized presence data should normally be used. Certainly in natural data, rather than very simple simulated data sets, ecological interpretation is little affected by centering or by standardization (Looney and Kershaw, 1985 *a* and *b*). Their data for the salt marsh are given in Fig. 11.10. It strongly reflects the minimal changes of ecological interpretation achieved by either species-centering or by species-standardization, in even very simple field data. The quadrat-centered and standardized ordination only

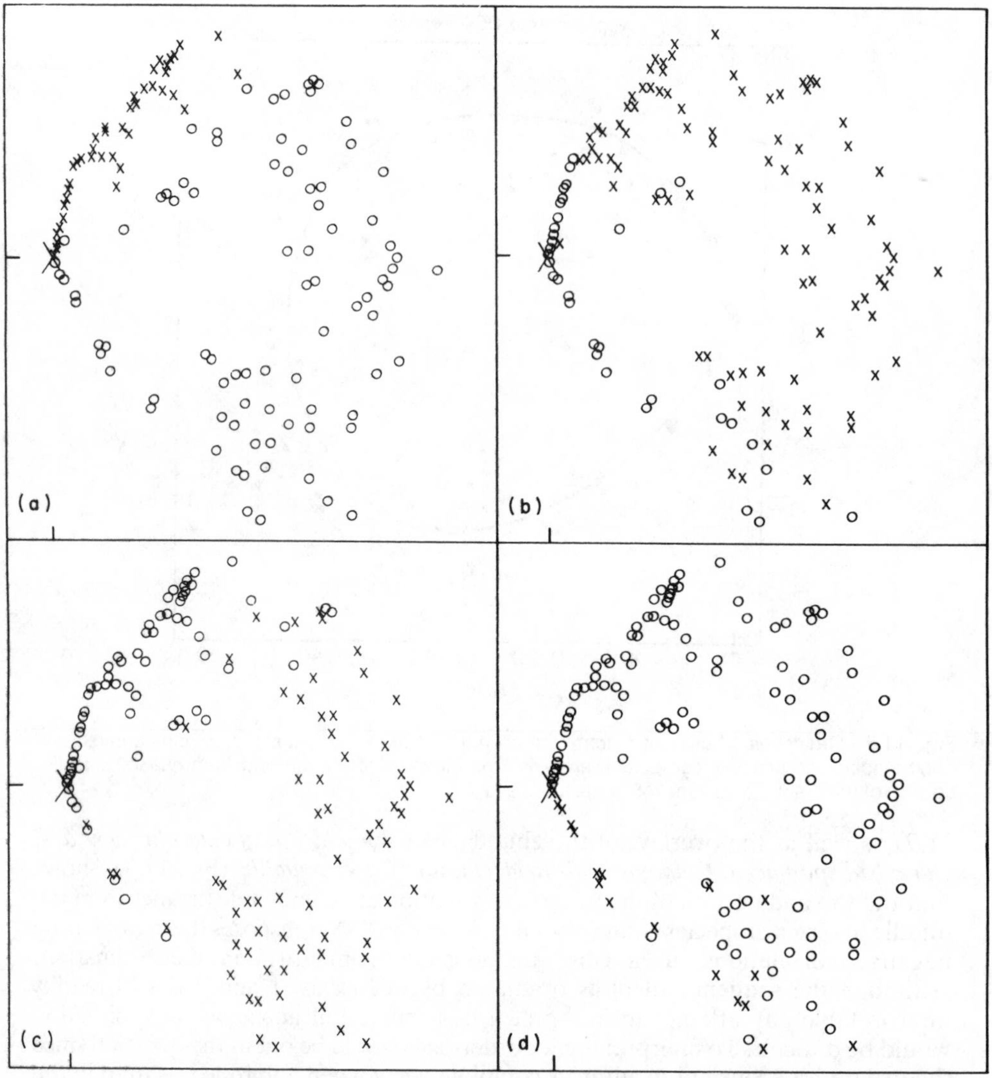

Fig. 11.9 Overlays for species presence (*) on the salt marsh PCA ordination for (**a**) *Puccinellia phryganodes*; (**b**) *Carex subspathacea*; (**c**) *Calamagrostis neglecta*; (**d**) *Carex aquatilis*.

partially corrects the loop distortion and it is concluded that neither quadrat nor species based data transformations will consistently correct the severe distortion normally induced by PCA when used as an ordination technique.

The raised-beach data set

Pierce and Kershaw (1976) described the colonization of young raised-beach ridges and the subsequent development of lichen–heath vegetation on the older parts of the ridges adjacent to the Hudson Bay salt marshes. This is used here as our first data set. The detailed morphology of the beach system together with the location of the sample plots has been presented above (Fig. 11.6). The leading beach ridge has a fairly uniform morphology along its length and represents a continuous time sequence. Thus the development of a complete raised-beach system is a complex

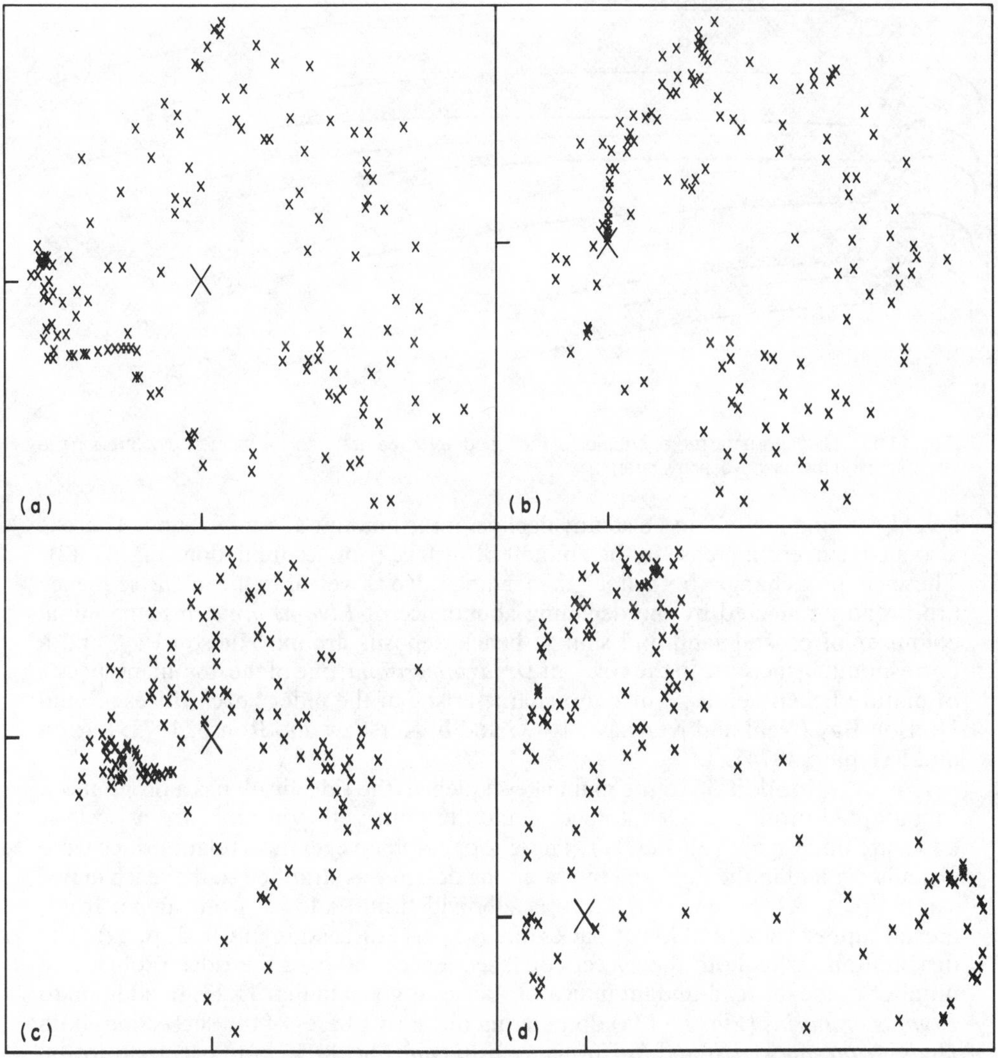

Fig. 11.10 The effect on the PCA ordination of the salt marsh data of: (**a**) centering by species; (**b**) centering by stands; (**c**) centering and standardizing by species; (**d**) centering and standardizing by stands.

interplay between depositional and isostatic uplift processes. An initial sand bar forms a locus from which centrifugal accretion coupled to uplift extends the length of the ridge, but with additional ridges incremented sequentially over time by high energy waves during successive storms. As a result the leading ridge tip (A1 in Fig. 11.6) is still being actively formed and colonized, and there is a resultant marked successional sequence back along the frontal ridge and this is shown diagrammatically in Fig. 11.11. The sample plots (A1–K) were located at regular intervals along the succession sequence. In addition however there is an identical but spatially compressed time sequence along the tips of each ridge running directly inland, and an additional set of plots (A2–A5) were also sampled on the end of each of these ridges (Fig. 11.11).

The major edaphic changes during the successional sequence are a regular increase in elevation along the frontal ridge with corresponding increases of ridge

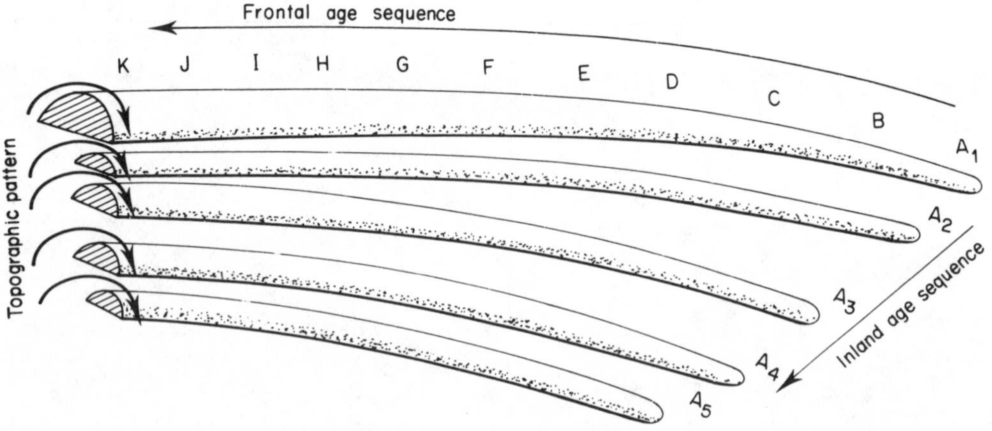

Fig. 11.11 Diagrammatic representation of the topographic features and age sequence controlling the vegetation on the raised-beach system.

height running inland, and a steady decline in the amount of bare ground. There is also a concurrent increase in the amount of surface peat accumulation (Fig. 11.12). These edaphic changes both along the leading ridge as well as in the inland sequence are broadly reflected by the declining abundance of *Elymus arenarius*, the initial colonizer of coastal sand and shingle beach deposits around Hudson Bay, and a corresponding increase in the cover of *Dryas integrifolia*, one of the dominant species of mature lichen heaths, which are characteristic of the older beach ridges around Hudson Bay (Neal and Kershaw, 1973*a* and *b*; Kershaw and Rouse, 1973; Larson and Kershaw, 1974).

However, in addition to the main age sequence, the ridge itself has a pronounced topographic profile which interacts markedly both with soil moisture as well as exposure and aspect (Fig. 11.11). These topographic-vegetation relationships were visually evident in the field and the sampling design was stratified so that each major sample plot, A1–K, and A1–A5, was subdivided into a lower front, upper front, crestal, upper back, and lower back zone (see also discussion in Ch. 2, p. 29). The distribution throughout the successional sequence, and over the ridge profile, of a number of the more abundant indicator species is given in Fig. 11.13. In addition to *Elymus arenarius* (Fig. 11.13a) dominating the early phases of the succession, both *Hedysarum mackenzii* and *Saxifraga tricuspidata* (Fig. 11.13 b–c) occupy a crestal position and are very abundant in the early stages of colonisation along the frontal beach ridge. Similarly, they dominate the second ridge but decline rapidly further inland. *Dryas integrifolia* is a dominant species of the crestal zone of the older ridges together with *Alectoria ochroleuca* and a number of other lichen species (Fig. 11.13 d and f). In marked contrast *Salix reticulata* is dominant on the lower slopes of the beach ridge particularly in the later stages of succession (Fig. 11.13e).

Thus the second data set we use here (see also Appendix 8) has two clearly defined major sources of variation which can be identified. Alternatively they can also be expressed by a simple graphical presentation. Again the data set thus constitutes an ideal test for examining the efficiency and capabilities of different ordination methods, utilizing field data with a natural level of variability rather than a simulated data set as generally has been used previously (Austin and Noy-Meir, 1971; Gauch and Whittaker, 1976; Hill and Gauch, 1980; Gauch and Whittaker, 1981; Gauch *et al.*, 1981).

Pierce and Kershaw (1976) used PCA to ordinate the primary samples at the plot level and showed a good correlation between the major successional sequence on the

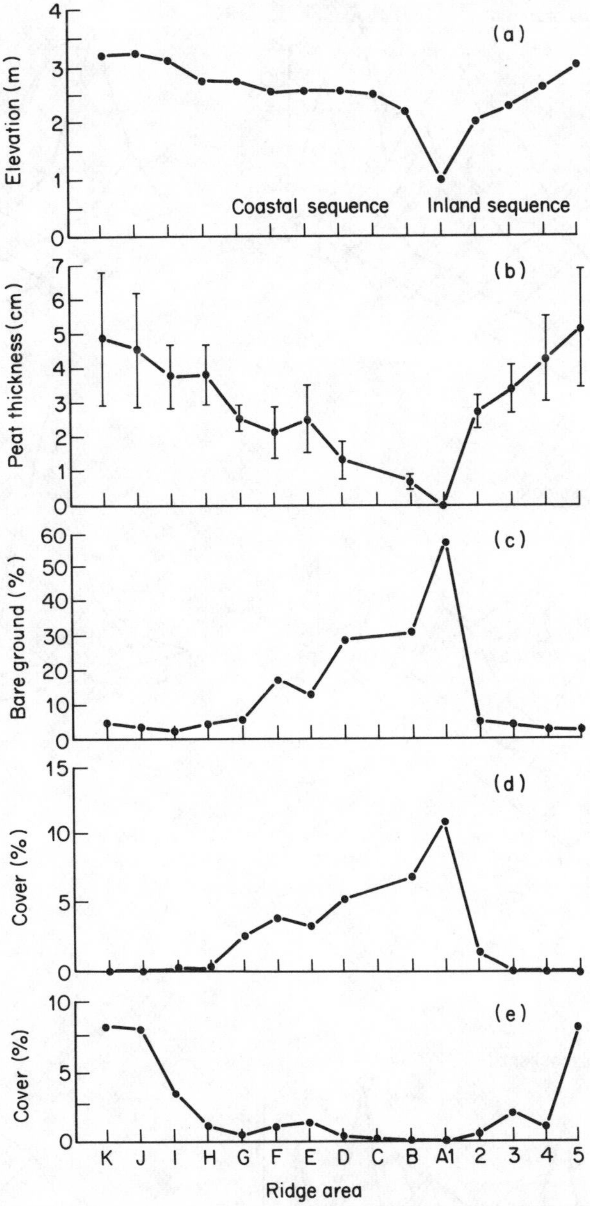

Fig. 11.12 The major vegetational and edaphic characteristics of the successional sequence of the raised-beach. As the ridge becomes older the elevation increases. (**a**) The ground becomes increasingly covered by vegetation with a gradual accumulation of humus (**b** and **c**); and the percentage cover of *Elymus* decreases whilst that of *Dryas* increases (**d** and **e**). (From Pierce and Kershaw, 1976; courtesy of *Can.J.Bot.*)

front ridge with the equivalent sequence running inland. This in turn closely matched the corresponding elevation data (Fig. 11.14). However even at this basic plot level there is a pronounced arch to the ordination sequence. The full data set at the quadrat or at the transect level in fact was ordinated also using PCA but the information was so obscured by distortion that the analysis was abandoned. The full ordination of the transect data is however reproduced here (Fig. 11.15), as percentage cover overlays on the plot ordination, for four species characteristic of both

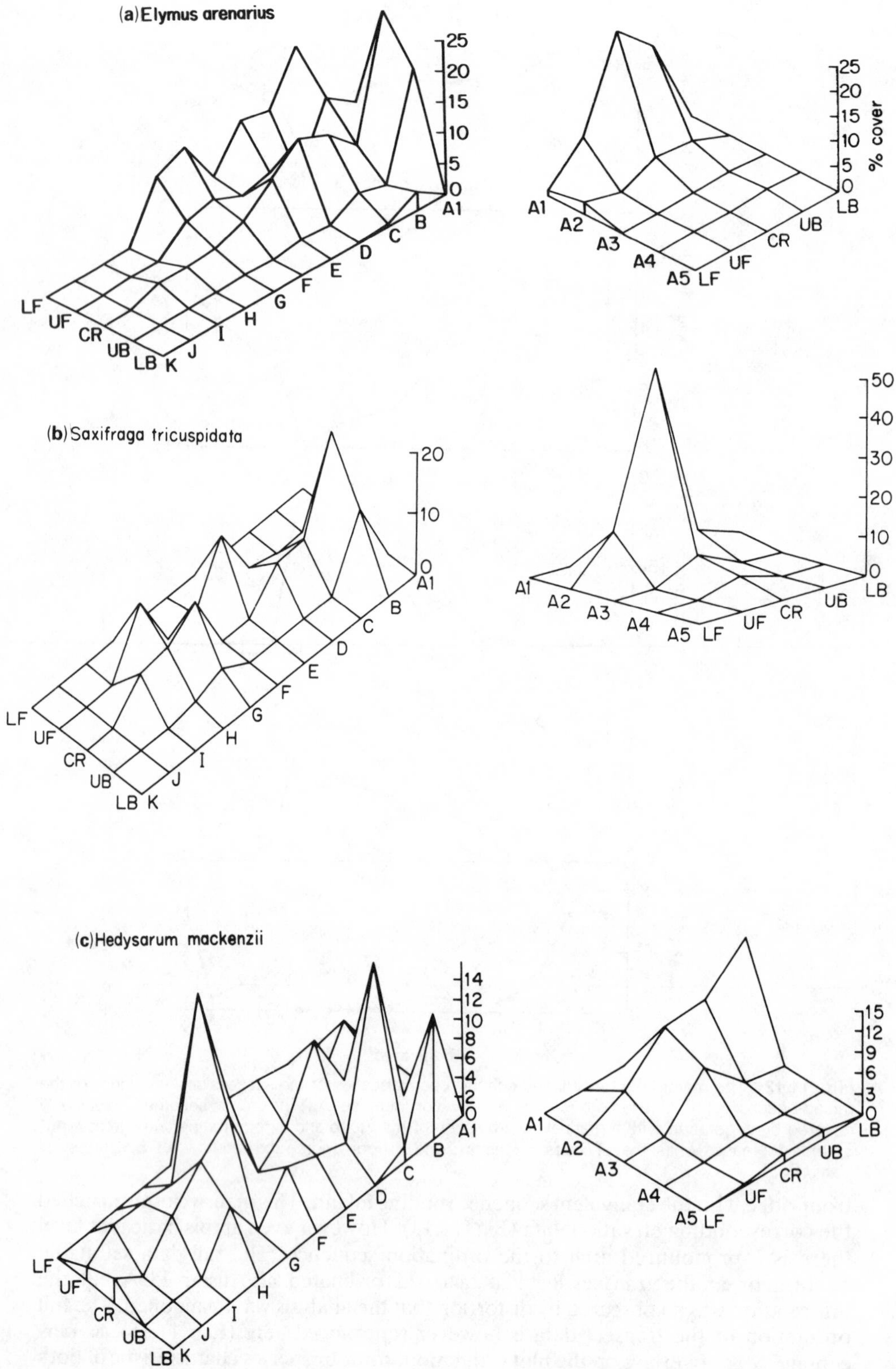
(a) Elymus arenarius
25
20
15
10
5
0
% cover
LF
UF
CR
UB
LB
K
J
I
H
G
F
E
D
C
B
A1
A2
A3
A4
A5
(b) Saxifraga tricuspidata
20
10
50
40
30
(c) Hedysarum mackenzii
14
12
8
6
4
2
9
3

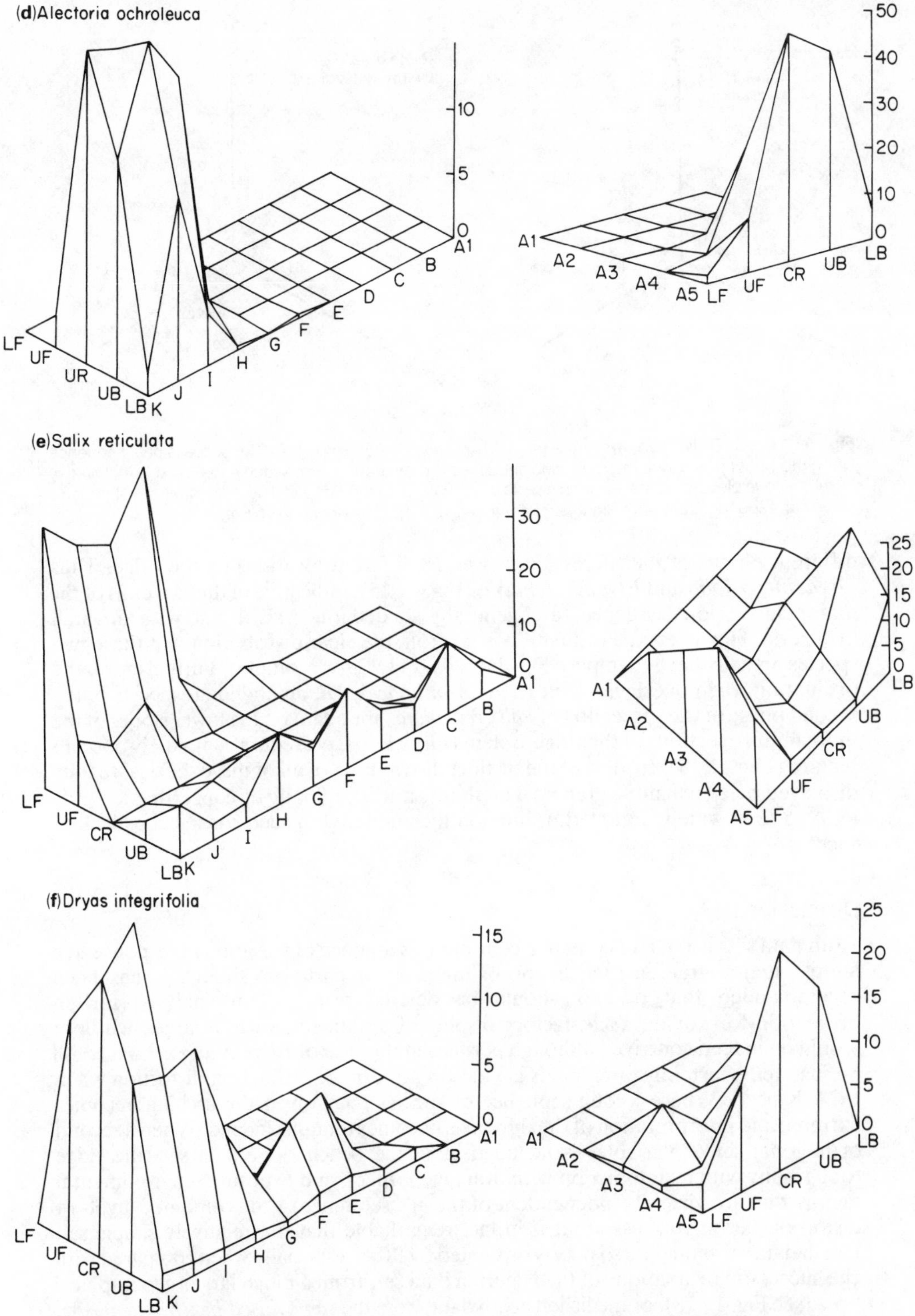

Fig. 11.13 The pattern of percentage cover in response to ridge topography and successional stage during the colonization of raised beaches. (**a**) *Elymus arenarius*; (**b**) *Saxifraga tricuspidata*; (**c**) *Hedysarum mackenzii*; (**d**) *Alectoria ochroleuca*; (**e**) *Salix reticulata*; (**f**) *Dryas integrifolia*. (See text.)

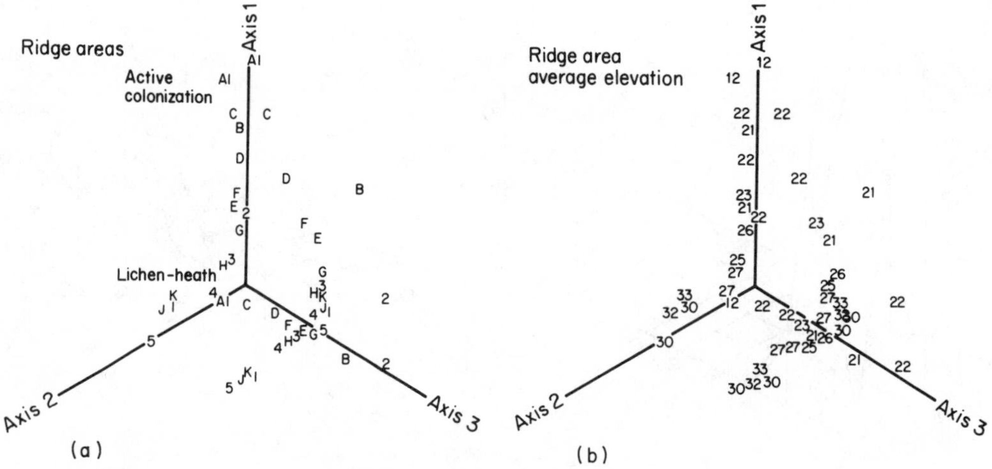

Fig. 11.14 (**a**) Ordination of sample plots A1–K and A1–A5 using PCA. The successional sequence along the front ridge as well as the inland sequence is ordinated well between Axes 1 and 2. (**b**) Overlay of the relative elevations of the sample plots again with both the ridge and inland sequences showing a close correspondence. (From Pierce and Kershaw, 1976; courtesy of *Can. J. Bot.*)

different phases of the succession as well as different locations on the ridge. Thus *Dryas integrifolia* and *Elymus arenarius* (Fig. 11.15), indicative of the two ends of the major succession sequence, are generally loaded on Axis 1 but with an arch distortion. However, the ordination is far from the clear distribution that these two species actually have (compare Figs 11.12 and 11.13 for example). Furthermore, the ordination when overlayed with *Hedysarum mackenzii*, an abundant species of the second phase of the succession of *Salix reticulata*, abundant on the lower slopes of the more mature sections of the ridge system (Fig. 11.13), presents a completely obscure picture (Fig. 11.15). In view of the distinct distribution of all of these species, readily displayed by even simple graphical methods, it was orginally disconcerting to apply PCA, a then widely accepted ordination technique, with such conspicuous lack of success.

Discussion

Both data sets used here contain a continuous sequence of variation in response to a simple major gradient which approximates to linearity. Although it has been recommended that outliers should be deleted prior to an analysis, for an ordination to provide a satisfactory display of a data set, such an approach here would be indeed contrived although perhaps in the case of *Elymus arenarius* it could be justified. Certainly 'arching' is a widespread form of ordination distortion when PCA is used. Where a long sequence of species is involved the arching becomes extreme and interpretation of the third axis becomes tenuous indeed. When a second orthogonal set of variables is included as in the beach ridge data set (the ridge topography with its interaction with drainage, aspect, and exposure; environmental factors that are virtually independent of the age sequence) – the combined levels of distortion become excessive. It is in fact remarkable that the relatively simple salt marsh data set should be so poorly ordinated by PCA, especially when compared with the successful ordinations of the Nigerian data set, from a much larger geographical area (see Fig. 10.18), or the lichen heath data, from the *same raised-beach system* (Fig. 10.12). Certainly the beta diversity seems to be implicated but unfortunately this does not correlate necessarily with the size of the sampled area in question. Neither is it

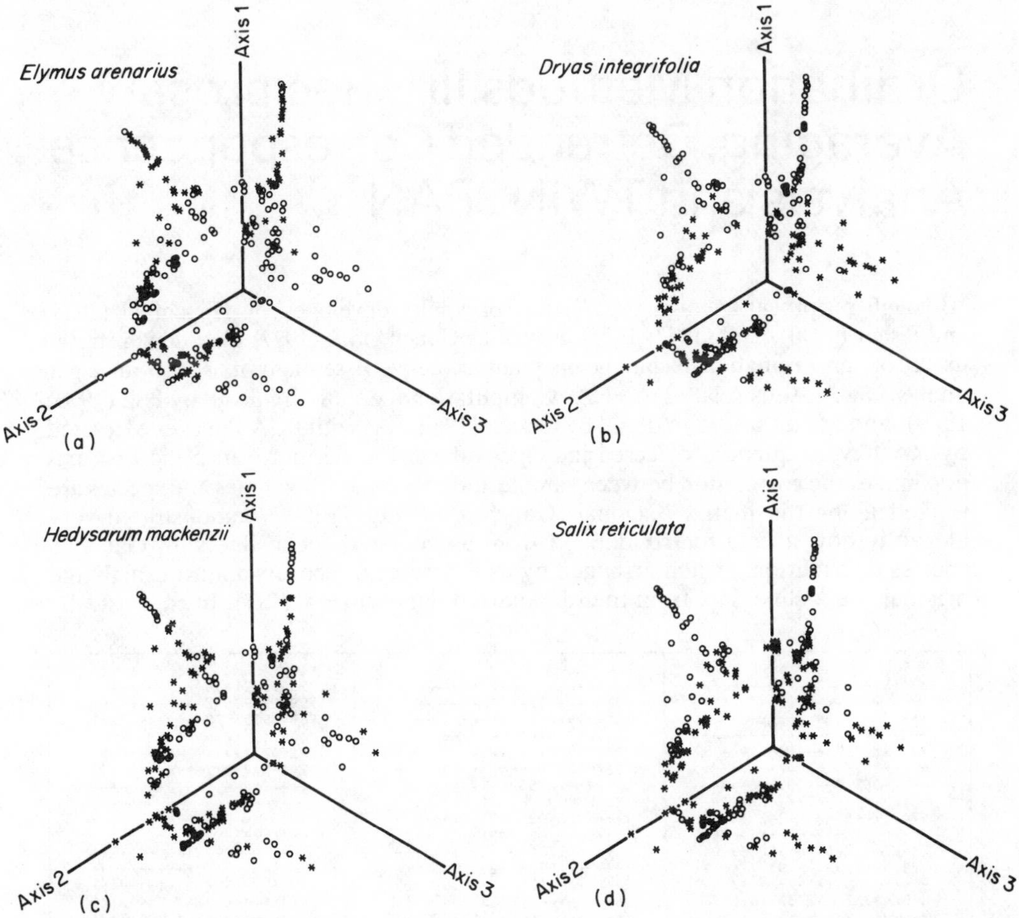

Fig. 11.15 Ordination using PCA of the individual topographic plots with the abundance values of (**a**) *Elymus arenarius*, (**b**) *Dryas integrifolia*, (**c**) *Hedysarum mackenzii*, and (**d**) *Salix reticulata*; overlayed as presence (*) or absence (o) values.

always apparent prior to an analysis particularly when the field personnel are inexperienced. As a result a modification of the ordination approach is desirable if not essential.

12

Ordination Methods III Reciprocal Averaging, Detrended Correspondence Analysis and TWINSPAN

Although reciprocal averaging (RA) was originally developed by Hirschfeld (1935) and Fisher (1940), Hill (1973; 1974) as well as Gauch *et al.* (1977) then demonstrated its use as an ordination technique in plant ecology. Essentially it is also an eigen analysis and is thus related to PCA. Computational details are given by Hill (1973; 1979), and are discussed in detail by Gauch (1982). As with PCA the axes extracted by RA have sequentially decreasing eigenvalues. Furthermore since the first axis maximizes the correlation between sample and species, both samples and species are ranked along the matrix diagonal. Gauch *et al.* (1977) have demonstrated very elegantly how a data matrix derived from a simulated coenocline with Gaussian species distributions, when arranged by its RA scores, recovers almost exactly the original coenocline data from the randomized data set (Fig. 12.1). In contrast, the

```
993+31+29612+++58++46+79
--+--+++-+--42+++-1--1--
+++73---++64---++9-21-++
---+-------+------+------
758++51719++++398-1+++13
++1-+352++-+9841+-6++7++
1+7--988+3--12952-6++4++
--+--+++----21++--+--+--
67137+++9346+++242+89+98
---++------++-----+-++-+-
23+27+--7+35--++11-99-95
428--2+5+9---+197-+++++1
---++------++-----+-++-+-
649++97928++23897+6+1413
------+-----++----+--+--
++1--362++--995++-7-+8++
--+--+++----21++--+--+--
+++63++++++54--+++7-22-1+
429-+527+8--++396-+++++1
++-97---1+99---++8-53-2+
```

(a)

```
+++++-------------------
21++++++++--------------
21++++++++--------------
4211++++++++------------
998765321+++++++++------
9876543211++++++++++++---
124689987 5321++++++-----
23467899999876432 11++++++
++++23579986421+++++----
-++++1258997421++++-----
+++11357899875311++++++-
++++++123568999976432 1++
+++++++++12346789998764 32
-----++++++1235799975321
---------++++++123579998
----++++++++++++12234567
---------+++++++++1234679
-------------------++++++++
-------------------++++++++
-----------------------+++
```

(b)

```
795993821+++---+---+-+--
759329197532---1--++++++
768439299876+++4+++3+2+1
3426719+++++++--+---+-+-+
34256271++++---+-+++++++
3241+6+79999+++8+++61523
+++++++++-----------------
+++++++++-----------------
---++-+------------------
--------------+++-++------
--------------21+-+++-++++
--------------21+-+++-++++
--------------421+1+++++++
+++--+-+++++987+654+3121
-----+-+++++998+765+3+21
--+--+-++124-++7++192958
++++++-11357+++811395978
+-+--+-++124+++6+2385979
--+--+-++++112426893958 7
+++++1+1234623476788 9999
```

(c)

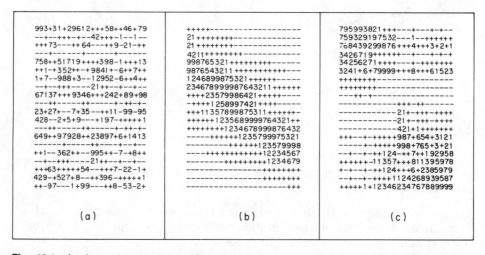

Fig. 12.1 A primary data matrix of species-by-sample scores arranged at random (**a**), according to species and sample ordination values from reciprocal averaging (**b**), and values from centered, non-standardized principal components analysis (**c**). Note that RA arranges the matrix to concentrate higher values along the matrix diagonal. The data matrix is a simulated coenocline with Gaussian species distributions and a beta diversity of 5 HC having twenty species and twenty-four samples. Absence is indicated by a dash (–) and scores are rounded into ten equal classes marked '+' and '1' to '9'. (From Gauch *et al.*, 1977; courtesy of *The Journal of Ecology*.)

ordering achieved using PCA although revealing some of the structure of the coenocline is far from satisfactory (Fig. 12.1c).

Gauch *et al.* (1977), and Gauch (1982) have concluded that in most cases RA is superior to PCA and although some arch-distortion is always present, the marked involutions sometimes induced by PCA are absent, and as a result interpretation at the ecological level is considerably more straightforward. The most serious problem associated with arch distortion however is that the extent of arching may exceed the eigenvalue of a secondary environmental gradient. Put more simply, the arch may be interpreted by the analysis as an additional set of ecological variations and as a result primary and secondary environmental gradients may be ordinated on axes 1 and 3 or 2 and 3, rather than on 1 and 2. With more complex data sets the relevant axes which are ecologically meaningfull may in fact be contained amongst even higher dimensions and it becomes difficult to distinguish mathematical artifacts from ecological meaning. An additional problem with RA, and direct consequence of the arch-distortion, is the compression of samples at the two ends of the first axis which should correspond to a simple linear coenocline. This is illustrated in Fig. 12.2 where

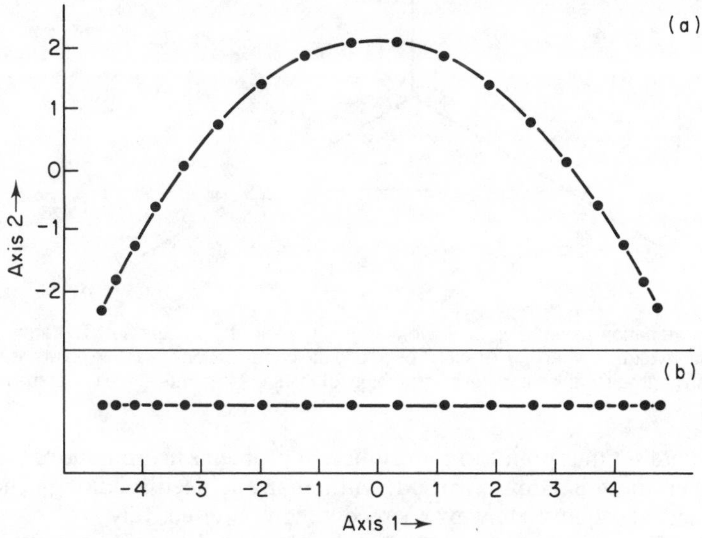

Fig. 12.2 Ordination of 18 regularly spaced samples using reciprocal averaging. The arch distortion is clearly evident in the ordination using axes 1 and 2 (**a**), whilst the compression of the two ends is quite marked in the ordination simply using axis 1 (**b**). (from Gauch, 1982; courtesy of Cambridge University Press.)

compression of the first axis ends coupled to the arch distortion, produces a marked degree of non-uniformity when the samples are ordinated along axis 1. As a result, RA can induce an uncontrolled degree of apparent sample dissimilarity along each axis.

The extent of these problems is very clearly illustrated in our two test data sets from the Pen Island salt marsh and raised-beach systems.

Ordination of the salt marsh and raised-beach data using reciprocal averaging

The full analysis of the salt marsh data using RA for the first three axes is given in Fig. 12.3. Although the sample sequence 1 to 145 is recovered very well and displayed along axis 1, the arch distortion is quite marked. Furthermore, the compression of samples ranges from extreme at the bottom end of the marsh with an apparent 'stretching' at the top of the marsh. However, and rather more significantly, there is evidence of an increasing degree of distortion of this arch leading to a complete involution of the sequence when displayed between axes 2 and 3. Although the sequence of eigenvalues (0.7730, 0.6260, 0.4669) does suggest that the major part of the sample variance is related to axis 2, it also points to potential interpretative problems when several higher order axes are essential for more complex data sets. Our simple but

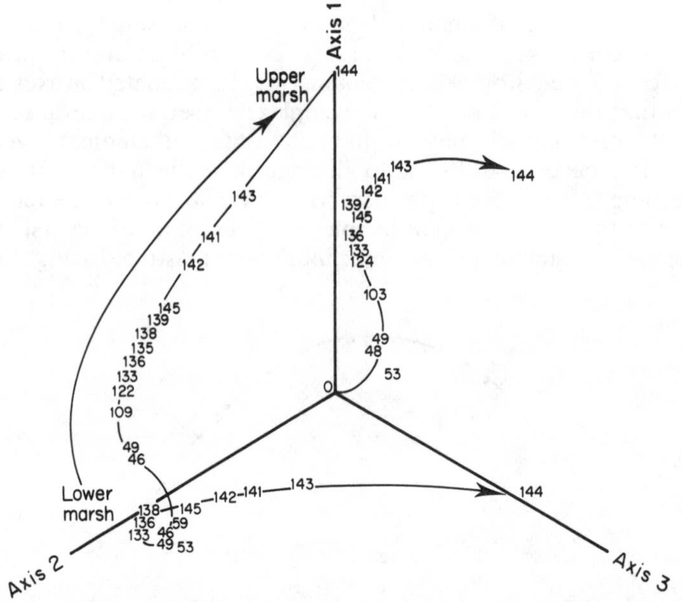

Fig. 12.3 Plot ordination using reciprocal averaging of the salt marsh data set. The arch is heavily distorted into a convolution when a third axis is used and the compression of sample points as the lower end of the marsh contrast with an apparent 'stretching' of the sample points at the top of the marsh.

natural field data set thus points to some inherent problems in ordinations using RA, especially when the full data set is used, without any subjective editing. The actual salt marsh quadrat sequence however is recovered very effectively.

The RA ordination of the raised-beach data set however does provide a considerable improvement over the corresponding analysis using PCA. The overlays of *Elymus arenarius, Hedysarum mackenzii, Salix reticulata* and *Dryas integrifolia* are given in Figs 12.4 and 12.5 and are summarized in Fig. 12.6. The major succession sequence is largely correlated with the first axis with *Elymus arenarius* having low and *Dryas integrifolia* high loadings on axis 1. The ordination on axes 1 and 3 reflects again the arch distortion, the true ordination of the secondary (topographic) ecological gradients being contained between axes 2 and 3. Thus the crestal samples dominated by *Dryas integrifolia* in the older and *Hedysarum mackenzii* in the younger parts of the ridge system, together with the plots sampled on the lower slopes dominated by *Salix reticulata* on the old and *Elymus arenarius* on the young ridge sections, are very clearly ordinated.

It is concluded that both the salt marsh and the beach ridge data support the findings of Gauch *et al.* (1977) which were subsequently amplified by Gauch (1982), pointing to a number of problems still inherent in RA as an ordination technique, although at the same time it does provide a method that is a considerable improvement on PCA. Thus, although when the salt marsh data were ordinated using PCA the results were severely distorted, the actual linear sequence was still reasonably clear. The corresponding ordination using RA however contained considerably less distortion and was as a result much more clear cut. In contrast, the beach-ridge data containing a single major successional sequence together with an orthogonal secondary topographic sequence was very poorly ordinated by PCA but very clearly using RA. Thus RA provided a much more reliable method of displaying the ecological information in the ridge data as well as recovering fairly exactly actual

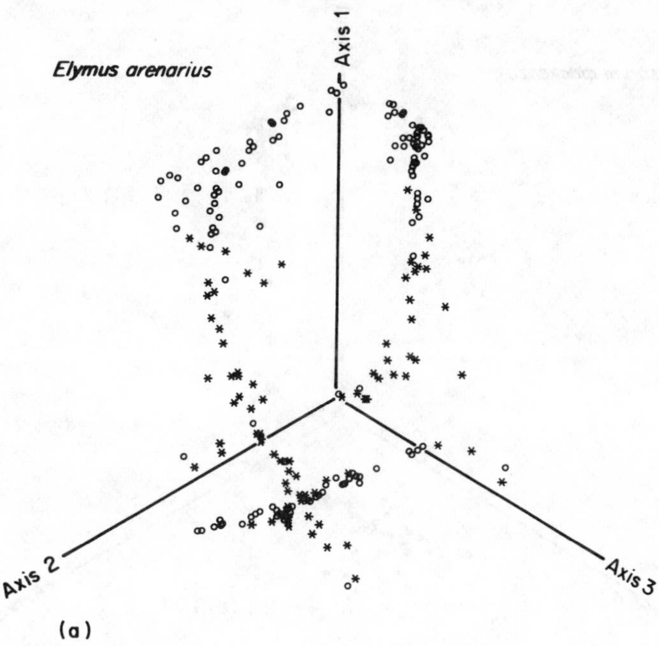

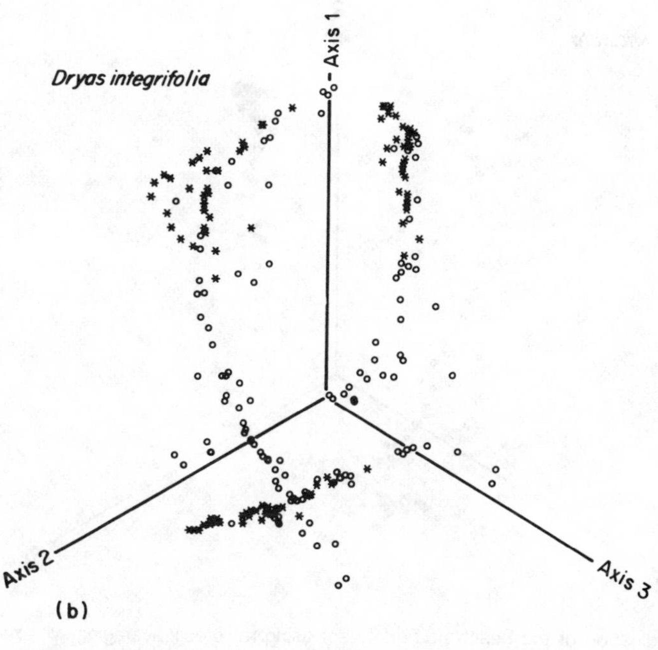

Fig. 12.4 Ordination of the beach ridge data set using reciprocal averaging with presence (*) and absence (o) overlayed for (**a**) *Elymus arenarius* and (**b**) *Dryas integrifolia*.

sample sequences in the linear gradient of the salt marsh. Although some residual distortion is still present, more seriously it is impossible to predict *which* particular axis pair or pairs carry meaningful ecological information.

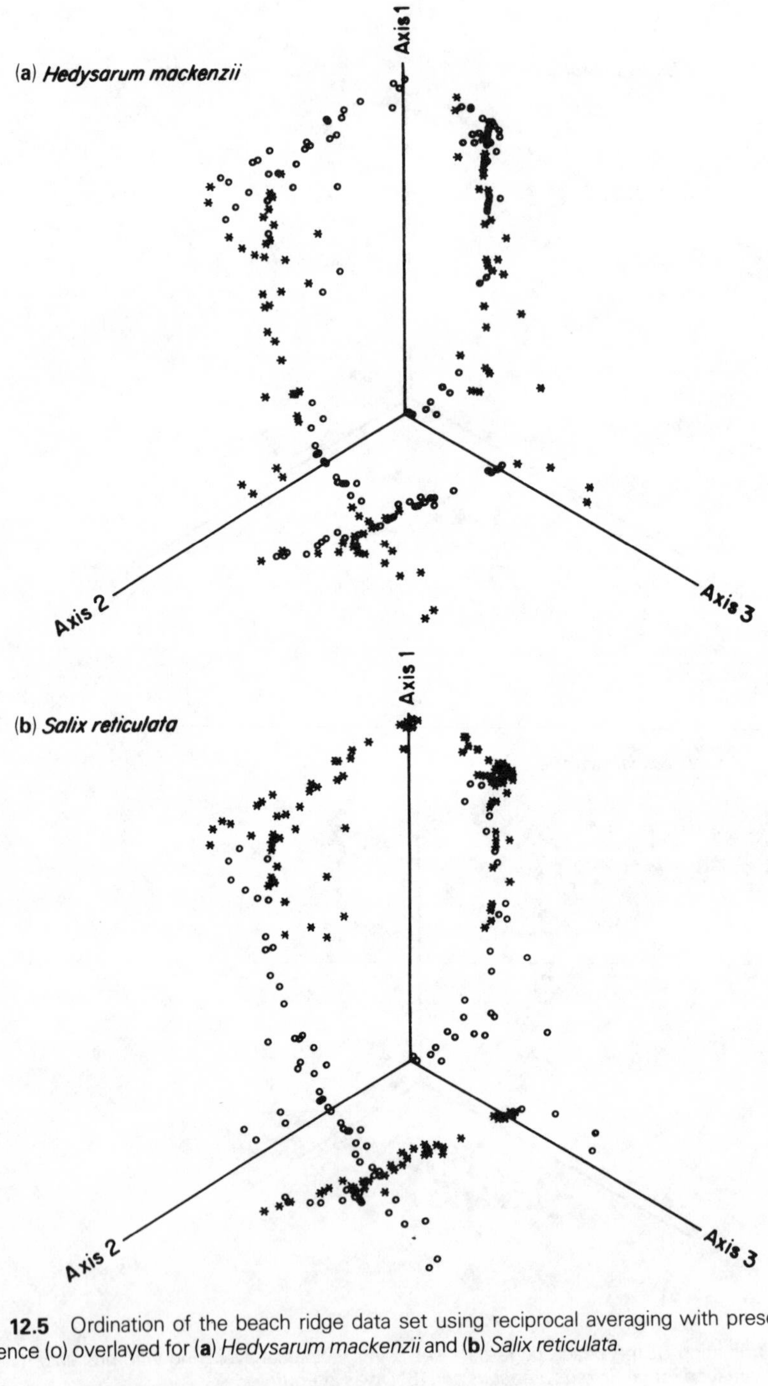

Fig. 12.5 Ordination of the beach ridge data set using reciprocal averaging with presence (*) and absence (o) overlayed for (**a**) *Hedysarum mackenzii* and (**b**) *Salix reticulata*.

Detrended correspondence analysis

In complex and extensive data sets it is usually difficult enough to interpret the axes of an ordination in terms of environmental criteria without the added difficulty of *first* identifying which specific axes actually carry the meaningful information. The disadvantage of arching and the potential postponement of real ecological

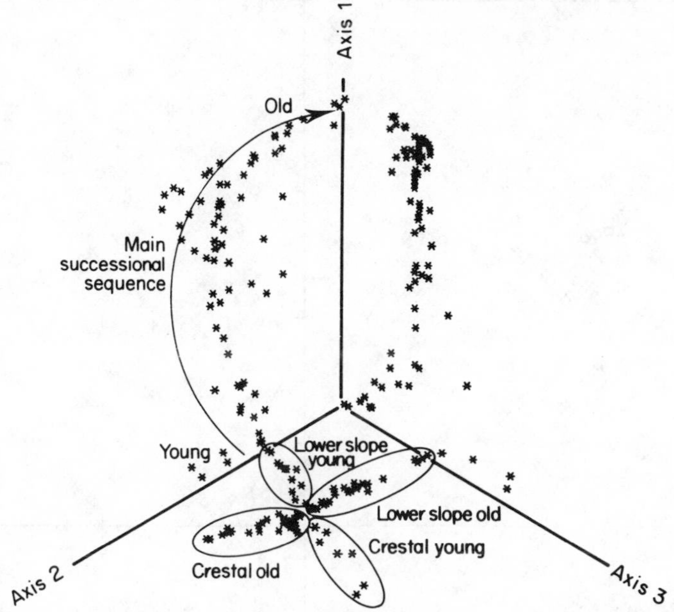

Fig. 12.6 Summary of the plot ordination of the raised beach data set using reciprocal averaging.

relationships to undefinable higher order axes, is accordingly quite a serious problem. Similarly, the inability of RA to preserve and recover ecological distance between adjacent pairs of closely similar quadrats in quite specific parts of an ordination, but not in others, also seriously detracts from its use in unknown ecological situations. Hill and Gauch (1980), Gauch (1982) have examined these problems and have proposed an alternative analysis, detrended correspondence analysis (DCA), which eliminates the arch effect by adjusting loadings on the second axis to a zero mean value within predefined segments of axis 1(Fig. 12.7). Although the actual method of computation is rather more complex than Fig. 12.7 might suggest, the details do not concern us here, but can be obtained from Hill (1979*a*) and Hill and Gauch (1980).

The results of the DCA analysis of the salt marsh data using the Fortran program provided by Hill (1979*a*, DECORANA) are given in Fig. 12.8. The ordination is

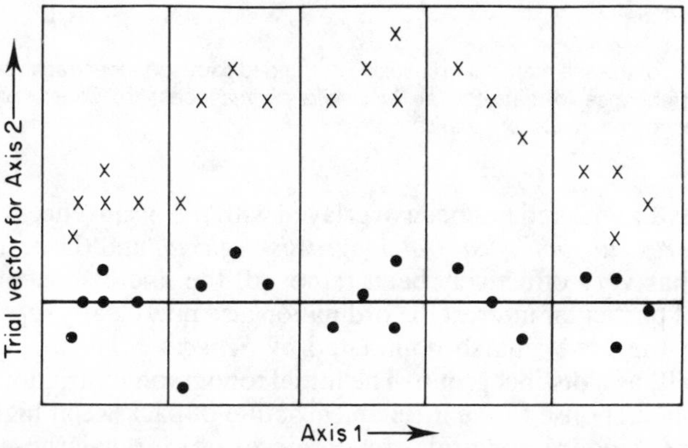

Fig. 12.7 Method of detrending used in detrended correspondence analysis. × = initial trial positions on second axis; ● = adjusted positions on second axis. (From Hill and Gauch, 1980; courtesy of *Vegetatio*.)

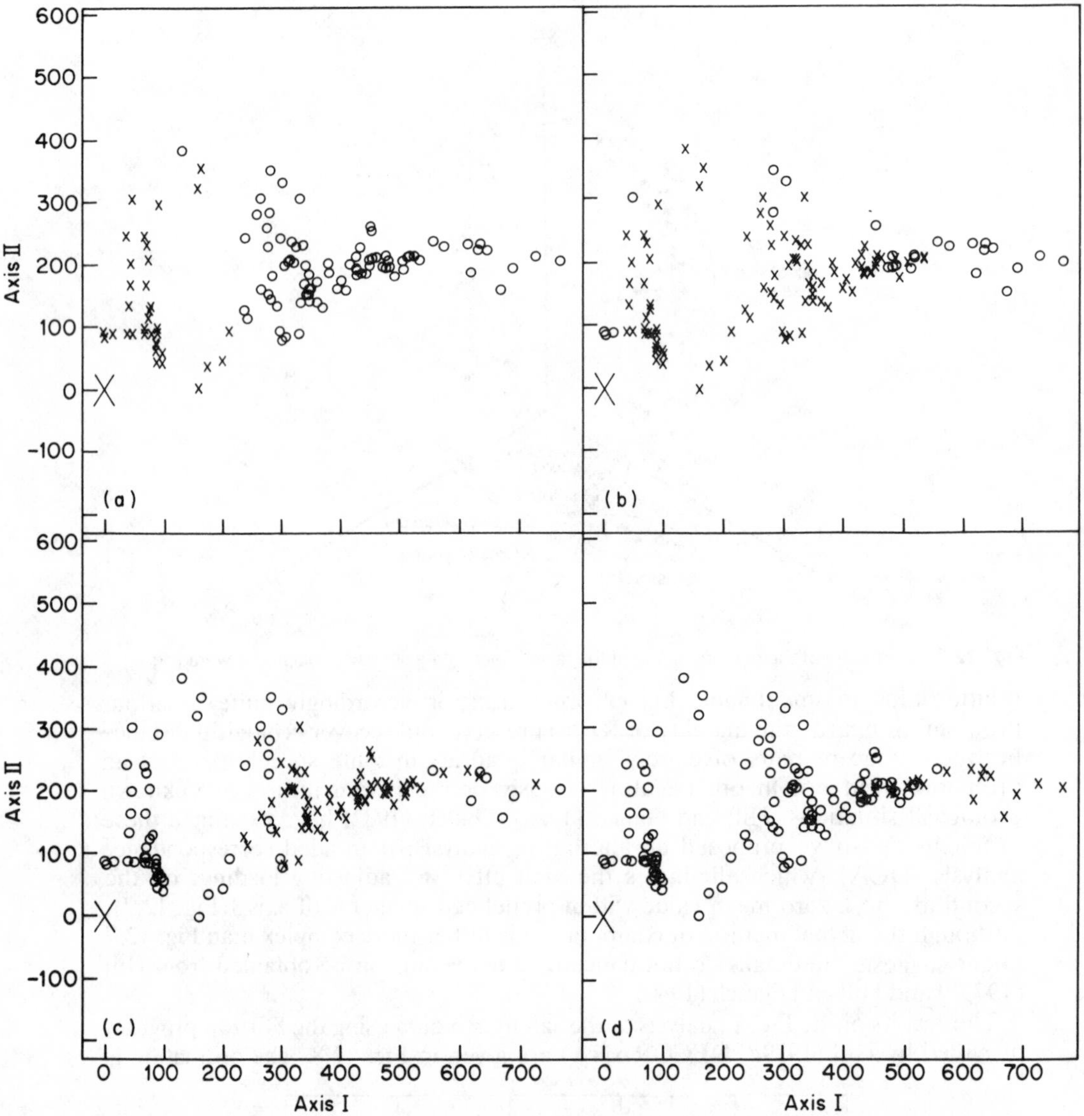

Fig. 12.8 Ordination of the salt marsh data set using detrended correspondence analysis with overlays of presence (*) and absence (o) data for (**a**) *Puccinellia phryganodes*; (**b**) *Carex subspathacea*; (**c**) *Calamagrostis neglecta*; and (**d**) *Carex aquatilis*.

displayed on axes 1 and 2 and has been overlayed with the occurrences of *Puccinellia phryganodes, Carex subspathacea, Calamagrostis neglecta*, and *Carex aquatilis*. The arch distortion has very effectively been removed, the linear sequence efficiently recovered, but of particular interest the ordination also now clearly separates the salt pan quadrats of the lower marsh dominated by *Senecio congestus* and *Hippurus vulgaris* (Fig. 12.9) as a distinct group. The initial topographic variation in the lower marsh develops in response to the mechanical action of pack-ice at high tide during the early summer. Under strong tidal action pieces of ice regularly scour the mud surface of the tidal flats creating shallow depressions or conversely 'bulldozing' mud into small mounds with the leading edge of the ice. The slight mounds are rapidly colonized by *Puccinellia phyryganodes* whilst the incipient pans are finally vegetated higher in the marsh by a number of characteristic species, notably *Senecio congestus* and *Hippurus vulgaris*. All the previous data analyses simply identify these species as typical plants of the lower marsh without any indication of their very specific ecology.

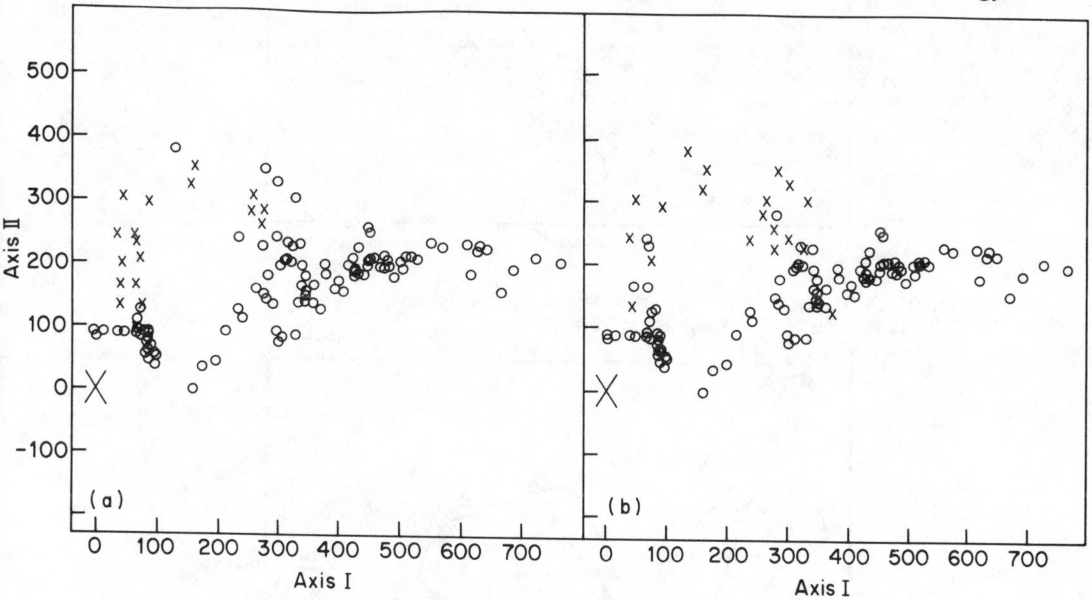

Fig. 12.9 Ordination of the salt marsh data set using detrended correspondence analysis with overlays of presence (*) and absence (o) data for (**a**) *Senecio congestus*; and (**b**) *Hippurus vulgaris*.

It is indicative of the powerful attributes of DCA that the analysis does ordinate these plots on a separate axis (see Looney and Kershaw, 1985, for further discussion and the classification of the data by TWINSPAN. Also see below and Fig. 12.10).

The corresponding DCA analysis of the raised-beach data set is equally satisfactory (see Fig. 12.11). The primary variance reflecting the age sequence of the beachridge system is ordinated almost linearly along axis 1, with the secondary variance due to the topographic ridge profile appearing correctly on axis 2. It is particularly satisfying to see that the full successional sequence is effectively displayed, *including* the quadrats dominated almost exclusively by *Elymus arenarius* which are floristically very distinct from the remaining quadrats. Thus Gauch (1982) has emphasized how both 'outliers' and 'noise' in a data set can limit the success of an ordination and equivalent comments have been made above in relation to component analysis (p. 200). It is perhaps reasonable to reject from a data set rare species which contribute largely to noise, the common species carrying the ecological information. However, in some situations, 'outliers' in a mathematical sense can be an integral part of the data set, as in our example here, and it is unfortunate if the quadrats in question are isolated from the remainder of the data. It is encouraging that DCA seems to be sufficiently robust for the *Elymus* plots in the ridge data to be satisfactorily incorporated in the ordination of the complete data set. It remains to be seen after more extensive use of DCA whether this conclusion is substantiated. DCA also ordinates both the age and topographic sequences very satisfactorily on the first two axes, eliminating the difficulty experienced with RA, in determining which axis pairs carry the required ecological information rather than its distortion.

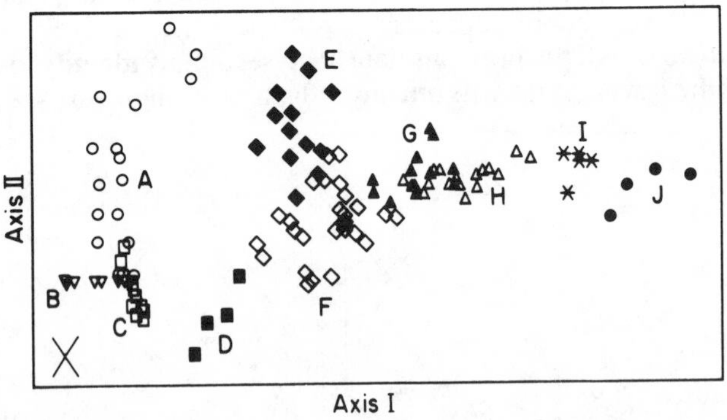

Fig. 12.10 Ordination of the salt marsh data set using detrended correspondence analysis with TWINSPAN Groupings A–J overlayed (see Table 12.1).

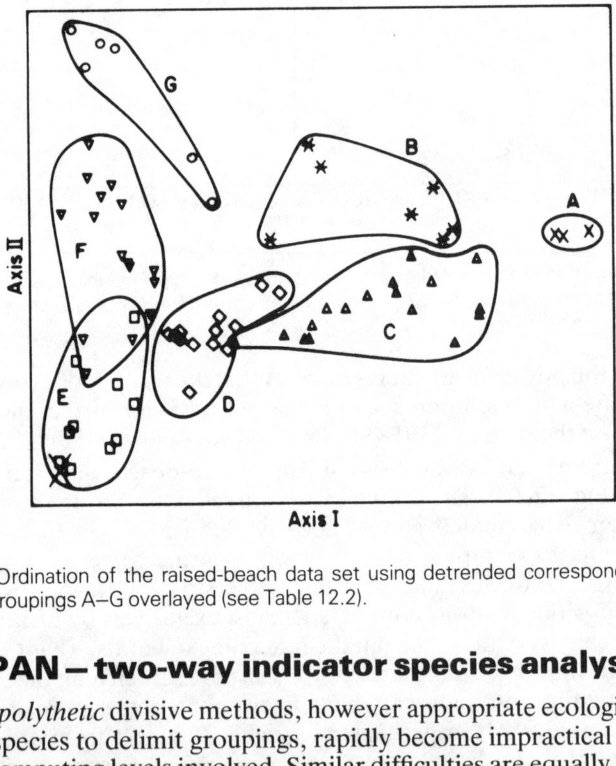

Fig. 12.11 Ordination of the raised-beach data set using detrended correspondence analysis with TWINSPAN groupings A–G overlayed (see Table 12.2).

TWINSPAN – two-way indicator species analysis

The use of *polythetic* divisive methods, however appropriate ecologically in their use of several species to delimit groupings, rapidly become impractical as a result of the excessive computing levels involved. Similar difficulties are equally inherent in some of the clustering methods that have been used where similarities *between* quadrats are emphasized. However, it is with these latter methods where the two apparently distinct approaches of classification on the one hand, contrasting with 'rigid' alternative, ordination, finally merge together. There has been considerable discussion over the last decade as to the most appropriate analysis in a specific situation, classification or ordination (see discussions for example in Greig-Smith (1980; 1983). Gauch (1982) very pungently focuses on the problem and emphasizes however useful classification units are, variation in vegetation is most usually, if not always, continuous. As a result, rather arbitrary divisions are imposed on essentially continuous data. In essence, however, an ordination displays also the samples contained in each plant association and this is quite evident in Figs 10.12 and 10.18 above (p. 202) where intermediates were largely eliminated during the sampling. Thus with the current conclusion that classification and ordination methods are quite compatable techniques we concur with Gauch (1982) that there is little to be gained in the use of complex polythetic agglomerative techniques except perhaps in special circumstances. Rather, we recommend either a direct Braun Blanquet phytosociological approach using computer assistance, for example TABORD (Maarel *et al.*, 1978; Maarel, 1982), or perhaps more constructively, ordination using DCA coupled to the ingenious approach Two-way Indicator Species Analysis (TWINSPAN), which delimits associations very successfully (Hill, 1979*b*).

TWINSPAN essentially recognises the power and efficiency of RA for establishing the major variation in a data set and utilises such an analysis to divide the data into two groups. Each of these two groups is then re-analysed and further divisions subsequently made in the same way. Recognition of the final levels of subdivision inevitably rest on the decision of the investigator. Hill has also recognized that a single species may in fact display quite different but very characteristic information at different abundance levels in a number of plant associations. As a result Hill (1979*b*) introduces the term 'pseudospecies' where a species at different levels of abundance is designated as having *differential indicator values*. This information is utilized in TWINSPAN and as a result very effectively overcomes the problem that *species abundances* as well as the *specific compostion* of an association is important. The reader is referred to Hill (1979*b*) and Gauch (1982) for a more detailed discussion of the actual method.

The analysis of the salt marsh data set (Appendix 7) using TWINSPAN is given in Table 12.1 in the form of the actual (slightly modified) computer print-out. The numbers in the body of the table represent the pseudospecies, scaled at levels 0 (blank), 3, 4 and 5 which are thus equivalent to 0, 5, 10 and 20% frequency respectively. At the bottom of thetable, the hierarchical divisions are based each time on RA, and are denoted by contrasted 0–1 groupings. Thus the primary split is between quadrats 74 and 41, numbered vertically along the top of the table, which essentially divides the lower marsh (low numbers) dominated by *Puccinellia phryganodes* from the middle and upper (fresh water) marsh (high numbers). Sequential division within the two primary groups are similarly indicated by blocks of 0s or 1s in successive lines at the bottom of the table. The divisions have been continued for 6 successive levels, the standard outer limits recommended for most data sets (Hill 1979*b*). However the final decision as to where to halt division or conversely what constitutes an ecologically meaningful level of sub-groupings, is usually entirely a subjective decision despite a number of attempts to develop objective tests. Numerous rules for halting hierarchical classification have been proposed (Popma *et al.*, 1983; Hill, 1979*b*; Gauch and Whittaker, 1981; Gates and Hansell, 1983) but in the final analysis the decision rests with the field observer as to what does or does not constitute ecological 'sense'. However, if the associations are then re-examined in the context of the corresponding DCA ordination, it becomes fairly straightforward to eliminate the doubtful groupings. Thus in the salt marsh data set, 10 plant associations (A–J) initially are recognized which also correlate with some topographic feature or position on the marsh. Furthermore when these 10 associations are then superimposed on the DCA ordination *the clarity of the groupings is essentially maintained*. The ordination showing the TWINSPAN groupings is given in Fig. 12.10 and *objectively substantiates* the reasonable nature of these groupings derived by the first 2, 3 or 4 hierarchical divisions. Thus the lower marsh salt pans are quite distinct entities and readily indentified by the presence of *Hippurus, Triglochin* and/or *Senecio* (Association A). Associations B and C, or the lower salt marsh, with discontinuous and continuous plant cover respectively are also

Table 12.1. The plant associations A–J delimited by TWINSPAN (see text). Species abbreviations are listed in Appendix 7.2.

		J	I	H	G	F	E	D	C	B	A
		1111	11111	11111111111111111111	111111111111	1111 1					
		4444	43334	22123133022222223333	010001011111	99000188899999088666667777	889967777888977	4444	2332442333444	22 11 111111222223333	5555634555555666
		1234	57890	45824956301367890123	825690734567	79024168901345135024780168	476892579012234	1569	7016488289037	032376968134578124593467	5678152024913356
PALU	SQUA	545	55 4								
SALI	CORD	445									
SALI	CAND	5554	5 3								
POTE	PALU	53	5			4 5					
CARE	AQUA	5553	55455	45 553 5 555							
CARE	RARI	4	555	5							
AULA	PALU	4 55		55							
RANU	PALA	5		4		3					
SAXI	HIRC	5 4	45555	4 5355 4		3					
SCOR	SCOR		555 3	45555545 55 5	4 55 5	5					
CARE	DIAN		45	3534354555 53 54535	4 3343 4						
ERIO	RUSS	5	5335	55535555555555555555	355455455555	3 455 33	5				
CALL	GIGA		55554	555455355555555555	3355 5554 5	5					
DREP	FLUI		3	5	3 53 4						
ARCT	FULV	3		43 555 55 553	5 5 5555	5 5 3	553 55				
CARE	SALI			5 55	553						
ERIO	ANGU			3	5 343						
RANU	GMEL				3						
UTRI	INTE				3 4 33						
PEDI	SP			3 5 53 553	55554						
DREP	UNCI				45 5	3 3					
CARD	PRAT			555555 555555555555	555555555555	555555455555535555	4 35				
CALA	NEGL	3	33	5555555 555555555555	555555555555	5555555555555555555454 45	5355 3 3 55				
EPIL	PALU	3		44343333555555445555	3 43 354	33 555545554 355 53555555	5 3 5 43 55				
DREP	REVO				5	3	4 4				
DREP	ADUN				5	3 3 44 3					

```
STEL   CRAS          33   3455 54555555544 33 535535553355 5553555555555 555555555555 5 3535555555  3  5
CAMP   STEL                                                 5   54 3 4  345    5535
CICU   VIRO                                         3        3433
GALI   SR                                         3          355 5     5533
CARE   GLAR                                     4 3 35535555 3  3  55554 55 55          555555555555555                                                     55
ATRI   GLAB             4                       55  3        555 55555555553       53 54                     4     3   3
DUPO   FISH                                                    3   5555533  55555555555 3   55 55554355  5                                              434         55
MNIU   CINC                                         3             4    3   3       3                     5
BRYU   PSEU      3 33   4335      5       3          5        5 5  4    333 3 55 5   3               353
DREP   EXAN                   5                   5         5555 55  44555555553 555 55555555555554  555
CARE   SUBS             55  5  53  5353533  5555455  355 555555555555555555555555555 5455554 55555  5555 5555555555555                   55555 5555555555533 53
FEST   RUBR                                                           5     555555555    5  3        5555 454
HIPP   VULG                                                                           3  5 5 3553355                                                     3    53555555
PUCC   PHRY                                                                                          5555 5555555555555 55333334455555555555555 555555555555 554
MNIU   PUNC      5                                                             3                     5544
STEL   HUMI                                                               553                        555  5554555535355 3
POTE   EGED                                                                5                         455  5
TRIG   PALU                                            4  3           33   33                34              534555                                      55535 355355 545
RANU   HYPE                                                                                                                                              3
SENE   CONG                                                  5                              3533                                                        553553 33455 555
             0000 00000 0000000000000000000 000000000000 0000000000000000000000000 00000000000000 1111 111111111111 111111111111111111111111 1111111111111111
             0000 00000 1111111111111111111 111111111111 1111111111111111111111111 11111111111111 0000 000000000000 000000000000000000000000 1111111111111111
             0000 11111 0000000000000000000 000000000000 1111111111111111111111111 11111111111111 0000 000000000000 111111111111111111111111 0000000000011111
                  01111 0000000000000000000 111111111111 0000000000000000000000000 11111111111111 0000 111111111111 000000000000000000011111 0000011111100011
                        0000011111111111111 000000111111 0000000000000001111111111 00001111111111      000111111111 00111111111111111111     00111000111
                        0011100011111111111 001111000011 0000001111111100111111111     00000000011        0001111111   00000111111111111
```

Table 12.2 The plant associations in the raised beach data set delineated by TWINSPAN. Species names are listed in Appendix 8.2.

```
                5667 3 1112211244556677 7226667712 5611156634451    553444452333 22 3453334     2
                616165 1561579005050505 2344893482 9323482794940 7937727382734889 0111266  34612
                   A   AAAAAAAA             AA     AA   AAA          A AA               A           AA AA
                HIJK2C 33455444DEFGHIJK K55IJJKK45 HI333HIJDEFG2 22GHDEFEFGBCCD2BB BEGCCDF  11B11
                LLLLLL LLLLLUULLLLLLLL UCUUCUCUCU UCUCUCUUUUUUL UUCUUUUCCUUCUCCCU LLLLULL  CULLU
                FFFFFB FBFFBFBBBBBBBBBB FRBBRBRBRF BRFRBRFFBBBBB FBRFFFFRRFFRBRRRB BFFFFFF  RBFFF

BART   ALPI      11111 2221211111111 1  1 1       1 1 1 1    1                            1111 11   21
DRYA   INTE         11 1212 121111111111 45445454 4 3231223413322
ALEC   OCHR       1    1  1311    1 1241 2554555454 24 1121      1  1
CORN   DIVE               1 1114 11111551 1555453545 3311111      1  1 11 1
CARE   AQUA     3232     11132  2132 1    1         1         1                              1
EMPE   NIGR          2  3213223432332323    31 1 232             1  1     1       1             1
VACC   ULIG       431 1 3131312122312444 1  2 1 2    1
ARCT   RUBR     1 1 33 3412243242433111 1222122213 2 33342224221  11                       2
CARE   CAPI       1  1    12111 1 11 11 2   1       111 1   1 1
PEDI   SP            1  111111 11   1 111 1      1 1 1 1   1    1
RHOD   LAPP          1    21111 1122212234 1 12 2 3 1    1
SALI   RETI     555544 5354432533434223 421212 212 3 22314123123    1                     223
TOFI   SP       122111 3221211121213222 1 11 1 2 1 2 1 1                                  111
PANA   SP       112211 1331 11212  2114 111 11 211 11111 12111 1                               11
EQUI   VARI     212244 2444424542435434 3 21 1 1 2 11  2      113               1   1 4
JUNC   SP       1   21 221112                    12    1      1   1 1                      2 11
POLY   VIVI     11  21 111111111111         1             1  1        1                     1
AULA   PALU     555442 211122 122 22 11 11 11  2 -1 3 1 11 112 12 11        2 1             214
MEES   TRIQ      2141   133111112 2122         1  11 1  1 11 11      1 1 1       1    1      1  1
TOME   SP       145521 425551 3112 4113 2        4      11        1      11       11 11 1 1  1
CETR   NIVA     111232 2515455235554553 5344443455 5555535511252 2135111113111       2
CLAD   GRAC          3  111212222111 111 1111111122 21213111 1132 1111     111           1
OCHR   UPSA       1    1 11111111211 1111  211111111 111111111111      1   1   1
STER   ALPI     33222   13 23211    1      4 1   111 1    1 1 21     1    12
DIC2   SP            1    1  21          13111  111 1 212        1          1 1                  1
CETR   I-E      1   213 4 34443341111 1  111 1 1 34 3 33241341342 22442 3 21 111 1           1
CETR   I-P      211122 5554455355554453 5334544455 5555545545555 5355455245 112 1  3 1  11
DIC1   SP        11 1  1 1  1 2  313423 1  2 3 21  11  11 1 11 1 21    1 1 1    111 2 1
DITR   FLEX       1   1 1  11    1111 212  11111  11 111  1 11  11 1 11  111  321        1
CARE   SP       11 121 11111 1311111111 1  111  11     11111 121     21111       1 11 11 1  1
CLAD   SP         1  11  1 1   111111111  1111111    111   22121     1 11  2            2 2   1
SALI   CORD     2 1134 1111222212241  1 111   2121  1332   31445 35331         241   35 4  2
CLAD   PYXI          113 11 1          11  1211111 11 1222111 21213 322 111241 1 121  3 3   1
PELT   CANI     111111 11111 11 1        1 1        1  11             1       1111    1  113   1
BRYU   SP       3312 2 1 1      21113221 1  11111 1 1       11    1 1   2         2 11 3552 41
SALI   CALL            3    1   1  1111         121       52   3333211     22 2  221    2  423 1 1
SHEP   CANA     22                21     1  11 1 11 2  1 22224531     1 2233   1   2
CETR   CUCU           1 1   1 1121111121231 3334334443 443325321 13  12441 1112 1  1
ASTR   ALPI     1    1         1            12111 1111 1111121211111 111121112 3133111 2121111
EPIL   ANGU            1                           1  11    2212 1   1 12111 11        1 121 1
HEDY   MACK           11            1         1111112 11 2213132212214 23322332142243321 4    1
OXYT   HUDS            1                    1       11 1 11    111 1      1 21 11 1221 32
SAXI   TRIC                  1    1      1 1 1    1      21 13 313      22 4333131  1 3 3544 1
POA    GLAU       1   11 11111    11   11                11   11    1 2 11  111 1 2111321211    12111
CIRR   SP          11 41            1   11              1 1      1    11 1 1211  1 11411312  1 3  11
HYLO   SP             221                   1     1     12    1  1    1 21   2   3 1 413
FEST   RUBR         1                          1  1                11111   1    211 33121       1
ANEM   PARV                                                        11 1    2  22313   1       1 1 1
TARA   SP                                                          11 1    1 11321111 1        1 11
ARTE   BORE                                     1         11             11 111111 1221223 1          1
FEST   BRAC           1    1      111 1    1    1 1 1  21   111 22221  223221455533324142 111    1 1
POA    ALPI           112                                 1 1    1      22111  3111213321421212  1113333 1
TORT   RURA                                                           2 45          5   555 1        12
ACHI   MILL      11   1         1  11          1    1  11 111 11111   112111122321213  1113111    1
DRAB   SP              1                                11          11     11 1 1 1 111  11     1111 11
AREN   SP                        1                                              2 11 11  1 1          2 154
ELYM   AREN             1        11                  1   2 1111111 12 222122332244241 31 2244144 55413
STEL   CRAS       1  11        1  1  1               1       11   11 11 1111111111221122 1112111 22  1

                 000000 0000000000000000 0000000000 0000000000000 00000000000000000 0000000 11111
                 000000 0000000000000000 0000000000 0000000000000 11111111111111111 1111111 00011
                 000000 0000000000000000 1111111111 1111111111111 00000000000000000 1111111
                 000000 1111111111111111 0000000000 1111111111111 00000000001111111 0001111
                 000011 0000000000000111 0000000011 0000000011111 00001111110000111
                        0000011111111     01111111   0011111100001      000111

                    G          F              E            D                C             B      A
```

clear cut and call for little comment. Association D however forms a distinct grouping which is particularly evident in DCA ordination, but which had previously been entirely overlooked both in the original discussion of the salt marsh data (Kershaw, 1976) as well as by the PCA and RA ordinations. These quadrats occur on the slight mounds left on the salt marsh by ice scour with *Festuca* and *Potentilla* present as characteristic species. Associations F and E represent the general middle marsh containing old salt pans (E) with *Hippurus*, and notably *Carex glareosa* interspersed through the *Calamagrositis, Stellaria, Carex subspathacea* association (F). G–I forms the fresh-water marsh groupings where the permafrost melt-water emerges below the first beach ridge and which are dominated sequentially by *Carex aquatilis* and *Saxifraga* (I); *Carex diandra, Calamagrostis* and *Eriophorum* (H); and *Carex glareosum, Eriophorum, Calamagrostis* (G). The final association J is indicative of the moss/permafrost hummocks immediately below the raised beach and characterized by *Salix* and *Aulacomnium*.

The analysis of the raised beach data set by TWINSPAN with the corresponding overlay on the DCA ordination of the groupings is equally successful (Table 12.2, Fig. 12.11). Seven groupings are evident in the first three levels of the hierarchy and these groupings also remain reasonably distinct when overlayed on the ordination, despite the fairly continuous nature of the vegetational succession and the original sampling design. It is important to recognize here that TWINSPAN and DCA provide two independent but complementary analyses each of which substantiates the other. It also emphasizes the complementary nature of ordination and classification and how indeed they can be very successfully merged together.

Ecological interpretation of both DCA ordinations or TWINSPAN hierarchies demands information on the distribution of environmental parameters or a fairly detailed knowledge of the ecology of at least some of the species located in

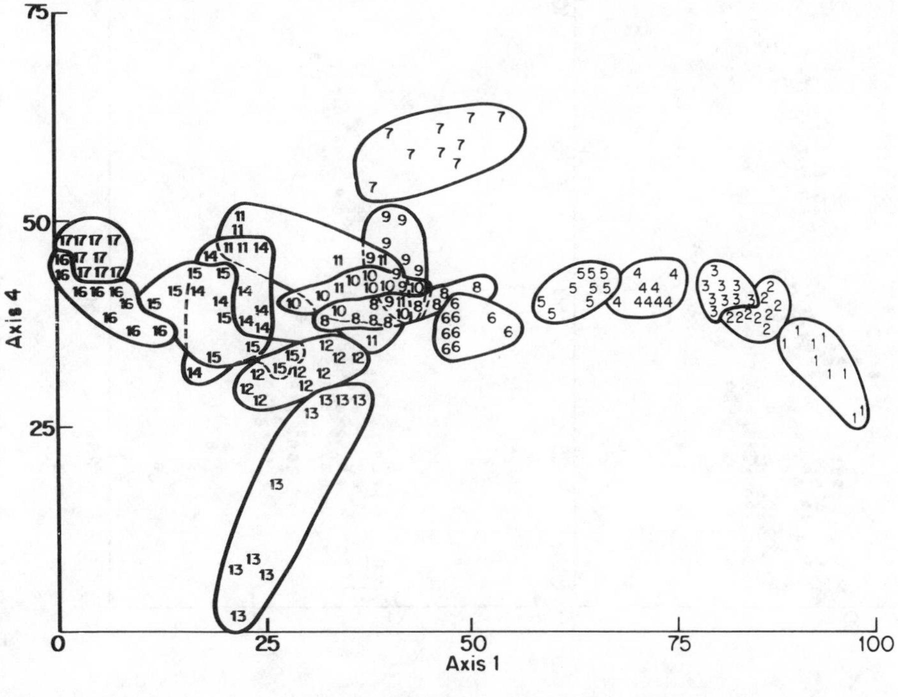

Fig. 12.12 Ordination of the individual sample quadrats located in 17 sample plots in a partially wooded meadow in S. Sweden. (From Persson, 1981; courtesy of *J. Ecol.*)

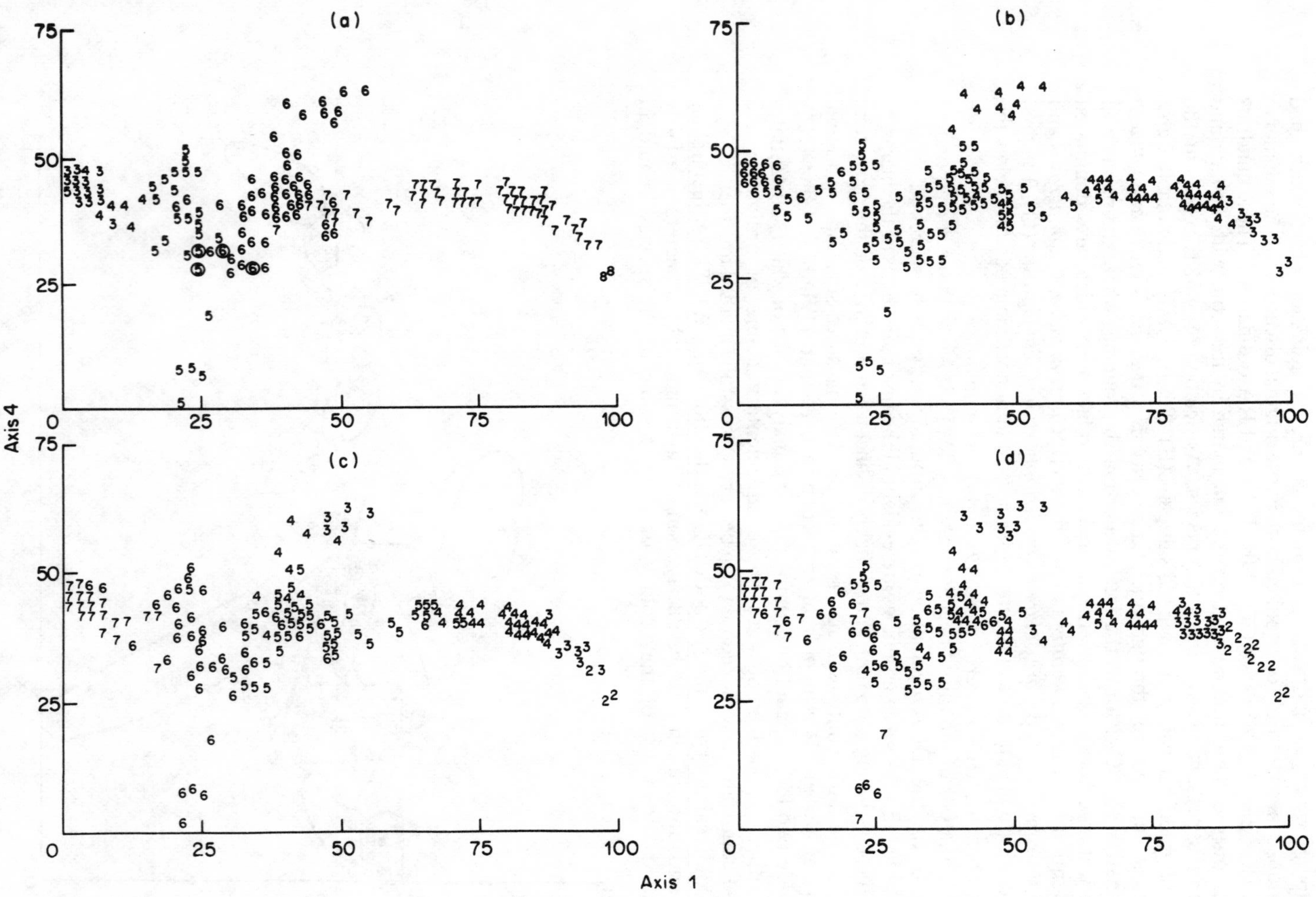

Fig. 12.13 Overlay of the characteristic indicator values for: (**a**) light, (**b**) moisture, (**c**) pH, and (**d**) nitrogen on the ordination given in Fig. 12.12. (From Persson, 1981; courtesy of *J. Ecol.*)

characteristic segments of the ordination. Persson (1981) has pointed out that there is often a lack of information on environmental variables, particularly where the ordination is largely used to generate hypotheses or dimensionality reduction (Dale, 1975). He does however use very effectively the scaled values provided by Ellenberg (1979) for six major environmental factors; light, temperature (derived from latitudinal and altitudinal ranges); soil moisture, soil pH and Ca^{2+} content; and nitrogen (as combined nitrate and ammonia). The RA ordination on axes 1 and 4 of Persson's data taken from a wooded-meadow area in S. Sweden, is given in Fig. 12.12. Within each plot the nine quadrats form fairly distinct groupings and show a more or less continuous sequence of variation. For each quadrat a characteristic indicator value was developed from the cover value of each species coupled to Ellenberg's indicator values for the range of environmental factors he examined. These characteristic indicator values for 4 factors are overlaid on the ordination in Fig. 12.13 in standard fashion. Although the visual correlation is quite good this can be readily quantified by a correlation coefficient calculated between the loading on an axis and the level of the indicator value. Scatter diagrams for light and moisture for this data (Fig. 12.14) show clearly the usefulness of this approach enabling the rigorous identification of ordination axes when adequate environmental data are available. This is true for either scaled quadrat values or equally for plot mean or median values.

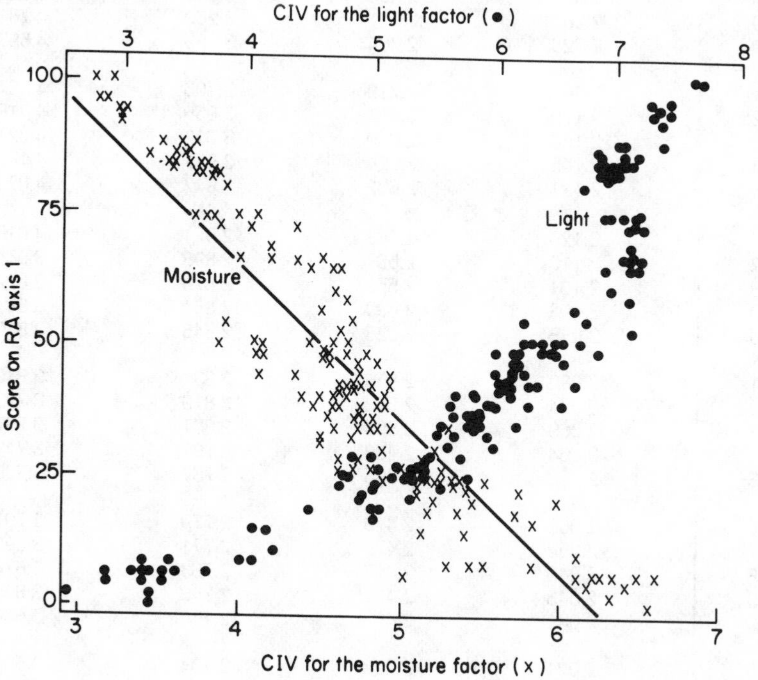

Fig. 12.14 The relation between position on RA ordination axis 1 and characteristic indicator value for light (0) and for moisture (x) for all quadrats. The linear regression between RA axis 1 score (y) and moisture CIV (x) is $y = -29.7x + 184$; $r^2 = 0.85$. (From Persson, 1981; courtesy of *J. Ecol.*)

We conclude that plant ecology has at last achieved a level of numerical *sufficiency* with DCA, TWINSPAN or TABORD providing extremely powerful and effective methods for organizing even large data sets. Future developments may 'fine tune' these methods but it seems evident that the *use* of these techniques to organize field data has now greater ecological merit than the development of a plethora of analagous techniques.

Appendix 1*
Distribution of *t*

	Probability			
n	.05	.02	.01	.001
1	12.706	31.821	63.657	636.619
2	4.303	6.965	9.925	31.598
3	3.182	4.541	5.841	12.941
4	2.776	3.747	4.604	8.610
5	2.571	3.365	4.032	6.859
6	2.447	3.143	3.707	5.959
7	2.365	2.998	3.499	5.405
8	2.306	2.896	3.355	5.041
9	2.262	2.821	3.250	4.781
10	2.228	2.764	3.169	4.587
11	2.201	2.718	3.106	4.437
12	2.179	2.681	3.055	4.318
13	2.160	2.650	3.012	4.221
14	2.145	2.624	2.977	4.140
15	2.131	2.602	2.947	4.073
16	2.120	2.583	2.921	4.015
17	2.110	2.567	2.898	3.965
18	2.101	2.552	2.878	3.922
19	2.093	2.539	2.861	3.883
20	2.086	2.528	2.845	3.850
21	2.080	2.518	2.831	3.819
22	2.074	2.508	2.819	3.792
23	2.069	2.500	2.807	3.767
24	2.064	2.492	2.797	3.745
25	2.060	2.485	2.787	3.725
26	2.056	2.479	2.779	3.707
27	2.052	2.473	2.771	3.690
28	2.048	2.467	2.763	3.674
29	2.045	2.462	2.756	3.659
30	2.042	2.457	2.750	3.646
40	2.021	2.423	2.704	3.551
60	2.000	2.390	2.660	3.460
120	1.980	2.358	2.617	3.373
∞	1.960	2.326	2.576	3.291

From Fisher and Yates; courtesy of Longman Group Ltd., London.

* The table on pages 239–50 are taken from Fisher and Yates *Statistical Table for Biological, Agricultureal and Medical Research*, published by Longman Group Ltd., London (previously published by Oliver and Boyd, Edinburgh), and by permission of the authors and publishers.

Appendix 2
Distribution of χ^2

	Probability			
n	.05	.02	.01	.001
1	3.841	5.412	6.635	10.827
2	5.991	7.824	9.210	13.815
3	7.815	9.837	11.345	16.268
4	9.488	11.668	13.277	18.465
5	11.070	13.388	15.086	20.517
6	12.592	15.033	16.812	22.457
7	14.067	16.622	18.475	24.322
8	15.507	18.168	20.090	26.125
9	16.919	19.679	21.666	27.877
10	18.307	21.161	23.209	29.588
11	19.675	22.618	24.725	31.264
12	21.026	24.054	26.217	32.909
13	22.362	25.472	27.688	34.528
14	23.685	26.873	29.141	36.123
15	24.996	28.259	30.578	37.697
16	26.296	29.633	32.000	39.252
17	27.587	30.995	33.409	40.790
18	28.869	32.346	34.805	42.312
19	30.144	33.687	36.191	43.820
20	31.410	35.020	37.566	45.315
21	32.671	36.343	38.932	46.797
22	33.924	37.659	40.289	48.268
23	35.172	38.968	41.638	49.728
24	36.415	40.270	42.980	51.179
25	37.652	41.566	44.314	52.620
26	38.885	42.856	45.642	54.052
27	40.113	44.140	46.963	55.476
28	41.337	45.419	48.278	56.893
29	42.557	46.693	49.588	58.302
30	43.773	47.962	50.892	59.703

From Fisher and Yates; courtesy of Longman Group Ltd., London.

Appendix 3
Variance Ratio

5 per cent points of e^{2z}

n_2 \ n_1	1	2	3	4	5	6	8	12	24	∞
1	161.4	199.5	215.7	224.6	230.2	234.0	238.9	243.9	249.0	254.3
2	18.51	19.00	19.16	19.25	19.30	19.33	19.37	19.41	19.45	19.50
3	10.13	9.55	9.28	9.12	9.01	8.94	8.84	8.74	8.64	8.53
4	7.71	6.94	6.59	6.39	6.26	6.16	6.04	5.91	5.77	5.63
5	6.61	5.79	5.41	5.19	5.05	4.95	4.82	4.68	4.53	4.36
6	5.99	5.14	4.76	4.53	4.39	4.28	4.15	4.00	3.84	3.67
7	5.59	4.74	4.35	4.12	3.97	3.87	3.73	3.57	3.41	3.23
8	5.32	4.46	4.07	3.84	3.69	3.58	3.44	3.28	3.12	2.93
9	5.12	4.26	3.86	3.63	3.48	3.37	3.23	3.07	2.90	2.71
10	4.96	4.10	3.71	3.48	3.33	3.22	3.07	2.91	2.74	2.54
11	4.84	3.98	3.59	3.36	3.20	3.09	2.95	2.79	2.61	2.40
12	4.75	3.88	3.49	3.26	3.11	3.00	2.85	2.69	2.50	2.30
13	4.67	3.80	3.41	3.18	3.02	2.92	2.77	2.60	2.42	2.21
14	4.60	3.74	3.34	3.11	2.96	2.85	2.70	2.53	2.35	2.13
15	4.54	3.68	3.29	3.06	2.90	2.79	2.64	2.48	2.29	2.07
16	4.49	3.63	3.24	3.01	2.85	2.74	2.59	2.42	2.24	2.01
17	4.45	3.59	3.20	2.96	2.81	2.70	2.55	2.38	2.19	1.96
18	4.41	3.55	3.16	2.93	2.77	2.66	2.51	2.34	2.15	1.92
19	4.38	3.52	3.13	2.90	2.74	2.63	2.48	2.31	2.11	1.88
20	4.35	3.49	3.10	2.87	2.71	2.60	2.45	2.28	2.08	1.84
21	4.32	3.47	3.07	2.84	2.68	2.57	2.42	2.25	2.05	1.81
22	4.30	3.44	3.05	2.82	2.66	2.55	2.40	2.23	2.03	1.78
23	4.28	3.42	3.03	2.80	2.64	2.53	2.38	2.20	2.00	1.76
24	4.26	3.40	3.01	2.78	2.62	2.51	2.36	2.18	1.98	1.73
25	4.24	3.38	2.99	2.76	2.60	2.49	2.34	2.26	1.96	1.71
26	4.22	3.37	2.98	2.74	2.59	2.47	2.32	2.15	1.95	1.69
27	4.21	3.35	2.96	2.73	2.57	2.46	2.30	2.13	1.93	1.67
28	4.20	3.34	2.95	2.71	2.56	2.44	2.29	2.12	1.91	1.65
29	4.18	3.33	2.93	2.70	2.54	2.43	2.28	2.10	1.90	1.64
30	4.17	3.32	2.92	2.69	2.53	2.42	2.27	2.09	1.89	1.62
40	4.08	3.23	2.84	2.61	2.45	2.34	2.18	2.00	1.79	1.51
60	4.00	3.15	2.76	2.52	2.37	2.25	2.10	1.92	1.70	1.39
120	3.92	3.07	2.68	2.45	2.29	2.17	2.02	1.83	1.61	1.25
∞	3.84	2.99	2.60	2.37	2.21	2.09	1.94	1.75	1.52	1.00

From Fisher and Yates; courtesy of Longman Group Ltd., London.

Variance Ratio – *continued*

1 per cent points of e^{2z}

n_2 \ n_1	1	2	3	4	5	6	8	12	24	∝
1	4052	4999	5403	5625	5764	5859	5981	6106	6234	6366
2	98.49	99.00	99.17	99.25	99.30	99.33	99.36	99.42	99.46	99.50
3	34.12	30.81	29.46	28.71	28.24	27.91	27.49	27.05	26.60	26.12
4	21.20	18.00	16.69	15.98	15.52	15.21	14.80	14.37	13.93	13.46
5	16.26	13.27	12.06	11.39	10.97	10.67	10.29	9.89	9.47	9.02
6	13.74	10.92	9.78	9.15	8.75	8.47	8.10	7.72	7.31	6.88
7	12.25	9.55	8.45	7.85	7.46	7.19	6.84	6.47	6.07	5.65
8	11.26	8.65	7.59	7.01	6.63	6.37	6.03	5.67	5.28	4.86
9	10.56	8.02	6.99	6.42	6.06	5.80	5.47	5.11	4.73	4.31
10	10.04	7.56	6.55	5.99	5.64	5.39	5.06	4.71	4.33	3.91
11	9.65	7.20	6.22	5.67	5.32	5.07	4.74	4.40	4.02	3.60
12	9.33	6.93	5.95	5.41	5.06	4.82	4.50	4.16	3.78	3.36
13	9.07	6.70	5.74	5.20	4.86	4.62	4.30	3.96	3.59	3.16
14	8.86	6.51	5.56	5.03	4.69	4.46	4.14	3.80	3.43	3.00
15	8.68	6.36	5.42	4.89	4.56	4.32	4.00	3.67	3.29	2.87
16	8.53	6.23	5.29	4.77	4.44	4.20	3.89	3.55	3.18	2.75
17	8.40	6.11	5.18	4.67	4.34	4.10	3.79	3.45	3.08	2.65
18	8.28	6.01	5.09	4.58	4.25	4.01	3.71	3.37	3.00	2.57
19	8.18	5.93	5.01	4.50	4.17	3.94	3.63	3.30	2.92	2.49
20	8.10	5.85	4.94	4.43	4.10	3.87	3.56	3.23	2.86	2.42
21	8.02	5.78	4.87	4.37	4.04	3.81	3.51	3.17	2.80	2.36
22	7.94	5.72	4.82	4.31	3.99	3.76	3.45	3.12	2.75	2.31
23	7.88	5.66	4.76	4.26	3.94	3.71	3.41	3.07	2.70	2.26
24	7.82	5.61	4.72	4.22	3.90	3.67	3.36	3.03	2.66	2.21
25	7.77	5.57	4.68	4.18	3.86	3.63	3.32	2.99	2.62	2.17
26	7.72	5.53	4.64	4.14	3.82	3.59	3.29	2.96	2.58	2.13
27	7.68	5.49	4.60	4.11	3.78	3.56	3.26	2.93	2.55	2.10
28	7.64	5.45	4.57	4.07	3.75	3.53	3.23	2.90	2.52	2.06
29	7.60	5.42	4.54	4.04	3.73	3.50	3.20	2.87	2.49	2.03
30	7.56	5.39	4.51	4.02	3.70	3.47	3.17	2.84	2.47	2.01
40	7.31	5.18	4.31	3.83	3.51	3.29	2.99	2.66	2.29	1.80
60	7.08	4.98	4.13	3.65	3.34	3.12	2.82	2.50	2.12	1.60
120	6.85	4.79	3.95	3.48	3.17	2.96	2.66	2.34	1.95	1.38
∝	6.64	4.60	3.78	3.32	3.02	2.80	2.51	2.18	1.79	1.00

From Fisher and Yates; courtesy of Longman Group Ltd., London.

Variance Ratio – *continued*

0.1 per cent points of e^{2z}

n_2 \ n_1	1	2	3	4	5	6	8	12	24	∝
1	405284	500000	540379	562500	576405	585937	598144	610667	623497	636619
2	998.5	999.0	999.2	999.2	999.3	999.3	999.4	999.4	999.5	999.5
3	167.5	148.5	141.1	137.1	134.6	132.8	130.6	128.3	125.9	123.5
4	74.14	61.25	56.18	53.44	51.71	50.53	49.00	47.41	45.77	44.05
5	47.04	36.61	33.20	31.09	29.75	28.84	27.64	26.42	25.14	23.78
6	35.51	27.00	23.70	21.90	20.81	20.03	19.03	17.99	16.89	15.75
7	29.22	21.69	18.77	17.19	16.21	15.52	14.63	13.71	12.73	11.69
8	25.42	18.49	15.83	14.39	13.49	12.86	12.04	11.19	10.30	9.34
9	22.86	16.39	13.90	12.56	11.71	11.13	10.37	9.57	8.72	7.81
10	21.04	14.91	12.55	11.28	10.48	9.92	9.20	8.45	7.64	6.76
11	19.69	13.81	11.56	10.35	9.58	9.05	8.35	7.63	6.85	6.00
12	18.64	12.97	10.80	9.63	8.89	8.38	7.71	7.00	6.25	5.42
13	17.81	12.31	10.21	9.07	8.35	7.86	7.21	6.52	5.78	4.97
14	17.14	11.78	9.73	8.62	7.92	7.43	6.80	6.13	5.41	4.60
15	16.59	11.34	9.34	8.25	7.57	7.09	6.47	5.81	5.10	4.31
16	16.12	10.97	9.00	7.94	7.27	6.81	6.19	5.55	4.85	4.06
17	15.72	10.66	8.73	7.68	7.02	6.56	5.96	5.32	4.63	3.85
18	15.38	10.39	8.49	7.46	6.81	6.35	5.76	5.13	4.45	3.67
19	15.08	10.16	8.28	7.26	6.61	6.18	5.59	4.97	4.29	3.52
20	14.82	9.95	8.10	7.10	6.46	6.02	5.44	4.82	4.15	3.38
21	14.59	9.77	7.94	6.95	6.32	5.88	5.31	4.70	4.03	3.26
22	14.38	9.61	7.80	6.81	6.19	5.76	5.19	4.58	3.92	3.15
23	14.19	9.47	7.67	6.69	6.08	5.65	5.09	4.48	3.82	3.05
24	14.03	9.34	7.55	6.59	5.98	5.55	4.99	4.39	3.74	2.97
25	13.88	9.22	7.45	6.49	5.88	5.46	4.91	4.31	3.66	2.89
26	13.74	9.12	7.36	6.41	5.80	5.38	4.83	4.24	3.59	2.82
27	13.61	9.02	7.27	6.33	5.73	5.31	4.76	4.17	3.52	2.75
28	13.50	8.93	7.19	6.25	5.66	5.24	4.69	4.11	3.46	2.70
29	13.39	8.85	7.12	6.19	5.59	5.18	4.64	4.05	3.41	2.64
30	13.29	8.77	7.05	6.12	5.53	5.12	4.58	4.00	3.36	2.59
40	12.61	8.25	6.60	5.70	5.13	4.73	4.21	3.64	3.01	2.23
60	11.97	7.76	6.17	5.31	4.76	4.37	3.87	3.31	2.69	1.90
120	11.38	7.31	5.79	4.95	4.42	4.04	3.55	3.02	2.40	1.56
∝	10.83	6.91	5.42	4.62	4.10	3.74	3.27	2.74	2.13	1.00

From Fisher and Yates; courtesy of Longman Group Ltd., London.

Appendix 4

Values of the correlation coefficient for different levels of significance

n	.1	.05	.02	.01	.001
1	.98769	.99692	.999507	.999877	.9999988
2	.90000	.95000	.98000	.990000	.99900
3	.8054	.8783	.93433	.95873	.99116
4	.7293	.8114	.8822	.91720	.97406
5	.6694	.7545	.8329	.8745	.95074
6	.6215	.7067	.7887	.8343	.92493
7	.5822	.6664	.7498	.7977	.8982
8	.5494	.6319	.7155	.7646	8721
9	.5214	.6021	.6851	.7348	.8471
10	.4973	.5760	.6581	.7079	.8233
11	.4762	.5529	.6339	.6835	.8010
12	.4575	.5324	.6120	.6614	.7800
13	.4409	.5139	.5923	.6411	.7603
14	.4259	.4973	.5742	.6226	.7420
15	.4124	.4821	.5577	.6055	.7246
16	.4000	.4683	.5425	.5897	.7084
17	.3887	.4555	.5285	.5751	.6932
18	.3783	.4438	.5155	.5614	.6787
19	.3687	.4329	.5034	.5487	.6652
20	.3598	.4227	.4921	.5368	.6524
25	.3233	.3809	.4451	.4869	.5974
30	.2960	.3494	.4093	.4487	.5541
35	.2746	.3246	.3810	.4182	.5189
40	.2573	.3044	.3578	.3932	.4896
45	.2428	.2875	.3384	.3721	.4648
50	.2306	.2732	.3218	.3541	.4433
60	.2108	.2500	.2948	.3248	.4078
70	.1954	.2319	.2727	.3017	.3799
80	.1829	.2172	.2565	.2830	.3568
90	.1726	.2050	.2427	.2673	.3375
100	.1638	.1946	.2301	.2540	.3211

From Fisher and Yates; courtesy of Longman Group Ltd., London.

Appendix 5

Random numbers (I)

03 47 43 73 86	36 96 47 36 61	46 98 63 71 62	33 26 16 80 45	60 11 14 10 95
97 74 24 67 62	42 81 14 57 20	42 53 32 37 32	27 07 36 07 51	24 51 79 89 73
16 76 62 27 66	56 50 26 71 07	32 90 79 78 53	13 55 38 58 59	88 97 54 14 10
12 56 85 99 26	96 96 68 27 31	05 03 72 93 15	57 12 10 14 21	88 26 49 81 76
55 59 56 35 64	38 54 82 46 22	31 62 43 09 90	06 18 44 32 53	23 83 01 30 30
16 22 77 94 39	49 54 43 54 82	17 37 93 23 78	87 35 20 96 43	84 26 34 91 64
84 42 17 53 31	57 24 55 06 88	77 04 74 47 67	21 76 33 50 25	83 92 12 06 76
63 01 63 78 59	16 95 55 67 19	98 10 50 71 75	12 86 73 58 07	44 39 52 38 79
33 21 12 34 29	78 64 56 07 82	52 42 07 44 38	15 51 00 13 42	99 66 02 79 54
57 60 86 32 44	09 47 27 96 54	49 17 46 09 62	90 52 84 77 27	08 02 73 43 28
18 18 07 92 46	44 17 16 58 09	79 83 86 19 62	06 76 50 03 10	55 23 64 05 05
26 62 38 97 75	84 16 07 44 99	83 11 46 32 24	20 14 85 88 45	10 93 72 88 71
23 42 40 64 74	82 97 77 77 81	07 45 32 14 08	32 98 94 07 72	93 85 79 10 75
52 36 28 19 95	50 92 26 11 97	00 56 76 31 38	80 22 02 53 53	86 60 42 04 53
34 85 94 35 12	83 39 50 08 30	42 34 07 96 88	54 42 06 87 98	35 85 29 48 39
70 29 17 12 13	40 33 20 38 26	13 89 51 03 74	17 76 37 13 04	07 74 21 19 30
56 62 18 37 35	96 83 50 87 75	97 12 25 93 47	70 33 24 03 54	97 77 46 44 80
99 49 57 22 77	88 42 95 45 72	16 64 36 16 00	04 43 18 66 79	94 77 24 21 90
16 08 15 04 72	33 27 14 34 09	45 59 34 68 49	12 72 07 34 45	99 27 72 95 14
31 16 93 32 43	50 27 89 87 19	20 15 37 00 49	52 85 66 60 44	38 68 88 11 80
68 34 30 13 70	55 74 30 77 40	44 22 78 84 26	04 33 46 09 52	68 07 97 06 57
74 57 25 65 76	59 29 97 68 60	71 91 38 67 54	13 58 18 24 76	15 54 55 95 52
27 42 37 86 53	48 55 90 65 72	96 57 69 36 10	96 46 92 42 45	97 60 49 04 91
00 39 68 29 61	66 37 32 20 30	77 84 57 03 29	10 45 65 04 26	11 04 96 67 24
29 94 98 94 24	68 49 69 10 82	53 75 91 93 30	34 25 20 57 27	40 48 73 51 92
16 90 82 66 59	83 62 64 11 12	67 19 00 71 74	60 47 21 29 68	02 02 37 03 31
11 27 94 75 06	06 09 19 74 66	02 94 37 34 02	76 70 90 30 86	38 45 94 30 38
35 24 10 16 20	33 32 51 26 38	79 78 45 04 91	16 92 53 56 16	02 75 50 95 98
38 23 16 86 38	42 38 97 01 50	87 75 66 81 41	40 01 74 91 62	48 51 84 08 32
31 96 25 91 47	96 44 33 49 13	34 86 82 53 91	00 52 43 48 85	27 55 26 89 62
66 67 40 67 14	64 05 71 95 86	11 05 65 09 68	76 83 20 37 90	57 16 00 11 66
14 90 84 45 11	75 73 88 05 90	52 27 41 14 86	22 98 12 22 08	07 52 74 95 80
68 05 51 18 00	33 96 02 75 19	07 60 62 93 55	59 33 82 43 90	49 37 38 44 59
20 46 78 73 90	97 51 40 14 02	04 02 33 31 08	39 54 16 49 36	47 95 93 13 30
64 19 58 97 79	15 06 15 93 20	01 90 10 75 06	40 78 78 89 62	02 67 74 17 33
05 26 93 70 60	22 35 85 15 13	92 03 51 59 77	59 56 78 06 83	52 91 05 70 74
07 97 10 88 23	09 98 42 99 64	61 71 62 99 15	06 51 29 16 93	58 05 77 09 51
68 71 86 85 85	54 87 66 47 54	73 32 08 11 12	44 95 92 63 16	29 56 24 29 48
26 99 61 65 53	58 37 78 80 70	42 10 50 67 42	32 17 55 85 74	94 44 67 16 94
14 65 52 68 75	87 59 36 22 41	26 78 63 06 55	13 08 27 01 50	15 29 39 39 43
17 53 77 58 71	71 41 61 50 72	12 41 94 96 26	44 95 27 36 99	02 96 74 30 83
90 26 59 21 19	23 52 23 33 12	96 93 02 18 39	07 02 18 36 07	25 99 32 70 23
41 23 52 55 99	31 04 49 69 96	10 47 48 45 88	13 41 43 89 20	97 17 14 49 17
60 20 50 81 69	31 99 73 68 68	35 81 33 03 76	24 30 12 48 60	18 99 10 72 34
91 25 38 05 90	94 58 28 41 36	45 37 59 03 09	90 35 57 29 12	82 62 54 65 60
34 50 57 74 37	98 80 33 00 91	09 77 93 19 82	74 94 80 04 04	45 07 31 66 49
85 22 04 39 43	73 81 53 94 79	33 62 46 86 28	08 31 54 46 31	53 94 13 38 47
09 79 13 77 48	73 82 97 22 21	05 03 27 24 83	72 89 44 05 60	35 80 39 94 88
88 75 80 18 14	22 95 75 42 49	39 32 82 22 49	02 48 07 70 37	16 04 61 67 87
90 96 23 70 00	39 00 03 06 90	58 85 78 38 36	94 37 30 69 32	94 89 00 76 33

Fisher and Yates; courtesy of Longman Group Ltd., London.

Random numbers (II)

53 74 23 99 67	61 32 28 69 84	94 62 67 86 24	98 33 41 19 95	47 53 53 38 09
63 38 06 86 54	99 00 65 26 94	02 82 90 23 07	79 62 67 80 60	75 91 12 81 19
35 30 58 21 46	06 72 17 10 94	25 21 31 75 96	49 28 24 00 49	55 65 79 78 07
63 43 36 82 69	65 51 18 37 88	61 38 44 12 45	32 92 85 88 65	54 34 81 85 35
98 25 37 55 26	01 91 82 81 46	74 71 12 94 97	24 02 71 37 07	03 92 18 66 75
02 63 21 17 69	71 50 80 89 56	38 15 70 11 48	43 40 45 86 98	00 83 26 91 03
64 55 22 21 82	48 22 28 06 00	61 54 13 43 91	82 78 12 23 29	06 66 24 12 27
85 07 26 13 89	01 10 07 82 04	59 63 69 36 03	69 11 15 83 80	13 29 54 19 28
58 54 16 24 15	51 54 44 82 00	62 61 65 04 69	38 18 65 18 97	85 72 13 49 21
34 85 27 84 87	61 48 64 56 26	90 18 48 13 26	37 70 15 42 57	65 65 80 39 07
03 92 18 27 46	57 99 16 96 56	30 33 72 85 22	84 64 38 56 98	99 01 30 98 64
62 95 30 27 59	37 75 41 66 48	86 97 80 61 45	23 53 04 01 63	45 76 08 64 27
08 45 93 15 22	60 21 75 46 91	98 77 27 85 42	28 88 61 08 84	69 62 03 42 73
07 08 55 18 40	45 44 75 13 90	24 94 96 61 02	57 55 66 83 15	73 42 37 11 61
01 85 89 95 66	51 10 19 34 88	15 84 97 19 75	12 76 39 43 78	64 63 91 08 25
72 84 71 14 35	19 11 58 49 26	50 11 17 17 76	86 31 57 20 18	95 60 78 46 75
88 78 28 16 84	13 52 53 94 53	75 45 69 30 96	73 89 65 70 31	99 17 43 48 76
45 17 75 65 57	28 40 19 72 12	25 12 74 75 67	60 40 60 81 19	24 62 01 61 16
96 76 28 12 54	22 01 11 94 25	71 96 16 16 88	68 64 36 74 45	19 59 50 88 92
43 31 67 72 30	24 02 94 08 63	38 32 36 66 02	69 36 38 25 39	48 03 45 15 22
50 44 66 44 21	66 06 58 05 62	68 15 54 35 02	42 35 48 96 32	14 52 41 52 48
22 66 22 15 86	26 63 75 41 99	58 42 36 72 24	58 37 52 18 51	03 37 18 39 11
96 24 40 14 51	23 22 30 88 57	95 67 47 29 83	94 69 40 06 07	18 16 36 78 86
31 73 91 61 19	60 20 72 93 48	98 57 07 23 69	65 95 39 69 58	56 80 30 19 44
78 60 73 99 84	43 89 94 36 45	56 69 47 07 41	90 22 91 07 12	78 35 34 08 72
84 37 90 61 56	70 10 23 98 05	85 11 34 76 60	76 48 45 34 60	01 64 18 39 96
36 67 10 08 23	98 93 35 08 86	99 29 76 29 81	33 34 91 58 93	63 14 52 32 52
07 28 59 07 48	89 64 58 89 75	83 85 62 27 89	30 14 78 56 27	86 63 59 80 02
10 15 83 87 60	79 24 31 66 56	21 48 24 06 93	91 98 94 05 49	01 47 59 38 00
55 19 68 97 65	03 73 52 16 56	00 53 55 90 27	33 42 29 38 87	22 18 88 83 34
53 81 29 13 39	35 01 20 71 34	62 33 74 82 14	53 73 19 09 03	56 54 29 56 93
51 86 32 68 92	33 98 74 66 99	40 14 71 94 58	45 94 19 38 81	14 44 99 81 07
35 91 70 29 13	80 03 54 07 27	96 94 78 32 66	50 95 52 74 33	13 80 55 62 54
37 71 67 95 13	20 02 44 95 94	64 85 04 05 72	01 32 90 76 14	53 89 74 60 41
93 66 13 83 27	92 79 64 64 72	28 54 96 53 84	48 14 52 98 94	56 07 93 89 30
02 96 08 45 65	13 05 00 41 84	93 07 54 72 59	21 45 57 09 77	19 48 56 27 44
49 83 43 48 35	82 88 33 69 96	72 36 04 19 76	47 45 15 18 60	82 11 08 95 97
84 60 71 62 46	40 80 81 30 37	34 39 23 05 38	25 15 35 71 30	88 12 57 21 77
18 17 30 88 71	44 91 14 88 47	89 23 30 63 15	56 34 20 47 89	99 82 93 24 98
79 69 10 61 78	71 32 76 95 62	87 00 22 58 40	92 54 01 75 25	43 11 71 99 31
75 93 36 57 83	56 20 14 82 11	74 21 97 90 65	96 42 68 63 86	74 54 13 26 94
38 30 92 29 03	06 28 81 39 38	62 25 06 84 63	61 29 08 93 67	04 32 92 08 09
51 29 50 10 34	31 57 75 95 80	51 97 02 74 77	76 15 48 49 44	18 55 63 77 09
21 31 38 86 24	37 79 81 53 74	73 24 16 10 33	52 83 90 94 76	70 47 14 54 36
29 01 23 87 88	58 02 39 37 67	42 10 14 20 92	16 55 23 42 45	54 96 09 11 06
95 33 95 22 00	18 74 72 00 18	38 79 58 69 32	81 76 80 26 92	82 80 84 25 39
90 84 60 79 80	24 36 59 87 38	82 07 53 89 35	96 35 23 79 18	05 98 90 07 35
46 40 62 98 82	54 97 20 56 95	15 74 80 08 32	16 46 70 50 80	67 72 16 42 79
20 31 89 03 43	38 46 82 68 72	32 14 82 99 70	80 60 47 18 97	63 49 30 21 30
71 59 73 05 50	08 22 23 71 77	91 01 93 20 49	82 96 59 26 94	66 39 67 98 60

From Fisher and Yates; courtesy of Longman Group Ltd., London.

Random numbers (III)

22 17 68 65 84	68 95 23 92 35	87 02 22 57 51	61 09 43 95 06	58 24 82 03 47
19 36 27 59 46	13 79 93 37 55	39 77 32 77 09	85 52 05 30 62	47 83 51 62 74
16 77 23 02 77	09 61 87 25 21	28 06 24 25 93	16 71 13 59 78	23 05 47 47 25
78 43 76 71 61	20 44 90 32 64	97 67 63 99 61	46 38 03 93 22	69 81 21 99 21
03 28 28 26 08	73 37 32 04 05	69 30 16 09 05	88 69 58 28 99	35 07 44 73 47
93 22 53 64 39	07 10 63 76 35	87 03 04 79 88	08 13 13 85 51	55 34 57 72 69
78 76 58 54 74	92 38 70 96 92	52 06 79 79 45	82 63 18 27 44	69 66 92 19 09
23 68 35 26 00	99 53 93 61 28	52 70 05 48 34	56 65 05 61 86	90 92 10 70 80
15 39 25 70 99	93 86 52 77 65	15 33 59 05 28	22 87 26 07 47	86 96 98 29 06
58 71 96 30 24	18 46 23 34 27	85 13 99 24 44	49 18 09 79 49	74 16 32 23 02
57 35 27 33 72	24 53 63 94 09	41 10 76 47 91	44 04 95 49 66	39 60 04 59 81
48 50 86 54 48	22 06 34 72 52	82 21 15 65 20	33 29 94 71 11	15 91 29 12 03
61 96 48 95 03	07 16 39 33 66	98 56 10 56 79	77 21 30 27 12	90 49 22 23 62
36 93 89 41 26	29 70 83 63 51	99 74 20 52 36	87 09 41 15 09	98 60 16 03 03
18 87 00 42 31	57 90 12 02 07	23 47 37 17 31	54 08 01 88 63	39 41 88 92 10
88 56 53 27 59	33 35 72 67 47	77 34 55 45 70	08 18 27 38 90	16 95 86 70 75
09 72 95 84 29	49 41 31 06 70	42 38 06 45 18	64 84 73 31 65	52 53 37 97 15
12 96 88 17 31	65 19 69 02 83	60 75 86 90 68	24 64 19 35 51	56 61 87 39 12
85 94 57 24 16	92 09 84 38 76	22 00 27 69 85	29 81 94 78 70	21 94 47 90 12
38 64 43 59 98	98 77 87 68 07	91 51 67 62 44	40 98 05 93 78	23 32 65 41 18
53 44 09 42 72	00 41 86 79 79	68 47 22 00 20	35 55 31 51 51	00 83 63 22 55
40 76 66 26 84	57 99 99 90 37	36 63 32 08 58	37 40 13 68 97	87 64 81 07 83
02 17 79 18 05	12 59 52 57 02	22 07 90 47 03	28 14 11 30 79	20 69 22 40 98
95 17 82 06 53	31 51 10 96 46	92 06 88 07 77	56 11 50 81 69	40 23 72 51 39
35 76 22 42 92	96 11 83 44 80	34 68 35 48 77	33 42 40 90 60	73 96 53 97 86
26 29 13 56 41	85 47 04 66 08	34 72 57 59 13	82 43 80 46 15	38 26 61 70 04
77 80 20 75 82	72 82 32 99 90	63 95 73 76 63	89 73 44 99 05	48 67 26 43 18
46 40 66 44 52	91 36 74 43 53	30 82 13 54 00	78 45 63 98 35	55 03 36 67 68
37 56 08 18 09	77 53 84 46 47	31 91 18 95 58	24 16 74 11 53	44 10 13 85 57
61 65 61 68 66	37 27 47 39 19	84 83 70 07 48	53 21 40 60 71	95 06 79 88 54
93 43 69 64 07	34 18 04 52 35	56 27 09 24 86	61 85 53 83 45	19 90 70 99 00
21 96 60 12 99	11 20 99 45 18	48 13 93 55 34	18 37 79 49 90	65 97 38 20 46
95 20 47 97 97	27 37 83 28 71	00 06 41 41 74	45 89 09 39 84	51 67 11 52 49
97 86 21 78 73	10 65 81 92 59	58 76 17 14 97	04 76 62 16 17	17 95 70 45 80
69 92 06 34 13	59 71 74 17 32	27 55 10 24 19	23 71 82 13 74	63 52 52 01 41
04 31 17 21 56	33 73 99 19 87	26 72 39 27 67	53 77 57 68 93	60 61 97 22 61
61 06 98 03 91	87 14 77 43 06	43 00 65 98 50	45 60 33 01 07	98 99 46 50 47
85 93 85 86 88	72 87 08 62 40	16 06 10 89 20	23 21 34 74 97	76 38 03 29 63
21 74 32 47 45	73 96 07 94 52	09 65 90 77 47	25 76 16 19 33	53 05 70 53 30
15 69 53 82 80	79 96 23 53 10	65 39 07 16 29	45 33 02 43 70	02 87 40 41 45
02 89 08 04 49	20 21 14 68 86	87 63 93 95 17	11 29 01 95 80	35 14 97 35 33
87 18 15 89 79	85 43 01 72 73	08 61 74 51 69	89 74 39 82 15	94 51 33 41 67
98 83 71 94 22	59 97 50 99 52	08 52 85 08 40	87 80 61 65 31	91 51 80 32 44
10 08 58 21 66	72 68 49 29 31	89 85 84 46 06	59 73 19 85 23	65 09 29 75 63
47 90 56 10 08	88 02 84 27 83	42 29 72 23 19	66 56 45 65 79	20 71 53 20 25
22 85 61 68 90	49 64 92 85 44	16 40 12 89 88	50 14 49 81 06	01 82 77 45 12
67 80 43 79 33	12 83 11 41 16	25 58 19 68 70	77 02 54 00 52	53 43 37 15 26
27 62 50 96 72	79 44 61 40 15	14 53 40 65 39	27 31 58 50 28	11 39 03 34 25
33 78 80 87 15	38 30 06 38 21	14 47 47 07 26	54 96 87 53 32	40 36 40 96 76
13 13 92 66 99	47 24 49 57 74	32 25 43 62 17	10 97 11 69 84	99 63 22 32 98

From Fisher and Yates; courtesy of Longman Group Ltd., London.

Random numbers (IV)

10 27 53 96 23	71 50 54 36 23	54 31 04 82 98	04 14 12 15 09	26 78 25 47 47
28 41 50 61 88	64 85 27 20 18	83 36 36 05 56	39 71 65 09 62	94 76 62 11 89
34 21 42 57 02	59 19 18 97 48	80 30 03 30 98	05 24 67 70 07	84 97 50 87 46
61 81 77 23 23	82 82 11 54 08	53 28 70 58 96	44 07 39 55 43	42 34 43 39 28
61 15 18 13 54	16 86 20 26 88	90 74 80 55 09	14 53 90 51 17	52 01 63 01 59
91 76 21 64 64	44 91 13 32 97	75 31 62 66 54	84 80 32 75 77	56 08 25 70 29
00 97 79 08 06	37 30 28 59 85	53 56 68 53 40	01 74 39 59 73	30 19 99 85 48
36 46 18 34 94	75 20 80 27 77	78 91 69 16 00	08 43 18 73 68	67 69 61 34 25
88 98 99 60 50	65 95 79 42 94	93 62 40 89 96	43 56 47 71 66	46 76 29 67 02
04 37 59 87 21	05 02 03 24 17	47 97 81 56 51	92 34 86 01 82	55 51 33 12 91
63 62 06 34 41	94 21 78 55 09	72 76 45 16 94	29 95 81 83 83	79 88 01 97 30
78 47 23 53 90	34 41 92 45 71	09 23 70 70 07	12 38 92 79 43	14 85 11 47 23
87 68 62 15 43	53 14 36 59 25	54 47 33 70 15	59 24 48 40 35	50 03 42 99 36
47 60 92 10 77	88 59 53 11 52	66 25 69 07 04	48 68 64 71 06	61 65 70 22 12
56 88 87 59 41	65 28 04 67 53	95 79 88 37 31	50 41 06 94 76	81 83 17 16 33
02 57 45 86 67	73 43 07 34 48	44 26 87 93 29	77 09 61 67 84	06 69 44 77 75
31 54 14 13 17	48 62 11 90 60	68 12 93 64 28	46 24 79 16 76	14 60 25 51 01
28 50 16 43 36	28 97 85 58 99	67 22 52 76 23	24 70 36 54 54	59 28 61 71 96
63 29 62 66 50	02 63 45 52 38	67 63 47 54 75	83 24 78 43 20	92 63 13 47 48
45 65 58 26 51	76 96 59 38 72	86 57 45 71 46	44 67 76 14 55	44 88 01 62 12
39 65 36 63 70	77 45 85 50 51	74 13 39 35 22	30 53 36 02 95	49 34 88 73 61
73 71 98 16 04	29 18 94 51 23	76 51 94 84 86	79 93 96 38 63	08 58 25 58 94
72 20 56 20 11	72 65 71 08 86	79 57 95 13 91	97 48 72 66 48	09 71 17 24 89
75 17 26 99 76	89 37 20 70 01	77 31 61 95 46	26 97 05 73 51	53 33 18 72 87
37 48 60 82 29	81 30 15 39 14	48 38 75 93 29	06 87 37 78 48	45 56 00 84 47
68 08 02 80 72	83 71 46 30 49	89 17 95 88 29	02 39 56 03 46	97 74 06 56 17
14 23 98 61 67	70 52 85 01 50	01 84 02 78 43	10 62 98 19 41	18 83 99 47 99
49 08 96 21 44	25 27 99 41 28	07 41 08 34 66	19 42 74 39 91	41 96 53 78 72
78 37 06 08 43	63 61 62 42 29	39 68 95 10 96	09 24 23 00 62	56 12 80 73 16
37 21 34 17 68	68 96 83 23 56	32 84 60 15 31	44 73 67 34 77	91 15 79 74 58
14 29 09 34 04	87 83 07 55 07	76 58 30 83 64	87 29 25 58 84	86 50 60 00 25
58 43 28 06 36	49 52 83 51 14	47 56 91 29 34	05 87 31 06 95	12 45 57 09 09
10 43 67 29 70	80 62 80 03 42	10 80 21 38 84	90 56 35 03 09	43 12 74 49 14
44 38 88 39 54	86 97 37 44 22	00 95 01 31 76	17 16 29 56 63	38 78 94 49 81
90 69 59 19 51	85 39 52 85 13	07 28 37 07 61	11 16 36 27 03	78 86 72 04 95
41 47 10 25 62	97 05 31 03 61	20 26 36 31 62	68 69 86 95 44	84 95 48 46 45
91 94 14 63 19	75 89 11 47 11	31 56 34 19 09	79 57 92 36 59	14 93 87 81 40
80 06 54 18 86	09 18 94 06 19	98 40 07 17 81	22 45 44 84 11	24 62 20 42 31
67 72 77 63 48	84 08 31 55 58	24 33 45 77 58	80 45 67 93 82	75 70 16 08 24
59 40 24 13 27	79 26 88 86 30	01 31 60 10 39	53 58 47 70 93	85 81 56 39 38
05 90 35 89 95	01 61 16 96 94	50 78 13 69 36	37 68 53 37 31	71 26 35 03 71
44 43 80 69 98	46 68 05 14 82	90 78 50 05 62	77 79 13 57 44	59 60 10 39 66
61 81 31 96 82	00 57 25 60 59	46 72 60 18 77	55 66 12 62 11	08 99 55 64 57
42 88 07 10 05	24 98 65 63 21	47 21 61 88 32	27 80 30 21 60	10 92 35 36 12
77 94 30 05 39	28 10 99 00 27	12 73 73 99 12	49 99 57 94 82	96 88 57 17 91
78 83 19 76 16	94 11 68 84 26	23 54 20 86 85	23 86 66 99 07	36 37 34 92 09
87 76 59 61 81	43 63 64 61 61	65 76 36 95 90	18 48 27 45 68	27 23 65 30 72
91 43 05 96 47	55 78 99 95 24	37 55 85 78 78	01 48 41 19 10	35 19 54 07 73
84 97 77 72 73	09 62 06 65 72	87 12 49 03 60	41 15 20 76 27	50 47 02 29 16
87 41 60 76 83	44 88 96 07 80	83 05 83 38 96	73 70 66 81 90	30 56 10 48 59

From Fisher and Yates; courtesy of Longman Group Ltd., London.

Random numbers (V)

28 89 65 87 08	13 50 63 04 23	25 47 57 91 13	52 62 24 19 94	91 67 48 57 10
30 29 43 65 42	78 66 28 55 80	47 46 41 90 08	55 98 78 10 70	49 92 05 12 07
95 74 62 60 53	51 57 32 22 27	12 72 72 27 77	44 67 32 23 13	67 95 07 76 30
01 85 54 96 72	66 86 65 64 60	56 59 75 36 75	46 44 33 63 71	54 50 06 44 75
10 91 46 96 86	19 83 52 47 53	65 00 51 93 51	30 80 05 19 29	56 23 27 19 03
05 33 18 08 51	51 78 57 26 17	34 87 96 23 95	89 99 93 39 79	11 28 94 15 52
04 43 13 37 00	79 68 96 26 60	70 39 83 66 56	62 03 55 86 57	77 55 33 62 02
05 85 40 25 24	73 52 93 70 50	48 21 47 74 63	17 27 27 51 26	35 96 29 00 45
84 90 90 65 77	63 99 25 69 02	09 04 03 35 78	19 79 95 07 21	02 84 48 51 97
28 55 53 09 48	86 28 30 02 35	71 30 32 06 47	93 74 21 86 33	49 90 21 69 74
89 83 40 69 80	97 96 47 59 97	56 33 24 87 36	17 18 16 90 46	75 27 28 52 13
73 20 96 05 68	93 41 69 96 07	97 50 81 79 59	42 37 13 81 83	92 42 85 04 31
10 89 07 76 21	40 24 74 36 42	40 33 04 46 24	35 63 02 31 61	34 59 43 36 96
91 50 27 78 37	06 06 16 25 98	17 78 80 36 85	26 41 77 63 37	71 63 94 94 33
03 45 44 66 88	97 81 26 03 89	39 46 67 21 17	98 10 39 33 15	61 63 00 25 92
89 41 58 91 63	65 99 59 97 84	90 14 79 61 55	56 16 88 87 60	32 15 99 67 43
13 43 00 97 26	16 91 21 32 41	60 22 66 72 17	31 85 33 69 07	68 49 20 43 29
71 71 00 51 72	62 03 89 26 32	35 27 99 18 25	78 12 03 09 70	50 93 19 35 56
19 28 15 00 41	92 27 73 40 38	37 11 05 75 16	98 81 99 37 29	92 20 32 39 67
56 38 30 92 30	45 51 94 69 04	00 84 14 36 37	95 66 39 01 09	21 68 40 95 79
39 27 52 89 11	00 81 06 28 48	12 08 05 75 26	03 35 63 05 77	13 81 20 67 58
73 13 28 58 01	05 06 42 42 07	60 60 29 99 93	72 93 78 04 36	25 76 01 54 03
81 60 84 51 57	12 68 46 55 89	60 09 71 87 89	70 81 10 95 91	83 79 68 20 66
05 62 98 07 85	07 79 26 69 61	67 85 72 37 41	85 79 76 48 23	61 58 87 08 05
62 97 16 29 18	52 16 16 23 56	62 95 80 97 63	32 25 34 03 36	48 84 60 37 65
31 13 63 21 08	16 01 92 58 21	48 79 74 73 72	08 64 80 91 38	07 28 66 61 59
97 38 35 34 19	89 84 05 34 47	88 09 31 54 88	97 96 86 01 69	46 13 95 65 96
32 11 78 33 82	51 99 98 44 39	12 75 10 60 36	80 66 39 94 97	42 36 31 16 59
81 99 13 37 05	08 12 60 39 23	61 73 84 89 18	26 02 04 37 95	96 18 69 06 30
45 74 00 03 05	69 99 47 26 52	48 06 30 00 18	03 30 28 55 59	66 10 71 44 05
11 84 13 69 01	88 91 28 79 50	71 42 14 96 55	98 59 96 01 36	88 77 90 45 59
14 66 12 87 22	59 45 27 08 51	85 64 23 85 41	64 72 08 59 44	67 98 36 65 56
40 25 67 87 82	84 27 17 30 37	48 69 49 02 58	98 02 50 58 11	95 39 06 35 63
44 48 97 49 43	65 45 53 41 07	14 83 46 74 11	76 66 63 60 08	90 54 33 65 84
41 94 54 06 57	48 28 01 83 84	09 11 21 91 73	97 28 44 74 06	22 30 95 69 72
07 12 15 58 84	93 18 31 83 45	54 52 62 29 91	53 58 54 66 05	47 19 63 92 75
64 27 90 43 52	18 26 32 96 83	50 58 45 27 57	14 96 39 64 85	73 87 96 76 23
80 71 86 41 03	45 62 63 40 88	35 69 34 10 94	32 22 52 04 74	69 63 21 83 41
27 06 08 09 92	26 22 59 28 27	38 58 22 14 79	24 32 12 38 42	33 56 90 92 57
54 68 97 20 54	33 26 74 03 30	74 22 19 13 48	30 28 01 92 49	58 61 52 27 03
02 92 65 68 99	05 53 15 26 70	04 69 22 64 07	04 73 25 74 82	78 35 22 21 88
83 52 57 78 62	98 61 70 48 22	68 50 64 55 75	42 70 32 09 60	58 70 61 43 97
82 82 76 31 33	85 13 41 38 10	16 47 61 43 77	83 27 19 70 41	34 78 77 60 25
38 61 34 09 49	04 41 66 09 76	20 50 73 40 95	24 77 95 73 20	47 42 80 61 03
01 01 11 88 38	03 10 16 82 24	39 58 20 12 39	82 77 02 18 88	33 11 49 15 16
21 66 14 38 28	54 08 18 07 04	92 17 63 36 75	33 14 11 11 78	97 30 53 62 38
32 29 30 69 59	68 50 33 31 47	15 64 88 75 27	04 51 41 61 96	86 62 93 66 71
04 59 21 65 47	39 90 89 86 77	46 86 86 88 86	50 09 13 24 91	54 80 67 78 66
38 64 50 07 36	56 50 45 94 25	48 28 48 30 51	60 73 73 03 87	68 47 37 10 84
48 33 50 83 53	59 77 64 59 90	58 92 62 50 18	93 09 45 89 06	13 26 98 86 29

From Fisher and Yates; courtesy of Longman Group Ltd., London.

Random numbers (VI)

25 19 64 82 84	62 74 29 92 24	61 03 91 22 48	64 94 63 15 07	66 85 12 00 27
23 02 41 46 04	44 31 52 43 07	44 06 03 09 34	19 83 94 62 94	48 28 01 51 92
55 85 66 96 28	28 30 62 58 83	65 68 62 42 45	13 08 60 46 28	95 68 45 52 43
68 45 19 69 59	35 14 82 56 80	22 06 52 26 39	59 78 98 76 14	36 09 03 01 86
69 31 46 29 85	18 88 26 95 54	01 02 14 03 05	48 00 26 43 85	33 93 81 45 95
37 31 61 28 98	94 61 47 03 10	67 80 84 41 26	88 84 59 69 14	77 32 82 81 89
66 42 19 24 94	13 13 38 69 96	76 69 76 24 13	43 83 10 13 24	18 32 84 85 04
33 65 78 12 35	91 59 11 38 44	23 31 48 75 74	05 30 08 46 32	90 04 93 56 16
76 32 06 19 35	22 95 30 19 29	57 74 43 20 90	20 25 36 70 69	38 32 11 01 01
43 33 42 02 59	20 39 84 95 61	58 22 04 02 99	99 78 78 83 82	43 67 16 38 95
28 31 93 43 94	87 73 19 38 47	54 36 90 98 10	83 43 32 26 26	22 00 90 59 22
97 19 21 63 34	69 33 17 03 02	11 15 50 46 08	42 69 60 17 42	14 68 61 14 48
82 80 37 14 20	56 39 59 89 63	33 90 38 44 50	78 22 87 10 88	06 58 87 39 67
03 68 03 13 60	64 13 09 37 11	86 02 57 41 99	31 66 60 65 64	03 03 02 58 97
65 16 58 11 01	98 78 80 63 23	07 37 66 20 56	20 96 06 79 80	33 39 40 49 42
24 65 58 57 04	18 62 85 28 24	26 45 17 82 76	39 65 01 73 91	50 37 49 38 73
02 72 64 07 75	85 66 48 38 73	75 10 96 59 31	48 78 58 08 88	72 08 54 57 17
79 16 78 63 99	43 61 00 66 42	76 26 71 14 33	33 86 76 71 66	37 85 05 56 07
04 75 14 93 39	68 52 16 83 34	64 09 44 62 58	48 32 72 26 95	32 67 35 49 71
40 64 64 57 60	97 00 12 91 33	22 14 73 01 11	83 97 68 95 65	67 77 80 98 87
06 27 07 34 26	01 52 48 69 57	19 17 53 55 96	02 41 03 89 33	86 85 73 02 32
62 40 03 87 10	96 88 22 46 94	35 56 60 94 20	60 73 04 84 98	96 45 18 47 07
00 98 48 18 97	91 51 63 27 95	74 25 84 03 07	88 29 04 79 84	03 71 13 78 26
50 64 19 18 91	98 55 83 46 09	49 66 41 12 45	41 49 36 83 43	53 75 35 13 39
38 54 52 25 78	01 98 00 89 85	86 12 22 89 25	10 10 71 19 45	88 84 77 00 07
46 86 80 97 78	65 12 64 64 70	58 41 05 49 08	68 68 88 54 00	81 61 61 80 41
90 72 92 93 10	09 12 81 93 63	69 30 02 04 26	92 36 48 69 45	91 99 08 07 65
66 21 41 77 60	99 35 72 61 22	52 40 74 67 29	97 50 71 39 79	57 82 14 88 06
87 05 46 52 76	89 96 34 22 37	27 11 57 04 19	57 93 08 35 69	07 51 19 92 66
46 90 61 03 06	89 85 33 22 80	34 89 12 29 37	44 71 38 40 37	15 49 55 51 08
11 88 53 06 09	81 83 33 98 29	91 27 59 43 09	70 72 51 49 73	35 97 25 83 41
11 05 92 06 97	68 82 34 08 83	25 40 58 40 64	56 42 78 54 06	60 96 96 12 82
33 94 24 20 28	62 42 07 12 63	34 39 02 92 31	80 61 68 44 19	09 92 14 73 49
24 89 74 75 61	61 02 73 36 85	67 28 50 49 85	37 79 95 02 66	73 19 76 28 13
15 19 74 67 23	61 38 93 73 68	76 23 15 58 20	35 36 82 82 59	01 33 48 17 66
05 64 12 70 88	80 58 35 06 88	73 48 27 39 43	43 40 13 35 45	55 10 54 38 50
57 49 36 44 06	74 93 55 39 26	27 70 98 76 68	78 36 26 24 06	43 24 56 40 80
77 82 96 96 97	60 42 17 18 48	16 34 92 19 52	98 84 48 42 92	83 19 06 77 78
24 10 70 06 51	59 62 37 95 42	53 67 14 95 29	84 65 43 07 30	77 54 00 15 42
50 00 07 78 23	49 54 36 85 14	18 50 54 18 82	23 79 80 71 37	60 62 95 40 30
44 37 76 21 96	37 03 08 98 64	90 85 59 43 64	17 79 96 52 35	21 05 22 59 30
90 57 55 17 47	53 26 79 20 38	69 90 58 64 03	33 48 32 91 54	68 44 90 24 25
50 74 64 67 42	95 28 12 73 23	32 54 98 64 94	82 17 18 17 14	55 10 61 64 29
44 04 70 22 02	84 31 64 64 08	52 55 04 24 29	91 95 43 81 14	66 13 18 47 44
32 74 61 64 73	21 46 51 44 77	72 48 92 00 05	83 59 89 65 06	53 76 70 58 78
75 73 51 70 49	12 53 67 51 54	38 10 11 67 73	22 32 61 43 75	31 61 22 21 11
76 18 36 16 34	16 28 25 82 98	64 26 70 54 87	49 48 55 11 39	94 25 20 80 85
00 17 37 71 81	64 21 91 15 82	81 04 14 52 11	39 07 30 60 77	39 18 27 85 68
54 95 57 55 04	12 77 40 70 14	79 86 61 57 50	52 49 41 73 46	05 63 34 92 33
69 99 95 54 63	44 37 33 53 17	38 06 58 37 93	47 10 62 31 28	63 59 40 40 32

From Fisher and Yates; courtesy of Longman Group Ltd., London.

Appendix 6

Appendix 6.1 Contiguous Quadrat Density Data (pp. 254–5) for three Species with Pattern Scales and the Interactive Associations shown in Fig. 7.8, p. 134. The columns for a species within a transect are continued from the bottom of p. 254 to the top of p. 255.

Each transect provides quadrat density values for blocksize one Ns(1). Ns(2) is derived by grouping adjacent quadrat density values in pairs. for example, species 1 in transect 1 has the following sequence of density values for Ns (1):

	2 2	0 0	0 1	3 2	0 0	0 0	0 3	3 2	...
Block size 2 is then:	4	0	1	5	0	0	3	5	...
Block size 4:	4		6		0		8		...

and so on.

Σx^2 is then calculated for each block size ($\Sigma^1 x^2$, $\Sigma^2 x^2$, $\Sigma^4 x^2$, etc.), scaled, and the variance partitioned between the range of block size available. The analysis of variance table for species 1, 2 and 3 is given in Appendix 6.2 below.

A short-cut for the partitioning of the sum of squares is simply obtained by subtracting $\Sigma x^2/\text{Ns}(2)$ from $\Sigma x^2/\text{Ns}(1)$ to yield SS. Ns (1)
$\Sigma x^2/\text{Ns}(4)$ from $\Sigma x^2/\text{Ns}(2)$ to yield SS. Ns(2)
and so on.

SS Ns(128) is derived from $\Sigma x^2/\text{Ns}(128) - \dfrac{(\Sigma x)^2}{\text{N}}$

where N = 128 quadrats × 10 replicates = 1280.

The derivation of this short-cut is as follows: the variance at Ns(128), the maximum block size, is given by

$$\frac{\Sigma^{128} x^2}{\text{Ns}(128)} - \frac{(\Sigma x)^2}{\text{N}} \text{ /degrees of freedom}$$

The variance at Ns(64) however has to be corrected for the variance attributed to Ns(128) i.e:

$$\frac{\dfrac{\Sigma^{64} x^2}{\text{Ns}(64)} - \dfrac{(\Sigma x)^2}{\text{N}} - \left[\dfrac{\Sigma^{128} x^2}{\text{Ns}(128)} - \dfrac{(\Sigma x)^2}{\text{N}}\right]}{n^{64}}$$

which simply reduces to $\dfrac{\Sigma^{64} x^2}{\text{Ns}(64)} - \dfrac{\Sigma^{128} x^2}{\text{Ns}(128)}$

The same partitioning of variance follows throughout the remainder of the table with appropriate corrections for the variance extracted at successively higher block sizes, allowing the quick method of driving the partitioned sum of squares.

The degrees of freedom are also similarly partitioned. Thus, in the example here, there are 10 transects 128 units long and, accordingly, there are $(n-1)$, 9 degrees of freedom available at Ns(128). At Ns(64) there are 20 such blocks in the total data with 19 degrees of freedom available. However 9 of these have been allocated to Ns(128) and as a result only 10 are available to allocate to Ns(64). The sum of the degrees of freedom (640 + 320 + 160 + . . .) = 1279. Thus there are a total of 1280 quadrats available with $n-1$ total degrees of freedom for partitioning between the individual block size variances. It is always useful to run this simple check to ensure that the correct partitioning of the degrees of freedom has been achieved.

For covariance analysis the expected variance assuming no relationship between species 1 and 2, 1 and 3 or 2 and 3 is simply the addition of the variance at each block size of the pairs of individual pattern analyses. A separate data set is then derived by adding, for example, the transect data of species 1 to that of species 2. Thus for transect 1:

	2 2 0 0 0 1 3 2 0 0 0 0 0 3 3 2 . . .	Species 1
and	0 0 3 3 1 0 0 0 2 1 1 0 0 0 0 0 . . .	Species 2
becomes	2 2 3 3 1 1 3 2 2 1 1 0 0 3 3 2 . . .	Sum

This is then blocked into

Ns(2)	4	6	2	5	3	1	3	5	. . .
Ns(4)	10		7		4		8		. . .

and so on and a pattern analysis carried out (Var A + B) as before yielding the *observed* variance at each block size. The *covariance* is then obtained as the difference observed − expected.

Kershaw (1960*a*) points out that although the *trend* of association is given by the difference between observed and expected variance,

Var (A + B) when expanded

is

Var A + Var B + 2 Covar AB

$$\text{i.e. } V_{A+B} = V_A + V_B + 2\,C_{AB}$$

accordingly

$$C_{AB} = \frac{V_{A+B} - (V_A + V_B)}{2}$$

i.e. the covariance = (Observed − Expected)/2

TRANSECT

1			2			3			4			5			6			7			8			9			10		
2	0	0	0	1	1	0	0	0	0	3	3	0	0	0	0	0	0	0	0	0	2	0	0	1	0	0	1	0	0
2	0	0	0	1	2	0	0	0	0	3	2	0	0	0	0	0	0	0	0	0	2	0	0	1	0	0	1	0	0
0	3	3	0	0	0	0	0	0	3	0	0	0	0	0	0	0	0	0	0	0	1	2	2	0	4	4	0	4	2
0	3	2	0	0	0	0	0	0	1	0	0	0	0	0	0	0	0	0	0	0	0	1	2	0	4	4	0	4	3
0	1	0	0	0	0	0	0	0	1	4	3	0	0	0	0	0	0	0	0	0	1	0	0	3	0	0	2	1	4
1	0	0	0	0	0	0	0	0	0	1	1	0	0	0	0	0	0	0	0	0	3	0	0	2	0	0	1	0	0
3	0	0	0	0	0	0	0	0	0	1	2	0	0	0	0	0	0	0	0	0	0	3	2	0	3	2	0	1	2
2	0	0	0	0	0	0	0	0	0	1	1	0	0	0	0	0	0	0	0	0	0	3	4	0	2	3	0	4	5
0	2	2	0	0	0	0	0	0	2	0	0	0	0	0	0	0	0	0	0	0	2	0	0	4	0	0	3	0	0
0	1	2	1	0	0	0	0	0	2	0	0	0	0	0	0	0	0	0	0	0	2	0	0	1	0	0	2	0	0
0	1	3	1	0	0	0	0	0	0	3	1	0	0	0	0	0	0	0	0	0	0	1	2	0	1	2	0	3	4
0	0	0	0	2	3	0	0	0	0	1	1	0	0	0	0	0	0	0	0	0	0	1	1	0	5	3	0	1	2
0	0	0	0	1	1	0	0	0	4	0	0	0	0	0	0	0	0	0	0	0	3	0	0	0	4	3	1	0	0
3	0	0	0	1	1	0	0	0	3	0	0	0	0	0	0	0	0	0	0	0	1	0	0	1	0	0	3	0	0
3	0	0	0	1	0	0	0	0	0	2	2	0	0	0	0	0	0	0	0	0	0	3	4	2	0	0	0	4	2
2	0	0	0	0	0	0	0	0	0	2	3	0	0	0	0	0	0	0	0	0	0	3	2	2	1	0	0	2	4
0	0	0	0	0	0	3	0	0	0	0	0	0	0	0	4	0	0	4	0	0	1	0	0	0	1	1	0	0	0
0	0	0	0	0	0	2	1	2	0	0	0	0	0	0	4	0	0	1	0	0	1	0	0	0	0	1	0	0	0
0	0	0	0	0	0	0	3	4	0	0	0	0	0	0	0	2	2	0	3	3	0	0	0	0	0	0	0	0	0
0	0	0	0	0	0	1	0	0	0	0	0	0	0	0	0	2	2	0	1	3	0	0	0	0	0	0	0	0	0
0	0	0	0	0	0	2	0	0	0	0	0	0	0	0	2	0	1	2	0	0	0	0	0	0	0	0	0	0	0
0	0	0	0	0	0	1	3	2	0	0	0	0	0	0	1	0	0	2	0	0	0	0	0	0	0	0	0	0	0
0	0	0	0	0	0	0	2	2	0	0	0	0	0	0	0	4	1	0	2	2	0	0	0	0	0	0	0	0	0
0	0	0	0	0	0	3	0	0	0	0	0	3	0	0	0	1	3	1	2	2	0	0	0	0	3	1	0	0	0
0	0	0	0	0	0	2	1	0	0	0	0	2	0	1	2	0	1	2	0	0	0	0	0	0	2	1	0	0	0
0	0	0	0	0	0	0	2	2	0	0	0	0	1	2	2	0	0	3	0	0	0	0	0	0	1	0	0	0	0
0	0	0	0	0	0	0	2	2	0	0	0	0	2	0	0	2	2	1	1	1	0	0	0	0	0	0	0	0	0
0	0	0	0	0	0	0	1	2	0	0	0	0	1	1	1	2	2	0	1	1	0	0	0	0	0	0	0	0	0
0	0	0	0	0	0	1	0	0	0	0	0	1	0	0	3	0	0	1	0	0	0	0	0	0	0	0	0	0	0
0	0	0	0	0	0	3	0	0	0	0	0	2	0	0	3	1	0	1	0	0	0	0	0	0	0	0	0	0	0
0	0	0	0	0	0	3	0	0	0	0	0	1	0	0	0	2	3	1	2	4	0	1	2	0	0	0	0	0	0
0	0	0	2	0	0	0	0	0	0	0	0	0	0	0	0	2	3	0	3	1	0	1	2	5	0	0	0	0	0
0	1	2	3	0	0	0	0	1	3	0	0	0	1	1	0	0	0	0	0	1	1	0	0	1	0	1	2	0	0
0	1	2	0	3	1	0	1	2	1	0	0	0	2	2	0	0	0	0	1	0	4	0	0	0	2	3	3	0	0
0	4	2	0	3	2	0	2	1	0	3	4	0	3	1	0	0	0	0	0	1	0	2	2	0	2	4	0	3	2
0	0	0	1	0	0	0	1	0	0	3	3	0	2	1	0	0	0	0	0	0	0	2	2	1	0	0	0	3	4
3	0	0	1	0	0	0	0	0	1	1	0	3	0	0	0	0	0	0	0	0	3	0	0	1	3	0	4	0	0
2	0	0	3	0	0	0	0	0	1	0	0	4	0	0	0	0	0	0	0	0	3	0	0	0	2	4	1	0	0
2	0	0	0	1	2	0	0	0	0	2	2	1	0	0	0	0	0	0	0	0	1	3	2	0	1	4	0	0	0
0	0	0	0	2	2	0	0	0	0	2	3	0	0	0	0	0	0	0	0	0	0	3	2	2	0	0	1	3	2
1	0	0	0	0	0	0	0	0	4	0	0	0	0	0	0	0	0	0	0	0	2	0	0	1	0	0	3	2	1
2	0	0	2	0	0	0	0	0	3	0	0	0	0	0	0	0	0	0	0	0	1	0	0	0	2	3	1	0	0
2	0	0	2	0	0	0	0	0	0	1	4	0	0	0	0	0	0	0	0	0	0	2	3	0	2	1	1	0	0
0	2	3	0	0	0	0	0	0	0	3	2	0	0	0	0	0	0	0	0	0	0	2	2	4	0	0	0	3	4
0	3	1	1	0	0	0	0	0	0	1	0	0	0	0	0	0	0	0	0	0	1	0	0	0	1	0	0	1	1
0	3	3	2	0	0	0	0	0	4	0	0	0	0	0	0	0	0	0	0	0	1	0	0	0	3	2	0	0	1
0	0	0	2	0	0	0	1	1	2	0	0	0	0	0	0	0	0	0	0	0	0	1	1	0	3	1	0	0	1
0	0	0	0	0	0	0	1	2	0	0	0	0	0	0	0	0	0	3	0	0	0	5	4	0	0	0	0	1	0
0	0	0	0	0	0	0	1	1	0	0	0	0	0	0	1	0	0	1	0	0	0	0	0	0	0	0	0	0	0
0	0	0	0	0	0	0	0	0	0	0	0	0	0	0	1	0	0	1	0	0	0	0	0	0	0	0	0	0	0
0	0	0	0	0	0	0	0	0	0	0	0	0	0	0	0	2	3	1	1	1	0	0	0	0	0	0	0	0	0
0	0	0	0	0	0	3	0	0	0	0	0	0	0	0	0	4	3	2	1	1	0	0	0	0	0	0	0	0	0
0	0	0	0	0	0	3	0	0	0	0	0	0	0	0	1	1	0	1	2	3	0	0	0	0	0	0	0	0	0
0	0	0	0	0	0	0	2	3	0	0	0	0	0	0	3	0	0	0	1	3	0	0	0	0	0	0	0	0	0
0	0	0	0	0	0	0	3	3	0	0	0	0	0	0	0	1	4	1	0	0	0	0	0	0	0	0	0	0	0
0	0	0	0	0	0	3	0	0	0	0	0	0	0	0	0	1	2	1	0	0	0	0	0	0	0	0	0	0	0
0	0	0	1	0	0	1	0	0	0	0	0	0	0	0	1	0	0	0	3	3	0	0	0	0	0	0	0	0	0
0	0	0	1	0	0	0	4	3	0	0	0	0	0	0	2	0	0	0	3	4	0	0	0	0	0	0	0	0	0
0	0	0	0	0	0	0	4	4	0	0	0	0	0	0	0	4	3	1	0	0	0	0	0	0	0	0	0	0	0
0	0	0	0	0	0	0	1	2	0	0	0	0	0	0	2	2	3	3	0	0	0	0	0	0	0	0	0	0	0
0	0	0	0	0	0	2	0	0	0	0	0	0	0	0	0	1	0	0	3	1	0	0	0	0	0	0	0	0	0
0	0	0	0	0	0	2	0	0	0	0	0	0	0	0	1	0	0	0	3	4	0	0	0	0	0	0	0	0	0
0	0	0	0	0	0	0	3	3	0	0	0	0	0	0	3	0	0	2	0	0	0	0	0	0	0	0	0	0	0
0	0	0	0	0	0	0	2	2	0	0	0	0	0	0	3	0	0	2	0	0	0	0	0	0	0	0	0	0	0
1	**2**	**3**	**1**	**2**	**3**	**1**	**2**	**3**	**1**	**2**	**3**	**1**	**2**	**3**	**1**	**2**	**3**	**1**	**2**	**3**	**1**	**2**	**3**	**1**	**2**	**3**	**1**	**2**	**3**

SPECIES

TRANSECT

1			2			3			4			5			6			7			8			9			10		
3	0	0	2	0	0	0	0	0	4	0	0	2	0	0	1	0	0	0	0	0	0	2	3	1	0	0	0	0	0
3	0	0	2	0	0	0	0	0	4	0	0	1	0	0	2	0	0	0	0	0	0	2	1	3	0	0	0	0	0
0	3	3	0	1	1	0	0	0	0	2	3	0	2	3	0	0	0	0	0	0	1	1	0	0	4	4	0	0	0
0	3	3	0	2	2	0	0	0	0	3	3	0	2	1	0	0	0	0	0	0	4	0	0	0	1	2	0	0	0
0	0	1	0	2	4	0	0	0	0	1	0	3	1	0	0	0	0	0	0	0	0	4	3	2	0	0	0	0	0
0	0	0	1	0	0	0	0	0	0	0	0	3	0	1	0	0	0	0	0	0	0	4	4	2	0	0	0	0	0
2	0	0	3	0	0	0	0	0	3	0	0	0	4	3	0	0	0	0	0	0	4	0	0	0	3	2	0	0	0
1	0	0	2	0	0	0	0	0	2	0	0	0	4	3	0	0	0	0	0	0	3	0	0	0	2	2	0	0	0
0	3	2	0	3	3	0	0	0	1	2	3	1	0	0	0	0	0	0	0	0	0	4	1	1	0	0	0	0	0
0	2	2	0	3	3	0	0	0	0	4	3	1	0	0	0	0	0	0	0	0	0	1	1	2	0	0	0	0	0
2	0	0	1	0	0	0	0	0	1	0	0	0	1	1	0	0	0	0	0	0	1	0	0	0	2	3	0	0	0
2	0	0	2	0	0	0	0	0	3	0	0	0	4	1	0	0	0	0	0	0	3	0	0	1	2	2	0	0	0
0	2	1	0	2	1	0	0	0	0	4	2	4	0	0	0	0	0	0	0	0	0	1	1	3	0	0	0	0	0
0	1	1	0	4	3	0	0	0	0	2	3	0	0	0	0	0	0	0	0	0	0	2	2	3	1	0	0	0	0
3	0	0	3	0	0	0	0	0	1	0	1	0	5	2	0	0	0	0	0	0	2	0	0	0	4	1	0	0	0
1	0	0	3	0	0	0	0	0	1	0	0	0	1	2	0	0	0	0	0	0	1	0	0	0	2	4	0	0	0
0	0	0	0	0	0	2	0	0	0	0	0	0	0	0	2	0	0	1	0	0	0	0	0	0	0	0	0	0	0
0	0	0	0	0	0	3	0	0	0	0	0	0	0	0	1	0	0	1	0	0	0	0	0	0	0	0	0	0	0
0	0	0	0	0	0	1	0	0	0	0	0	0	0	0	0	1	3	0	2	3	0	0	0	0	1	1	4	0	0
0	0	0	0	0	0	0	1	0	0	0	0	0	0	0	0	3	3	0	2	1	0	0	0	0	1	0	1	1	3
0	0	0	0	0	0	0	3	3	0	0	0	0	0	0	1	0	0	2	0	0	0	0	0	0	0	0	0	4	3
0	0	0	0	0	0	0	2	3	0	0	0	0	0	0	1	1	0	3	0	0	0	0	0	0	0	0	0	1	0
0	0	0	0	0	0	3	0	0	0	0	0	0	0	0	0	4	2	1	3	2	0	0	0	0	0	0	3	0	0
0	0	0	0	0	0	2	0	1	0	0	0	0	0	0	0	1	2	0	1	2	0	0	0	0	0	0	3	0	0
0	0	0	0	0	0	0	4	2	0	0	0	0	0	0	3	0	0	2	0	0	0	0	0	0	0	0	0	2	2
0	0	0	0	0	0	0	4	4	0	0	0	0	0	0	3	1	0	2	1	4	0	0	0	0	0	0	0	2	3
0	0	0	0	0	0	2	0	0	0	0	0	0	0	0	0	2	2	0	2	1	0	0	0	0	0	0	2	0	0
0	0	0	0	0	0	2	0	0	0	0	0	0	0	0	0	2	2	0	4	0	0	0	0	0	0	0	2	0	0
0	0	0	0	0	0	0	3	2	0	0	0	0	0	0	2	0	0	2	0	0	0	0	0	0	0	0	1	0	4
0	0	0	0	0	0	0	3	2	0	0	0	0	0	0	1	0	0	3	0	0	0	0	0	0	0	0	0	6	4
0	0	0	0	0	0	1	1	0	0	0	0	0	0	0	0	3	1	0	4	4	0	0	0	0	0	0	3	0	0
0	0	0	0	0	0	4	0	0	0	0	0	0	0	0	0	1	1	0	1	4	0	0	0	0	0	0	3	0	0
0	0	0	2	0	0	0	0	0	0	0	0	3	0	0	0	0	0	1	0	0	0	4	3	4	0	0	0	1	5
0	0	0	2	0	0	0	0	0	0	0	0	3	0	0	0	0	0	2	0	0	0	4	4	5	0	0	0	5	5
0	0	0	0	4	4	0	0	0	0	0	0	0	3	3	0	0	0	0	4	3	3	1	0	0	1	1	4	0	0
0	0	0	0	3	3	0	0	0	0	0	0	0	3	3	0	0	0	0	1	4	2	0	0	0	1	2	1	1	0
0	0	0	1	0	0	0	0	0	0	0	0	1	0	0	0	0	0	3	0	0	0	2	1	0	1	3	0	4	2
0	0	0	3	0	0	0	0	0	0	0	0	4	1	0	0	0	0	3	0	0	0	2	3	0	2	1	0	1	3
0	0	0	0	3	2	0	0	0	0	1	0	0	1	2	0	0	0	0	2	2	4	0	0	2	0	0	3	0	0
0	1	0	0	3	2	0	0	0	0	1	2	0	2	3	0	0	0	1	3	2	4	0	0	2	0	0	3	0	0
0	1	1	0	1	2	0	0	0	0	3	1	1	0	0	0	1	0	2	0	0	0	3	1	0	4	4	0	2	3
0	1	1	1	0	0	0	0	0	0	2	1	4	0	0	0	1	0	2	0	0	0	4	3	0	4	4	0	2	2
0	1	0	2	0	0	0	0	0	0	0	0	1	3	2	0	0	0	0	4	1	3	0	0	3	0	0	2	0	0
0	0	0	3	0	0	0	0	0	1	0	0	0	2	2	2	0	0	0	1	1	3	0	0	3	0	0	2	0	0
0	0	0	2	0	0	0	0	0	1	0	0	3	0	0	1	0	0	0	3	2	0	1	0	0	2	2	0	4	1
0	0	0	0	0	0	0	0	0	0	0	0	2	0	0	0	2	0	0	1	2	0	2	2	0	1	2	1	3	1
0	0	0	1	0	0	0	0	0	0	0	0	0	4	4	0	2	1	2	0	0	0	1	3	1	0	0	4	0	0
0	0	0	1	0	0	0	0	0	0	0	0	0	4	3	0	0	1	3	0	0	0	0	2	1	0	0	4	0	0
0	3	4	0	2	1	4	0	0	2	0	0	0	0	0	2	0	0	0	0	0	0	0	0	0	0	0	0	0	0
0	4	4	0	2	3	3	0	0	1	0	0	0	0	0	2	1	0	0	0	0	0	1	1	0	0	0	0	0	0
2	0	0	0	2	2	0	4	2	0	1	3	0	0	0	0	1	2	0	0	0	0	0	3	0	0	0	0	0	0
2	0	0	0	1	1	1	3	3	0	2	1	0	0	0	0	4	1	0	0	0	0	1	2	0	0	0	0	0	0
0	1	3	0	1	0	3	0	0	1	1	2	0	0	0	0	2	1	0	0	0	0	1	1	0	0	0	0	0	0
0	2	1	0	0	0	3	0	0	3	0	0	0	0	0	0	0	4	0	0	0	0	0	0	0	0	0	0	0	0
0	2	1	0	0	0	0	0	0	0	3	2	0	0	0	3	0	0	0	0	0	0	0	0	0	0	0	0	0	0
0	1	0	0	0	0	0	1	1	0	2	2	0	0	0	4	0	0	0	0	0	0	0	0	0	0	0	0	0	0
3	0	0	0	0	0	0	2	1	2	0	0	0	0	0	0	2	3	0	0	0	0	0	0	0	0	0	0	0	0
3	0	0	0	0	0	0	2	2	1	0	0	0	0	0	0	4	3	0	0	0	0	0	0	0	0	0	0	0	0
1	1	0	0	0	0	3	0	0	0	4	4	0	0	0	2	0	0	0	0	0	0	0	0	0	0	0	0	0	0
0	3	3	0	0	0	2	0	0	0	3	2	0	0	0	3	0	0	0	0	0	0	0	0	0	0	0	0	0	0
0	2	2	0	0	0	0	3	3	4	0	0	0	0	0	0	2	3	0	0	0	0	0	0	0	0	0	0	0	0
0	0	1	0	0	0	0	2	3	3	0	0	0	0	0	0	4	1	0	0	0	0	0	0	0	0	0	0	0	0
1	0	0	0	0	0	1	1	0	0	1	2	0	0	0	1	0	0	0	0	0	0	0	0	0	0	0	0	0	0
1	0	0	0	0	0	1	0	0	0	1	1	0	0	0	1	0	0	0	0	0	0	0	0	0	0	0	0	0	0
1	**2**	**3**	**1**	**2**	**3**	**1**	**2**	**3**	**1**	**2**	**3**	**1**	**2**	**3**	**1**	**2**	**3**	**1**	**2**	**3**	**1**	**2**	**3**	**1**	**2**	**3**	**1**	**2**	**3**

SPECIES

Appendix 6.2 Block-size/variance pattern analysis for species 1, 2 and 3. (Values have been rounded introducing some error in comparison with 4-decimal computer output.)

Block size (Ns)	**Sum of squares** (Σx^2)	Σx^2/**Ns**	**Partitioned sum of squares** (SS)	**Degrees of freedom** (n)	**Variance** (mean square)
		SPECIES 1			
1.	1791.00	1791.00	223.50	640.	.35
2.	3135.00	1567.50	665.25	320.	2.08
4.	3609.00	902.25	98.38	160.	.62
8.	6431.00	803.88	43.19	80.	.54
16.	12171.00	760.69	359.41	40.	8.99
32.	12841.00	401.28	9.20	20.	.46
64.	25093.00	392.08	7.54	10.	.75
128.	49221.00	384.54	5.00	9.	.56
$\frac{(\Sigma x)^2}{N}$		379.54	–	–	–
		SPECIES 2			
1.	2046.00	2046.00	288.00	640.	.45
2.	3516.00	1758.00	754.00	320.	2.36
4.	4016.00	1004.00	106.50	160.	.67
8.	7180.00	897.50	64.38	80.	.80
16.	13330.00	833.13	361.13	40.	9.03
32.	15104.00	472.00	10.97	20.	.55
64.	29506.00	461.03	18.33	10.	1.83
128.	56666.00	442.70	10.25	9.	1.14
$\frac{(\Sigma x)^2}{N}$		432.45	–	–	–
		SPECIES 3			
1.	1971.00	1971.00	248.50	640.	.39
2.	3445.00	1722.50	744.75	320.	2.33
4.	3911.00	977.75	115.63	160.	.72
8.	6897.00	862.13	63.31	80.	.79
16.	12781.00	798.81	351.78	40.	8.79
32.	14305.00	447.03	10.80	20.	.54
64.	27919.00	436.23	13.43	10.	1.34
128.	54119.00	442.80	14.42	9.	1.60
$\frac{(\Sigma x)^2}{N}$		408.38	–	–	–

Appendix 6.3 Covariance analysis of the data for species 1, 2 and 3 given in Appendix 6.1. (Values have been rounded to two decimal places introducing some error in comparison with 4-decimal computer output.)

Block size (Ns)	**Variance species 1**	**Variance species 2**	**Observed variance**	**Expected variance**	**Covariance**
		SPECIES 1 × 2			
1.	.35	.45	.64	.80	−0.08
2.	2.08	2.36	.81	4.44	−1.81
4.	.62	.67	.99	1.28	−0.14
8.	.54	.80	1.09	1.34	−0.13
16.	8.99	9.03	35.28	18.01	8.64
32.	.46	.55	.99	1.01	−0.01
64.	.75	1.83	4.01	2.59	0.71
128.	.56	1.14	2.98	1.69	0.64
		SPECIES 1 × 3			
1.	.35	.39	.59	.74	−0.75
2.	2.08	2.33	.87	4.41	−1.77
4.	.62	.72	1.06	1.34	−0.14
8.	.54	.79	.91	1.33	−0.21
16.	8.99	8.79	34.79	17.78	8.51
32.	.46	.54	.86	1.00	−0.07
64.	.75	1.34	3.29	2.10	0.60
128.	.56	1.60	3.81	2.16	0.82
		SPECIES 2 × 3			
1.	.45	.39	1.15	.84	0.16
2.	2.36	2.33	8.93	4.68	2.12
4.	.67	.72	2.36	1.39	0.49
8.	.80	.79	2.86	1.60	0.63
16.	9.03	8.79	35.44	17.82	8.81
32.	.55	.54	1.93	1.09	0.42
64.	1.83	1.34	6.03	3.18	1.43
128.	1.14	1.60	5.30	2.74	1.28

Appendix 7

Appendix 7.1 The salt marsh transect data are listed below by quadrat number (quadrats 1, 4, 5, 9, 10 and 12; all species are absent). Within each quadrat the species number is followed by the number of occurrences of that species. For example in quadrat 20 species 11, 25 occurrences; species 38, 1 occurrence. The species are listed alphabetically with their corresponding number in Appendix 7.2 below. (Note that MEES TRIQ, SALI CALC, and CHRY ARCT did not occur on this transect replicate.) The FORMAT used here corresponds that used as input to TWINSPAN and DECORANA and the name list and quadrat list, in correct input sequence, are also included (see Hill, 1979*a* and *b* for further details.) The full transect sequence is provided in quadrats 1 to 145. Quadrats 151 to 180 provide additional salt pan data sampled randomly across the marsh.

2	11 1							
3	11 2							
6	11 4							
7	11 1							
8	11 3							
11	11 8							
13	11 6							
14	11 16							
15	11 5							
16	11 1							
17	11 6							
18	11 15							
19	11 2							
20	11 25	38 1						
21	11 20							
22	11 25							
23	11 25	42 1						
24	11 19							
25	11 11							
26	1 19	11 25	27 6	38 3				
27	1 19	11 25	27 8	38 7	41 25	43 3		
28	1 25	11 25	38 17	42 4				
29	1 9	11 25						
30	1 25	11 25	27 1	38 18	43 10			
31	1 25	11 25	27 3	38 14	43 3			
32	1 25	11 25	38 13					
33	1 21	11 25						
34	1 6	11 21						
35	1 16	11 25	46 1					
36	1 25	11 25						
37	1 16	11 25						
38	1 17	11 24	38 1					
39	1 25	11 25	38 5					
40	1 25	11 25	38 1					
41	1 25	11 10	13 12	38 6	41 3	43 23		
42	1 25	11 25	27 1					
43	1 25	11 25	38 7	42 1				
44	1 25	11 25	27 12	38 5				
45	1 25	11 5	13 5	15 1	26 17	37 11	38 13	41 11
45	43 25							
46	1 25	11 14	13 4	15 11	26 17	38 19	41 14	43 22
47	1 25	11 17	38 11					
48	1 21	11 25	27 12	38 20				
49	1 25	10 5	11 5	13 3	14 12	15 2	26 12	43 6
50	1 25	11 25	27 7	46 2				
51	1 2	11 21	27 15	45 5	46 6			
52	1 5	11 25	27 9	45 7	46 2			
53	1 2	45 21						
54	1 6	11 23	27 1	45 1	46 3			
55	1 8	11 25	27 7	46 5				
56	1 17	11 12	27 21	33 2	46 7			
57	1 15	10 3	11 25	27 12	46 2			
58	1 14	10 1	11 19	27 2	45 2	46 16		
59	1 13	11 8	27 12	45 18	46 13			
60	1 25	10 18	12 3	15 2	26 20	37 5	38 5	43 17
61	1 8	10 4	11 14	27 18	46 8			

continued

62	1 25	10 25	12 5	26 24	37 17	38 6	41 25	43 6
62	47 9							
63	11 20	27 21	45 12	46 18				
64	1 24	10 23	12 4	15 1	26 25	37 15	38 1	43 10
64	47 1							
65	1 5	4 16	10 13	11 7	27 3	45 23	46 16	
66	1 2	4 23	10 11	11 3	27 5	45 24	46 23	
67	1 23	10 25	26 25	37 25	43 18	47 18		
68	1 19	10 22	15 5	17 6	26 1	37 25	42 17	43 25
68	47 20							
69	1 17	4 10	10 25	26 12	37 2	45 9		
70	1 5	10 19	13 1	14 1	15 7	17 16	37 22	42 2
70	43 25	47 12						
71	1 25	10 25	17 1	26 25	37 25	43 12	47 10	
72	1 22	4 10	10 25	12 1	24 3	26 25	37 23	43 9
72	47 8							
73	4 25	9 25	10 6	24 4	26 25	45 16		
74	4 25	9 21	10 5	26 3	37 2	45 24		
75	1 3	4 25	26 14	27 1	37 11	45 16	46 1	
76	1 16	10 11	12 4	15 9	17 8	26 5	37 24	42 24
76	43 25	47 9						
77	4 25	10 8	26 25	27 4	37 11	46 6	47 3	
78	1 23	10 14	12 19	26 21	37 21	42 4	43 22	47 7
79	1 20	4 25	10 17	26 25	37 14	43 1	45 1	47 1
80	1 9	4 25	10 21	12 1	26 25	37 5	45 12	46 2
81	1 12	4 25	10 5	26 25	37 7	45 8	46 1	47 6
82	1 13	4 25	10 4	12 25	26 25	37 9	45 1	47 11
83	1 25	4 25	10 10	12 5	15 1	17 4	26 25	37 19
83	40 6	47 14						
84	1 19	4 25	10 2	12 5	26 25	37 15	40 4	47 19
85	1 20	4 9	9 2	10 12	12 25	15 1	17 17	26 11
85	37 25	40 19	47 5					
86	1 25	4 5	10 13	12 25	17 6	22 1	25 3	37 25
86	40 3	42 25	47 8					
87	1 3	4 25	9 24	12 2	15 2	26 25	45 2	
88	1 21	4 25	9 11	10 8	12 15	15 8	17 4	26 6
88	37 16	40 23	42 17	47 12				
89	1 24	4 19	10 14	12 25	22 1	25 2	26 23	27 1
89	37 13	40 5	42 7	47 3				
90	1 23	4 5	10 9	12 25	17 1	24 1	27 1	37 15
90	40 25	42 25	47 5					
91	1 25	4 3	10 19	12 25	14 1	15 4	37 23	40 25
91	42 25	47 14	49 6					
92	1 19	4 25	10 2	12 21	26 25	45 1		
93	1 15	10 2	12 7	17 4	26 3	37 10	40 25	42 20
93	43 25	47 7	49 15					
94	1 24	4 14	10 2	12 25	26 4	37 14	40 22	42 25
94	47 4	49 2						
95	1 25	4 8	8 2	9 8	12 25	26 25	27 1	31 1
95	40 2	42 9	49 1					
96	1 22	4 16	9 12	12 25	26 25	37 1	40 1	
97	1 23	4 1	8 2	12 24	25 2	37 19	40 23	42 21
97	47 2	48 1	49 2					
98	1 17	4 23	8 8	9 2	12 25	26 25	37 5	40 6
98	47 2							
99	1 20	12 25	34 4	37 21	40 25	42 23	47 2	48 3
99	49 18							
100	1 21	10 1	12 25	17 7	25 2	26 14	37 17	40 25
100	42 10	46 6	48 2	49 21				
101	1 24	8 2	12 25	14 1	17 1	20 15	26 7	27 1
101	37 8	40 7	42 2	47 2				
102	1 18	4 1	8 4	12 25	26 25	37 2	40 14	48 1
103	1 12	3 8	8 9	12 25	20 2	26 25	37 8	40 9
103	47 5							
104	1 14	8 6	12 25	26 15	34 24	37 21	40 17	42 10
104	47 6	49 5						
105	1 22	4 2	8 5	12 25	20 25	37 19	40 21	
106	1 7	3 1	8 4	12 25	14 1	20 25	23 11	35 1

continued

106	37 17	40 20	49 1					
107	1 5	3 1	4 6	7 1	8 4	9 5	12 25	20 25
107	27 3	37 8	40 18	47 4				
108	1 15	4 3	7 5	8 2	9 10	12 25	20 2	22 3
108	23 1	24 5	37 15	40 20	42 12	47 2		
109	1 4	3 2	4 2	8 15	12 25	23 2	25 24	37 2
109	40 6	42 1						
110	1 11	3 3	4 7	8 9	12 25	20 25	37 18	40 23
110	48 1							
111	1 18	8 18	12 25	14 3	15 10	18 10	25 3	26 8
111	37 21	39 1	40 25	42 8	47 7			
112	1 7	3 4	8 18	12 25	18 3	20 1	22 5	37 1
112	40 12	42 12						
113	4 2	6 8	7 3	8 21	12 25	20 25	37 10	40 10
113	44 6	47 1						
114	3 3	4 15	6 5	7 1	8 22	9 13	12 18	18 9
114	20 4	26 6	35 4	37 1	40 7	44 9		
115	1 1	4 14	6 2	8 11	9 13	12 23	18 25	27 2
115	37 2	40 14	44 20	47 2				
116	1 8	4 6	8 21	9 12	12 25	15 10	22 17	35 1
116	37 21	40 25	44 10	47 6				
117	1 10	4 15	8 15	9 7	12 25	18 18	20 9	23 3
117	32 1	35 1	37 8	40 8	44 4	47 3		
118	1 6	3 2	8 6	9 4	12 24	15 2	18 10	20 8
118	37 22	40 23	47 1					
119	1 5	2 4	3 18	8 9	9 6	12 11	18 15	20 4
119	23 11	37 14	39 1	40 8	44 2	47 1		
120	1 1	3 8	8 13	9 18	12 15	20 25	31 3	37 21
120	40 25	44 7	47 7					
121	2 13	8 20	9 7	12 25	20 25	37 23	40 24	47 13
122	3 3	6 25	8 1	9 2	12 10	15 1	18 6	20 25
122	37 17	40 19	47 3					
123	2 7	8 6	12 22	15 7	18 13	20 11	37 6	40 25
123	44 7	47 5						
124	3 1	6 5	7 1	8 25	12 25	18 3	19 25	37 2
124	40 12	47 4						
125	1 8	3 16	8 19	12 17	15 4	18 11	19 24	37 3
125	39 3	40 13	42 4	47 4				
126	1 12	2 2	3 11	8 12	9 14	12 14	18 18	20 10
126	37 9	40 11	44 1	47 9				
127	1 2	3 2	8 23	9 19	12 25	20 25	37 11	40 20
127	47 15							
128	1 6	8 17	9 11	12 18	20 25	37 4	40 19	47 4
129	1 1	2 16	3 9	8 20	9 2	12 13	20 25	37 3
129	39 4	40 19	44 6	47 3				
130	1 10	3 4	8 10	12 21	20 25	40 15	44 11	47 10
131	1 1	2 9	3 15	8 21	12 16	20 25	37 1	40 19
131	44 1	47 6						
132	1 1	2 8	3 2	5 24	8 23	12 10	20 25	37 2
132	40 13	47 13						
133	2 6	3 11	8 24	12 14	15 1	18 25	20 25	40 25
133	47 5							
134	3 2	6 16	8 25	12 12	15 8	18 25	20 25	39 8
134	40 23	47 2						
135	2 15	3 3	8 10	9 5	12 8	18 4	20 25	37 3
135	39 23	47 2						
136	3 12	8 13	9 8	18 25	20 25	37 6	39 23	47 2
137	2 12	5 25	8 5	18 15	20 17	21 6	37 1	39 10
138	2 4	3 4	5 25	8 2	12 2	15 2	18 17	20 14
138	37 1	39 21						
139	2 20	3 5	5 14	8 2	12 2	15 2	20 18	21 4
139	23 2	39 17						
140	2 19	8 8	18 2	20 3	30 2	39 7		
141	2 21	8 5	12 1	19 3	21 25	29 4	30 14	39 17
141	47 2							
142	2 25	5 3	9 1	21 4	29 3	30 5	31 6	34 16
143	2 19	19 6	21 24	29 14	30 5	34 2	39 3	
144	2 2	19 15	30 4					

continued

```
145      2 21   13  6   15  1   18 25   20 24   21 12   30  7   34 25
145     39  3
151      1 11   11  7   27  4   45 21   46  4
152      1  2   11 25   27 24   45 19   46  6
153      1  8   11 25   27 23   45 21   46  6
154     11  1   27  5   45 10   46  1
155      1 19   11 22   27 18   45 22   46  4
156     11  6   27  5   45 25   46  5
157     11  2   27  6   45 25
158     27  5   45 25
159     11 11   27 17   45 25
160      1  7   11 25   27 16   45 25   46  8
161      1  4   11 25   27 22   45 23   46  1
162      1 10   11 25   27 18   45 19   46  4
163     11 25   27 18   45 24   46  5
164     11 15   27  8   45 22   46  2
165      1 11   11 25   27  4   45 23   46  3
166      1 23   11 25   27  3   45 24   46  6
167      1 10   11  9   27  3   45 15   46  4
168     11  4   27  2   45 22   46  2
169      1 12   11 25   27 24   45 22   46  2
170      1  4   11 25   27 16   45 25   46 11
171      1 11   10  7   11  1   27 11   45 25   46 18
172      1  4    4 25   27  7   45 23   46 12
173      1  2    4 23   11  1   27 10   45 25   46  5
174      1  4    4  8   27  4   45 22   46 18
175      1  3    4 21   11  2   27 12   45 24   46 19
176      1  2   11 19   27  9   45 25   46 24
177     27  4   45 25   46 10
178      4  2   27 13   45 15   46 21
179      1  3   11 19   27 24   45  6   46 19
180      1 15   11 25   27 19   45 24   46 24
  0
CARESUBSCAREAQUACAREDIANCAREGLARCARERARICARESALIERIOANGUERIORUSSARCTFULVDUPOFISC
PUCCPHRYCALANEGLMNIUPUNCMNIUCINCBRYUPSEUMEESTRIQCAMPSTELSCORSCORAULAPALUCALLGIGA
PALUSQUADREPUNCIDREPFLUIDREPREVODREPADUNDREPEXANTRIQPALUSALICALCSALICORDSALICAND
RANUPALARANUGMELRANUHYPEPOTEPALUUTRIINTECHRYARCTSTELCRASSTELHUMISAXAHIRCCARDPRAT
POTEEGEDATRIGLABFESTRUBRPEDI  SPHIPPVULGSENECONGEPILPALUCICUVIROGALI  SP
QUAD   1QUAD   2QUAD   3QUAD   4QUAD   5QUAD   6QUAD   7QUAD   8QUAD   9QUAD  10
QUAD  11QUAD  12QUAD  13QUAD  14QUAD  15QUAD  16QUAD  17QUAD  18QUAD  19QUAD  20
QUAD  21QUAD  22QUAD  23QUAD  24QUAD  25QUAD  26QUAD  27QUAD  28QUAD  29QUAD  30
QUAD  31QUAD  32QUAD  33QUAD  34QUAD  35QUAD  36QUAD  37QUAD  38QUAD  39QUAD  40
QUAD  41QUAD  42QUAD  43QUAD  44QUAD  45QUAD  46QUAD  47QUAD  48QUAD  49QUAD  50
QUAD  51QUAD  52QUAD  53QUAD  54QUAD  55QUAD  56QUAD  57QUAD  58QUAD  59QUAD  60
QUAD  61QUAD  62QUAD  63QUAD  64QUAD  65QUAD  66QUAD  67QUAD  68QUAD  69QUAD  70
QUAD  71QUAD  72QUAD  73QUAD  74QUAD  75QUAD  76QUAD  77QUAD  78QUAD  79QUAD  80
QUAD  81QUAD  82QUAD  83QUAD  84QUAD  85QUAD  86QUAD  87QUAD  88QUAD  89QUAD  90
QUAD  91QUAD  92QUAD  93QUAD  94QUAD  95QUAD  96QUAD  97QUAD  98QUAD  99QUAD 100
QUAD 101QUAD 102QUAD 103QUAD 104QUAD 105QUAD 106QUAD 107QUAD 108QUAD 109QUAD 110
QUAD 111QUAD 112QUAD 113QUAD 114QUAD 115QUAD 116QUAD 117QUAD 118QUAD 119QUAD 120
QUAD 121QUAD 122QUAD 123QUAD 124QUAD 125QUAD 126QUAD 127QUAD 128QUAD 129QUAD 130
QUAD 131QUAD 132QUAD 133QUAD 134QUAD 135QUAD 136QUAD 137QUAD 138QUAD 139QUAD 140
QUAD 141QUAD 142QUAD 143QUAD 144QUAD 145QUAD 146QUAD 147QUAD 148QUAD 149QUAD 150
QUAD 151QUAD 152QUAD 153QUAD 154QUAD 155QUAD 156QUAD 157QUAD 158QUAD 159QUAD 160
QUAD 161QUAD 162QUAD 163QUAD 164QUAD 165QUAD 166QUAD 167QUAD 168QUAD 169QUAD 170
QUAD 171QUAD 172QUAD 173QUAD 174QUAD 175QUAD 176QUAD 177QUAD 178QUAD 179QUAD 180
```

Appendix 7.2 Alphabetical list of the species names in the salt marsh data set and their reference (order) number used here.

9 *Arctophila fulva*	42 *Atriplex glabriuscula*	19 *Aulacomnium palustre*
15 *Bryum pseudotriquetrum*	12 *Calamagrostis neglecta*	20 *Calliergon giganteum*
17 *Campylium stellatum*	40 *Cardamine pratensis*	2 *Carex aquatilis*
3 *C. diandra*	4 *C. glareosa*	5 *C. rariflora*
6 *C. salina*	1 *C. subspathacea*	48 *Cicuta virosa*
25 *Drepanocladus aduncus*	26 *D. exannulatus*	23 *D. fluitans*
24 *D. revolvens*	22 *D. uncinatus*	10 *Dupontia fisheri*
47 *Epilobium palustre*	7 *Eriophorum angustifolium*	8 *E. russeolum*
43 *Festuca rubra*	49 *Galium* sp.	45 *Hippurus vulgaris*
14 *Mnium cinclidioides*	13 *M. punctatum*	21 *Paludella squarrosa*
44 *Pedicularis* sp.	41 *Potentilla egedii*	34 *P. palustris*
11 *Puccinellia phryganodes*	32 *Ranunculus gmelini*	33 *R. hyperboreus*
31 *R. palasii*	30 *Salix candida*	29 *S. cordifolia*
39 *Saxifraga hirculus*	18 *Scorpidium scorpioides*	46 *Senecio congestus*
37 *Stellaria crassipes*	38 *S. humifusa*	27 *Triglochin palustre*
35 *Utricularia intermedia*		

Appendix 8

Appendix 8.1 The ridge data set is listed by transect/plot sequence: AI, LF (lower front); AIUF (upper front); AICR (crest) etc., and the full sequence is listed at the end of the data string, together with the abbreviated species names. A full alphabetical list of species together with their code numbers is given below in Appendix 8.2. The data is entered as a species number followed by the mean percentage cover of that species in a particular plot and ridge location: FORMAT (I4, 2x, 8 (I3, F5.2)). N.B. 5 = AILB lower back completely devoid of vegetation.

```
1     4 21.6 13  1.3
2     4 15.4 13  8.6 30   .3
3     4  3.1  7  2.3 11   .1 13 27.7 16  1.4 22   .3 30  4.7 58  1.8
4     5   .2  7  1.3 11   .5 13 20.0 30  4.0 58  2.4
6     3  9.3  6  1.3  7  1.0  8  2.2 10   .5 11   .5 12   .1 14  3.0
6    16  1.9 18   .3 19  2.4 21   .2 22  1.4 23  3.6 26  5.5 27 12.9
6    30   .4 32   .4 34  1.5 37  1.5 38  3.6 39  8.6 40   .7 41  6.3
6    42  1.0 45   .7 46   .1 47  2.4 48 16.9 50 18.0 51   .2 54 17.0
6    55  4.2 56   .9 57  3.1
7     1  1.6  5   .4  6  1.3 13  2.9 15   .1 16  4.2 18  3.1 19   .4
7    22  8.1 23   .1 26  8.9 28 13.2 30   .1 34   .8 35   .3 36  1.0
7    37  3.0 38 42.4 39  3.3 41  1.4 42  7.2 43   .4 46   .1 48  1.0
7    49   .1 50  1.3 51  2.6 52   .6 53   .5 55  4.1 56   .1 57  1.5
7    58 14.0
8     1  1.9  5  2.6  6  1.3 13   .6 16  1.7 17  4.0 18  8.4 22   .1
8    26  1.3 28 56.2 30  1.3 34   .1 36   .1 42  2.8 50   .8 51  1.4
8    56   .6 58 32.1
9     1  1.9  3   .4  5   .5  6  1.7 11   .1 13  4.0 16  4.2 18  8.8
9    22  1.4 26 21.4 28  6.6 29   .5 30   .4 34  1.1 36  3.0 37  2.3
9    38  8.6 39  1.3 40   .4 41   .8 42  2.6 48   .1 50  2.0 51   .6
9    55  1.8 57   .3 58 29.6
10    3  1.2  6   .3 12  2.7 13  3.1 16  1.1 17   .1 18 13.9 19   .7
10   22   .5 26 20.5 27  5.4 28  2.6 30   .1 37  2.8 38 50.4 39  2.1
10   41  2.8 42  6.1 45   .1 47   .5 48  2.8 49   .4 50   .6 51  1.1
10   52   .1 53   .6 54  9.4 55   .5 56   .5 58  4.8
11    3  7.0  7  2.3  8  1.3  9   .2 12   .8 14  7.2 19  2.3 21   .3
11   23   .1 24  2.0 26   .8 27 20.4 32  6.7 33  5.3 34   .4 35   .1
11   37 12.3 38 31.8 39  4.2 41  1.1 42   .2 44   .9 45   .7 46   .5
11   47   .3 48  4.7 49   .7 51   .9 52   .4 53   .3 54  3.0 56   .7
11   57 17.9
12    3  9.5  6   .5  7   .2 12  5.6 18   .1 19   .5 23   .5 24   .5
12   26  7.9 27  4.8 32   .4 34   .1 36  5.2 37  7.1 38 49.6 39 29.6
12   40   .2 41  3.5 42  2.6 43   .3 44  1.2 45   .5 47  1.2 48   .1
12   53   .4 57   .2
```

continued

13	1	.1	3	5.4	6	1.2	8	.2	12	.7	13	.3	18	6.2	22	.1
13	26	8.4	27	3.4	28	6.2	29	4.9	35	1.1	36	6.8	37	6.1	38	36.5
13	39	28.4	41	.6	42	4.8	43	1.1	44	.7	45	.1	46	.1	56	.3
14	1	.1	3	8.6	6	.3	7	.6	8	.2	10	.1	12	4.5	13	.6
14	16	.3	18	1.0	21	.2	26	4.5	27	8.4	28	.9	29	3.1	30	.3
14	32	.1	35	.6	36	4.4	37	4.2	38	51.5	39	42.2	41	6.3	42	1.4
14	43	1.7	44	.1	45	.7	46	.1	47	.3	48	.7	51	.3	54	2.4
14	56	.5														
15	3	12.6	7	2.2	8	.8	9	.3	10	.5	12	2.8	14	3.5	16	.2
15	19	2.1	21	.2	23	.1	24	1.7	26	.4	27	8.6	28	.1	32	4.3
15	33	.9	34	.1	36	.2	38	39.0	39	25.5	40	.1	41	.4	42	1.2
15	43	.2	44	1.0	45	5.7	46	.3	47	6.6	48	.6	54	11.7	56	8.7
15	57	3.0														
16	3	.3	7	4.5	8	.2	9	.3	10	4.9	12	.8	14	.9	19	.5
16	21	.1	23	.3	24	.1	26	1.5	27	28.2	32	3.7	33	6.4	34	.2
16	37	8.6	38	23.8	39	.5	41	.5	44	.3	45	8.3	46	.1	48	1.4
16	49	.1	51	.8	54	18.6	56	6.3	57	40.3						
17	3	14.3	6	.3	7	.2	10	.2	12	.6	13	.2	14	4.5	19	4.5
17	21	.1	23	.3	26	3.0	27	7.6	28	.1	32	.4	33	.6	35	1.4
17	36	3.3	37	17.9	38	36.7	39	50.2	41	2.0	43	.5	44	1.7	45	.2
17	47	2.7	48	2.0	51	.5	54	4.6	56	.3	57	.5				
18	1	.1	3	1.7	6	.1	8	.3	10	.2	13	.1	14	5.5	15	.7
18	16	.1	18	.4	19	.3	26	4.6	27	.9	28	3.6	30	.1	34	.4
18	35	24.4	36	16.7	37	9.4	38	31.1	39	31.3	41	2.7	42	.1	43	15.8
18	44	.4	45	.1	51	1.1	52	.2	53	.6						
19	3	7.4	4	1.2	7	.3	8	.2	12	3.3	13	.7	14	8.3	23	.2
19	24	1.5	25	.1	26	3.4	27	4.7	32	.9	33	2.0	35	.5	36	.7
19	37	9.2	38	38.1	39	47.5	41	2.5	43	15.8	44	1.4	45	1.2	46	.2
19	47	.5	54	11.1	56	.6										
20	1	.1	3	3.0	7	1.6	8	6.1	9	3.1	10	1.6	12	.9	14	14.1
20	16	.1	21	.9	23	.9	24	1.4	26	3.9	27	24.4	30	.2	32	1.8
20	33	.7	34	.6	36	1.3	37	6.9	38	9.6	39	2.6	40	.1	41	2.7
20	44	1.7	45	4.0	46	.2	47	.3	48	.2	50	.7	51	2.2	54	21.8
20	56	1.6	57	5.1												
21	3	3.1	7	.7	8	.7	9	7.2	10	.7	12	3.5	14	5.6	19	.6
21	21	1.5	23	.7	24	.9	25	.1	26	1.0	27	19.8	32	1.1	33	1.8
21	34	.1	35	1.9	36	1.4	37	10.2	38	18.2	39	30.8	40	.2	41	2.1
21	42	.1	43	1.6	44	.2	45	.2	46	.6	47	2.2	48	1.0	52	3.1
21	53	1.5	54	13.0	56	1.5	57	31.3								
22	3	5.6	6	.1	7	.2	8	.1	9	.9	10	.1	12	13.2	14	2.8
22	18	.3	19	3.1	20	.1	21	.4	22	.3	24	.6	26	1.6	27	3.0
22	28	.3	29	3.1	32	.1	34	.1	35	12.0	36	5.1	37	10.0	38	26.6
22	39	39.0	41	2.8	42	.1	43	21.1	44	.7	45	.1	47	.2	48	.3
22	49	.4	50	.6	52	2.1	53	.1	54	2.8	56	.1				
23	3	2.0	6	3.4	12	26.5	17	.2	18	1.2	26	1.5	27	2.0	29	1.2
23	35	50.7	36	5.2	37	1.1	38	6.8	39	8.2	40	1.9	41	.6	42	2.0
23	43	21.1	44	4.0	45	.5	46	.3	48	.3	52	.6	53	.7		
24	3	2.0	5	.3	6	.9	7	.1	10	.1	12	19.6	14	5.5	18	1.1
24	23	.2	24	.9	26	.8	27	1.4	28	.3	29	1.0	32	.5	35	44.2
24	36	9.0	37	1.6	38	7.9	39	17.8	40	.4	41	.6	42	.1	43	37.5
24	44	1.2	45	.3	47	.1	52	.6	53	.4	54	2.5				
25	3	3.7	7	4.2	8	.3	9	3.4	10	.7	14	3.0	19	.3	21	.7
25	23	.6	24	.2	26	2.2	27	15.6	30	.2	32	3.0	33	5.2	34	.1
25	35	6.2	36	1.8	37	12.6	38	10.1	39	15.5	41	.7	43	.4	44	.9
25	46	.4	47	5.0	48	3.4	52	.1	53	.8	54	10.5	56	.1	57	45.2
26	1	.1	4	.8	13	19.1										
27	1	8.6	4	1.6	5	.6	6	5.2	11	.1	13	12.4	16	7.6	18	4.7
27	20	.8	22	17.6	25	.6	29	1.8	30	.4	31	.1	34	8.9	39	.1
27	46	.4	50	16.9	57	.5	58	46.6								
28	1	6.2	4	.3	5	3.0	6	1.8	8	.2	11	.5	13	6.4	15	11.2
28	17	.2	18	3.0	20	5.3	22	2.0	25	3.1	28	18.4	29	3.6	30	2.8
28	34	.2	37	.6	38	.1	42	.3	46	.3	49	.4	50	2.0	51	.6
28	57	.8	58	61.4												
29	5	5.9	6	.2	8	1.1	11	.9	13	1.9	16	2.1	18	1.8	20	4.3
29	28	10.2	30	3.7	49	.3	51	1.0	54	1.9	58	26.0				
30	1	.9	3	2.2	5	.5	6	2.4	7	.1	8	.3	13	3.4	15	.4
30	16	1.6	18	11.7	22	1.0	25	12.1	26	6.8	27	4.8	28	.4	30	.1
30	32	.4	38	7.6	39	4.1	40	2.0	41	.4	42	9.9	46	.2	48	2.6
30	49	8.8	50	.9	51	2.7	54	11.9	57	.3	58	.6				
31	1	6.0	6	.1	7	.1	8	.1	9	.3	11	.1	13	10.3	15	2.9
31	19	1.6	22	5.4	26	17.2	30	2.0	31	.2	34	3.4	49	2.8		

continued

32	1	1.4	2	.1	6	.4	11	.4	13	.8	15	.3	17	.9	19	.4
32	22	6.1	25	.3	30	.8	34	1.0								
33	1	3.0	5	3.2	6	.5	11	.4	13	11.3	14	.2	16	5.5	17	6.5
33	18	3.3	20	2.0	22	4.4	28	5.6	30	2.8	31	.3	34	2.0	36	.2
33	37	1.7	38	1.0	39	.4	42	.8	50	.1	53	6.6	55	10.9	57	.4
34	1	.1	5	3.8	6	6.0	11	.4	13	3.5	16	2.4	17	5.0	18	16.5
34	20	3.9	22	1.5	26	3.1	30	2.8	34	.1	37	.8	38	.5	39	.2
34	50	1.1	53	3.0	54	.2	55	.6								
35	1	1.9	3	5.5	7	.1	8	.2	12	1.2	13	.4	15	.4	18	1.3
35	19	1.8	20	.1	22	3.3	23	.1	25	9.1	26	10.3	27	12.9	30	1.0
35	32	.4	33	.1	34	.2	36	1.3	37	9.6	38	4.4	39	2.1	40	.7
35	42	5.7	44	.2	45	.2	46	.1	48	3.1	49	4.7	50	.3	53	1.6
35	54	17.5	55	.9	57	1.9										
36	1	1.2	6	.7	7	.3	11	.7	13	11.5	18	.2	22	7.4	30	1.0
36	31	.4	34	.2	38	.5	45	.1	49	12.2	50	.7	56	.5		
37	1	1.0	4	2.6	5	1.3	6	3.3	8	1.7	11	.1	13	4.6	15	.5
37	16	4.7	18	4.8	20	2.1	22	2.3	26	.2	28	1.1	29	3.5	30	.6
37	31	.2	34	.2	36	.1	37	2.6	38	13.2	39	.3	40	.3	42	.8
37	49	3.3	55	2.4												
38	1	3.3	2	.4	4	.9	5	.3	6	6.5	8	.2	13	13.5	16	10.0
38	17	.6	18	9.4	20	.1	22	2.8	26	14.9	28	5.3	30	.4	31	.5
38	34	2.8	37	.4	38	2.4	42	.2	49	4.2	50	6.3	53	.3	55	5.6
38	57	.2														
39	1	.2	2	.3	3	4.1	6	.4	7	.1	12	1.0	13	.4	15	4.4
39	16	3.8	17	.4	18	1.9	19	.1	22	3.0	25	5.8	26	7.1	27	2.0
39	29	22.1	30	.3	31	.6	34	4.1	36	.7	37	12.5	38	14.7	39	.9
39	40	.6	42	4.4	44	.4	45	.3	48	.7	49	.1	53	.6	55	.1
40	3	15.2	7	1.2	8	.3	9	.1	12	.8	14	7.4	16	.1	18	.1
40	21	.1	23	.2	24	3.8	26	.1	27	5.2	29	2.7	32	3.9	33	3.4
40	34	.2	36	.1	37	10.9	38	48.9	39	6.0	40	.1	41	3.5	43	.2
40	44	3.4	45	1.0	48	2.5	49	3.5	53	.2	54	15.4	56	2.0	57	.6
41	1	.2	6	.1	7	.2	8	.5	13	4.1	16	.6	19	2.9	22	.5
41	25	3.3	26	47.7	27	2.3	30	.7	32	.1	46	.2	48	1.8	49	24.8
42	1	1.6	2	2.0	5	.2	6	1.4	8	.2	13	3.3	15	1.1	16	1.2
42	18	5.7	20	.2	22	1.9	25	2.1	28	9.4	29	8.1	30	1.7	31	.1
42	38	37.0	39	.8	42	.4	43	.1	44	.1	51	1.3				
43	1	1.8	2	5.4	4	.2	5	.1	6	1.8	13	9.3	16	22.2	17	4.3
43	18	3.3	20	1.0	22	5.1	30	.9	31	3.9	34	.3	36	1.1	38	2.0
43	39	.1	40	3.6	41	.3	42	3.9	46	1.0	48	3.1	51	1.8	53	.1
43	55	6.1	57	.8												
44	1	.1	2	.3	3	11.8	6	.3	8	1.6	11	.1	12	6.9	13	.8
44	16	3.5	17	.3	18	2.9	20	.2	21	.1	22	2.8	25	3.4	26	.2
44	27	5.7	29	5.9	30	.3	31	.1	37	1.9	38	60.1	39	1.0	40	2.0
44	41	.8	42	1.0	44	.3	45	1.1	48	2.2	50	.5	51	.4	57	.2
45	3	4.6	7	.8	8	1.7	9	9.2	10	.1	12	1.7	14	3.9	16	.5
45	23	1.7	24	2.4	25	.7	26	4.3	27	7.5	29	.2	32	1.6	33	4.6
45	36	1.9	37	1.0	38	27.0	39	25.8	40	.2	41	.1	43	.5	44	1.8
45	45	2.4	46	.1	48	2.3	49	.6	53	1.5	54	3.0	57	.8		
46	1	1.4	2	.3	6	1.7	7	.6	8	.1	11	.8	13	12.4	14	.4
46	15	.3	16	.6	22	9.2	25	1.0	26	2.4	30	1.4	31	.8	34	.1
46	37	.5	38	.5	40	.9	42	.6	45	.1	46	.2	49	.9	50	1.5
47	1	1.8	2	4.0	4	.5	5	.4	6	1.0	8	.3	11	.2	13	5.7
47	15	1.1	16	12.4	18	6.2	22	5.1	28	.8	29	7.1	30	.8	31	5.0
47	34	2.5	36	1.2	37	8.0	38	52.4	39	.5	42	1.7	50	.9	53	1.1
48	1	3.2	2	1.2	5	.2	6	2.3	11	.1	13	4.3	15	.8	16	22.2
48	17	1.5	18	1.3	20	.2	22	3.7	25	3.5	30	.4	31	1.6	34	.7
48	36	.7	37	4.1	38	16.4	39	.7	41	.7	42	12.8	44	1.1	46	.3
48	50	.3	53	.3	57	.4										
49	1	.3	3	2.1	6	.7	8	2.9	11	.1	12	5.9	14	.1	15	.4
49	16	2.2	17	.5	18	4.1	22	1.0	25	1.4	26	11.7	27	1.4	29	1.3
49	34	.9	36	.6	37	7.4	38	60.8	39	3.2	40	.3	41	.2	42	2.1
49	44	1.2	45	.5	50	.4	51	.3	54	.8						
50	1	.3	3	10.8	7	1.0	8	1.2	9	2.1	10	1.7	12	1.5	14	6.1
50	23	.9	24	2.0	25	1.4	26	2.0	27	14.7	28	.1	32	2.7	33	5.4
50	35	.2	36	.4	37	.3	38	25.2	39	24.0	40	.2	41	1.3	43	.1
50	44	1.2	49	.8	50	.3	51	3.8	53	.6	54	15.3	56	2.9	57	2.6
51	1	1.7	2	.2	6	4.1	7	.3	13	10.1	15	.8	16	.2	22	1.3
51	23	.6	25	8.2	27	6.9	30	.1	32	.1	34	.5	38	.1	40	2.5

continued

51	42	3.1	46	5.8	48	13.8	49	23.1	50	7.4	51	.7	52	.1	53	1.1
51	56	1.1														
52	1	2.0	2	8.3	4	.4	13	4.6	15	1.1	16	20.0	17	1.5	18	18.1
52	22	1.8	25	2.3	28	.2	30	.6	31	.7	34	.4	36	4.2	37	1.5
52	38	49.7	39	7.5	41	.5	42	.7	46	.2	48	.9	50	1.0	51	.1
52	55	.5	56	.6												
53	1	2.4	2	2.4	3	.1	6	1.5	8	3.1	11	.2	13	2.2	15	1.4
53	16	6.4	17	1.2	18	8.9	20	.4	22	.3	25	2.9	26	8.7	28	9.4
53	30	.7	31	.1	34	.9	36	14.5	37	11.7	38	29.0	39	6.4	41	.6
53	42	2.7	43	.1	44	.5	47	.1	50	.5	53	.1	56	.2		
54	1	.5	2	.6	3	3.8	6	1.2	8	.8	12	2.5	13	.3	16	2.9
54	17	.1	18	.1	22	.2	25	1.6	26	14.1	27	2.8	28	2.5	30	.1
54	31	.4	34	.2	35	.1	36	9.0	37	10.7	38	37.0	39	20.8	41	5.0
54	42	1.1	43	.4	44	.2	48	.6	53	.4	54	.3				
55	1	.1	3	6.4	7	.6	8	.7	12	.5	14	7.5	16	.2	21	.1
55	23	.2	24	.3	25	.8	26	10.8	27	7.0	30	.1	32	.7	33	.3
55	34	.1	36	2.5	37	1.2	38	27.9	39	30.2	40	.5	41	.9	43	.7
55	47	.2	48	4.5	49	.5	50	.3	51	.7	52	.2	53	.2	54	8.5
55	56	1.7														
56	1	.1	3	.2	6	.1	8	.1	9	7.9	19	.5	23	.4	26	3.3
56	27	49.1	29	3.9	32	.6	37	.3	38	2.9	39	.2	45	.3	46	.2
56	47	7.5	48	32.0	49	9.6	54	3.6	57	.7						
57	1	.1	5	.2	6	1.4	8	.1	13	.2	15	4.2	16	2.3	18	2.4
57	22	.2	25	3.9	26	9.3	27	.8	28	9.2	29	2.9	30	.1	36	13.8
57	37	14.2	38	30.2	39	21.8	40	1.9	41	1.6	43	.2	47	3.0	50	.2
57	53	.1														
58	1	.2	3	11.8	6	3.0	8	.3	12	2.6	13	1.8	15	3.9	16	.4
58	18	6.8	20	.1	23	.3	25	6.4	27	.2	28	5.7	29	3.1	30	.1
58	35	3.6	36	20.0	37	11.6	38	14.2	39	5.7	41	1.6	42	1.0	43	1.0
58	44	.4	48	1.5	50	.2	51	.2	52	.7	53	1.2	55	1.2		
59	1	.7	3	4.4	5	1.0	6	.8	7	.7	10	.1	11	.2	12	7.5
59	15	.9	16	2.1	18	4.7	20	.1	21	.3	22	1.3	25	22.8	27	5.9
59	28	.8	32	2.4	33	.1	35	2.2	36	17.6	37	6.9	38	75.6	39	53.7
59	40	.2	41	2.9	42	.2	43	6.5	44	1.6	45	1.2	46	.8	48	8.4
59	49	.8	51	1.5	53	1.7	54	1.3	55	.2						
60	3	6.6	7	.1	8	.7	9	.6	10	1.1	12	.9	14	3.9	23	.3
60	24	2.8	25	1.0	26	1.9	27	12.7	28	.1	32	5.3	33	2.8	34	.2
60	35	.2	36	1.3	37	1.6	38	10.4	39	19.4	40	.3	43	1.0	44	.5
60	45	3.1	48	3.0	49	8.9	51	5.1	52	7.1	54	38.0	56	3.7	57	18.1
61	1	.3	7	.2	8	.2	9	3.9	10	.6	23	.4	27	38.0	29	4.0
61	30	.4	32	2.4	33	10.8	34	.6	38	1.5	39	.2	40	.4	44	.3
61	45	.9	46	.5	47	5.4	48	44.0	49	8.5	50	1.5	51	.3	53	.3
61	54	1.6	56	3.9	57	15.6										
62	3	3.2	6	1.5	8	.5	10	.4	12	5.4	13	.4	15	3.0	16	.3
62	18	3.0	20	.1	25	6.1	27	12.4	29	5.6	34	.2	35	.7	36	8.0
62	37	1.0	38	54.1	39	27.9	40	3.2	41	.7	42	.1	43	.1	44	.2
62	45	.6	47	4.6	56	.3										
63	1	.8	5	.1	6	.8	11	.1	12	3.3	13	2.0	15	.2	16	.2
63	18	3.9	20	.6	25	4.3	26	.3	28	5.0	29	.8	34	1.0	35	15.3
63	36	13.2	38	33.0	39	21.0	40	.4	41	.4	42	4.9	43	6.2	44	.5
63	45	.4	51	.2	53	1.1	54	.3	55	3.1	56	.5	57	.3		
64	3	4.6	6	.3	8	.1	12	16.0	14	1.8	18	.1	24	2.0	25	.1
64	27	4.3	32	.2	33	4.3	35	18.4	36	12.5	38	14.9	39	19.0	40	.2
64	41	.1	42	1.2	43	35.6	44	.9	46	.2	48	.6	49	.3	51	2.3
64	52	.6	53	.4	54	.2	56	.3								
65	3	.1	8	.1	10	.5	12	1.0	14	6.4	21	.1	24	3.3	27	3.6
65	32	2.9	33	10.7	35	3.5	36	4.2	38	10.6	39	33.0	40	.6	41	.9
65	42	.1	43	36.4	44	.1	45	.5	49	4.8	51	15.2	52	.2	53	2.9
65	54	15.9	56	3.1	57	1.1										
66	3	.3	7	.6	9	6.0	17	.3	26	.7	27	41.3	32	2.3	33	6.1
66	38	.4	39	.2	45	2.0	46	.7	47	2.9	48	33.2	49	1.5	50	.1
66	51	.3	54	3.3	56	.3	57	22.9								
67	1	.5	3	4.6	6	2.5	8	.6	9	1.0	12	12.9	13	.1	14	.1
67	15	.4	18	3.2	20	.3	25	7.5	27	.9	29	12.7	36	4.5	37	6.5
67	38	40.3	39	22.2	40	3.3	41	.2	44	.9	45	3.0	47	.3	48	.2
67	49	1.8	51	.2	53	.1	56	1.2								
68	1	.3	3	.7	6	1.3	8	.1	12	26.3	16	.3	17	1.6	18	.3
68	25	4.8	27	.2	29	1.9	35	20.3	36	7.7	37	.5	38	25.6	39	14.8

continued

```
68    40    .1 41    .3 42    .5 43 12.8 44    .3 45    .2 48    .2 49    .3
68    53    .5
69     3   2.0  8    .2 12 15.0 14  1.8 18    .2 21    .3 24   3.1 25  1.1
69    27   3.4 32   1.0 33    .8 35 21.9 36  9.3 38 10.9 39 17.3 40    .3
69    41    .1 42    .9 43 39.4 44  1.1 45    .2 47   1.3 49   1.5 51  9.6
69    52    .3 53    .6 54    .3
70     3    .4  7    .1  8    .1 12  1.2 14  3.6 21    .2 24   6.1 27  2.9
70    28    .1 32   3.1 33 12.0 35 12.4 36  6.5 37    .3 38 27.2 39 49.3
70    40   1.1 41    .1 42    .8 43 34.1 44    .4 45   1.8 48    .3 49  3.7
70    51   2.1 52    .6 53    .1 54  5.3 57    .6
71     7    .5  8   1.9  9   3.1 22    .2 24    .1 26   1.4 27 48.0 32  1.3
71    33    .7 35    .2 36    .1 37  2.7 38  1.3 39   2.5 42    .1 45  2.1
71    46    .2 47   2.6 48 17.3 49  4.2 52    .4 54   4.1 55   2.8 56 12.8
71    57  27.8
72     3    .8  6    .3  7    .2  8  1.0  9    .7 12 18.3 16    .1 18  1.3
72    20    .1 21    .2 24    .3 26  1.6 27 11.7 32    .9 33   1.6 35  2.5
72    36   8.4 37    .3 38 48.1 39 26.2 41  1.0 42    .3 43    .6 45    .1
72    47  10.0 48   1.2 49    .2 51  1.0 54  8.4 55    .6 57   3.5
73     3   2.1  6    .4 12 23.0 16    .1 18  2.4 20    .1 26   2.3 29    .8
73    35  21.3 36  17.4 37    .3 38 14.3 39  7.6 40   1.0 41    .4 42    .2
73    43   9.3 44    .5 47    .8 49    .9 55    .2 56    .1
74     3   2.4  6    .1 10    .3 12 15.7 14  2.0 20    .1 21    .3 24  5.4
74    26    .6 27   3.7 29    .1 32  2.9 33  3.3 35 15.0 36 11.9 38 18.6
74    39  19.7 40   1.0 41    .3 43 25.8 44  1.5 45   2.0 47    .8 48  4.5
74    49   1.2 50    .1 51   2.9 52  2.1 54  1.0 56    .5 57 11.6
75     3   1.6  8    .4 10   2.0 12  1.5 14  6.6 21    .4 24 12.1 26    .1
75    27   9.7 29    .1 32   3.7 33 18.9 35  1.5 36  1.4 38   5.6 39  5.2
75    40    .5 41    .3 43   1.1 44  1.0 45 10.8 48   1.5 49    .3 51  9.1
75    52    .4 53   3.0 54 12.7 57  6.6
 0
ACHIMILLANEMPARVARCTRUBRAREN  SPARTEBOREASTRALPIBARTALPICARE  SPCAREAQUACARECAPI
DRAB  SPDRYAINTEELYMARENEMPENIGREPILANGUFESTBRACFESTRUBRHEDYMACKJUNC  SPOXYTHUDS
PEDI  SPPOA ALPIPOLYVIVIRHODLAPPSALICALLSALICORDSALIRETISAXITRICSHEP  SPSTELCRAS
TARA  SPTOFI  SPVACCULIGPOA GLAUALECOCHRCETRCUCUCETRI-E CETRI-P CETRNIVACLAD  SP
CLADGRACCLADPYXICORNDIVEOCHRUPSAPANA  SPPELTCANISTERALPIAULAPALUBRYU  SPCIRR  SP
DIC1  SPDIC2  SPDITRFLEXEQUIVARIHYLO  SPMEESTRIQTOME  SPTORTRURA
A1    LF A1    UF A1    CR A1    UB A1    LB A2    LF A2    UF A2    CR A2    UB A2    LB
A3    LF A3    UF A3    CR A3    UB A3    LB A4    LF A4    UF A4    CR A4    UB A4    LB
A5    LF A5    UF A5    CR A5    UB A5    LB B     LF B     UF B     CR B     UB B     LB
C     LF C     UF C     CR C     UB C     LB D     LF D     UF D     CR D     UB D     LB
E     LF E     UF E     CR E     UB E     LB F     LF F     UF F     CR F     UB F     LB
G     LF G     UF G     CR G     UB G     LB H     LF H     UF H     CR H     UB H     LB
I     LF I     UF I     CR I     UB I     LB J     LF J     UF J     CR J     UB J     LB
K     LF K     UF K     CR K     UB K     LB
```

Appendix 8.2 Alphabetical list of the species names in the ridge data set and their reference (order) number used here.

1 *Achillea millefolium*	35 *Alectoria ochroleuca*	2 *Anemone parviflora*
3 *Arctostaphylos rubra*	4 *Arenaria* sp.	5 *Artemisia borealis*
6 *Astragalus alpinus*	48 *Aulacomnium palustre*	7 *Bartsia alpina*
49 *Bryum* sp.	9 *Carex aquatilis*	10 *C. capillaris*
8 *Carex* sp.	36 *Cetraria cucullata*	37 *C. islandica* (erect) (Cetr I–E)
38 *C. islandica* (prostrate)(CetrI–P)	39 *C. nivalis*	50 *Cirriphyllum* sp.
41 *Cladonia gracilis*	42 *C. pyxidata*	40 *Cladonia* sp.
43 *Cornicularia divergens*	51 *Dicranella* sp. (Dic 1 sp.)	52 *Dicranum* sp. (Dic 2 sp.)
53 *Ditrichum flexicaule*	11 *Draba* sp.	12 *Dryas integrifolia*
13 *Elymus arenarius*	14 *Empetrum nigrum*	15 *Epilobium angustifolium*
54 *Equisetum variegatum*	16 *Festuca brachyphylla*	17 *F. rubra*
18 *Hedysarum mackenzii*	55 *Hylocomium* sp.	19 *Juncus* sp.
56 *Meesia triquetra*	44 *Ochrolechia upsaliensis*	20 *Oxytropis hudsonica*
45 *Pannaria* sp.	21 *Pedicularis* sp.	46 *Peltigera canina*
22 *Poa alpina*	34 *P. glauca*	23 *Polygonum viviparum*
24 *Rhododendron lapponicum*	25 *Salix callicarpaea*	26 *S. cordifolia*
27 *S. reticulata*	28 *Saxifraga tricuspidata*	29 *Shepherdia canadensis*
30 *Stellaria crassipes*	47 *Stereocaulon alpinum*	31 *Taraxacum* sp.
32 *Tofieldia* sp.	57 *Tomenthypnum* sp.	58 *Tortula ruralis*
33 *Vaccinium uliginosum*		

References

Agnew, A. D. Q. (1961). The ecology of *Juncus effusus* L. in North Wales. *J. Ecol.*, **49**, 83–102.

Agnew, A. D. Q. and Haines, R. W. (1960). Studies on the plant ecology of the Jazira of Central Iraq. *I. Bull. Col. Sci. Baghdad*, **5**, 41–60.

Alvin, K. L. (1960). Observations on the lichen ecology of South Haven peninsula, Studland Heath, Dorset. *J. Ecol.*, **48**, 331–9.

Anaya, A. L. and Del Amo, S. (1978). Allelopathic potential of *Ambrosia cumanensis*. H. B. K. (Compositae) in a tropical zone of Mexico. *J. Chem. Ecol.*, **4**, 289–304.

Anderson, D. J. (1961*a*). The structure of some upland plant communities in Caernarvonshire. II – The pattern shown by *Vaccinium myrtillus* and *Calluna vulgaris*. *J. Ecol.*, **49**, 731–8.

Anderson, D. J. (1961*b*). The structure of some upland plant communities in Caernarvonshire. I – The pattern shown by *Pteridium aquilinum*. *J. Ecol.*, **49**, 369–77.

Angevine, M. W. (1983). Variations in the demography of natural populations of the wild strawberries *Fragaria vesca* and *F. virginiana*. *J. Ecol.*, **71**, 959–74.

Archibald, E. E. A. (1948). Plant populations – I. A new application of Neyman's contagious distribution. *Ann. Bot., Lond.*, N.S., **12**, 221–35.

Archibald, E. E. A. (1950). Plant populations – II. The estimation of the number of individuals per unit area of species in heterogenous plant populations. *Ann. Bot., Lond.*, N.S., **14**, 7–21.

Ashby, E. (1935). The quantitative analysis of vegetation. *Ann. Bot., Lond.*, **49**, 779–802.

Ashby, E. (1948). Statistical ecology, a re-assessment. *Bot.Rev.*, **14**, 222–34.

Ashby, E. and Pidgeon, I. M. (1942). A new quantitative method of analysis of plant communities. *Aust. J. Sci.*, **5**, 19.

Aspinall, D. and Milthorpe, F. L. (1959). An analysis of competition between barley and white persicaria. I. The effects on growth. *Ann. appl. Biol.*, **47**, 156–72.

Austin, M. P. (1968). Pattern in a *Zerna erecta* dominated community. *J. Ecol.*, **56**, 197–218.

Austin. M. P. (1980). Searching for a model for use in vegetation analysis. *Vegetatio*, **42**, 11–21.

Austin, M. P. (1981). Permanent quadrats: An interface for theory and practice. *Vegetatio*, **46**, 1–10.

Austin, M. P. and Greig-Smith, P. (1968). The application of quantitative methods to vegetation survey. II. Some methodological problems of data from rain forest. *J. Ecol.*, **56**, 827–44.

Austin, M. P. and Noy-Meir, I. (1971). The problem of non-linearity in ordination: Experiments with two-gradient models. *J. Ecol.*, **59**, 763–74.

Austin, M. P. and Orloci, L. (1966). Geometric models in ecology. II. An evaluation of some ordination techniques. *J. Ecol.*, **54**, 217–27.

Barclay-Estrup, P. and Gimingham, C. H. (1969). The description and interpretation of cyclical processes in a heath community. I. Vegetational change in relation to the *Calluna* cycle. *J. Ecol.*, **57**, 737–58.

Barclay-Estrup, P. (1971). The description and interpretation of cyclical processes in a heath community. III. Micro-climate in relation to the *Calluna* cycle. *J. Ecol.*, **59**, 143–66.

Barkman, J. J. (1949). Le Fabronietum pusillae et quelques autres associations épiphytiques du Tessin. (Suisse Meridionale.) *Vegetatio*, **2**, 309.

Barnes, H. and Stanbury, F. A. (1951). A statistical study of plant distribution during the colonization and early development of vegetation on china clay residues. *J. Ecol.*, **39**, 171–81.

Beard, J. S. (1949). The natural vegetation of the Windward and Leeward Islands. *Oxf. For. Mem.*, No. 21.

Bedford, Duke of, and Pickering, S. U. (1919). *Science and Fruit growing.* Macmillan, London.

Beschel, R. E. (1961). Dating rock surfaces by lichen growth and its application to glaciology and physiography (Lichenometry). *Geology of the Arctic.* 1044–62. Univ. Toronto Press.

Billings, W. D. and Mooney, H. A. (1959). An apparent frost hummock-sorted polygon cycle in the alpine tundra of Wyoming. *Ecology,* **40**, 16–20.

Bjorkman, O. and Holmgren, P. (1963). Adaptability of the photosynthetic apparatus to light intensity in ecotypes from exposed and shaded habitats. *Physiol. Plant.,* **16**, 889–914.

Black, J. N. (1955). The influence of depth of sowing and temperature on pre-emergence weight changes in subterranean clover (*Trifolium subterraneum* L.) *Aust.J.Agric.Res.*, **6**, 203.

Black, J. N. (1956). The influence of seed size and depth of sowing on pre-emergence and early vegetative growth of subterranean clover (*Trifolium subterraneum* L.) *Aust. J. Agric. Res.*, **7**, 98–109.

Black, J. N. (1958). Competition between plants of different initial seed sizes in swards of subterranean clover (*Trifolium subterraneum* L.) with particular reference to leaf area and the light microclimate. *Aust. J. Agric. Res.*, **9**, 299–318.

Black, J. N. (1960). The significance of petiole length, leaf area, and light interception in competition between strains of subterranean clover (*Trifolium subterraneum* L.) grown in swards. *Aust. J. Agric. Res.*, **11**, 277–91.

Blackman, G. E. (1935). A study by statistical methods of the distribution of species in grassland associations. *Ann. Bot., Lond.,* **49**, 749–78.

Blackman, G. E. (1942). Statistical and ecological studies in the distribution of species in plant communities. I. Dispersion as a factor in the study of changes in plant populations. *Ann. Bot., Lond.,* **6**, 351–70.

Braun-Blanquet, J. (1927). *Pflanzensoziologie.* Springer, Wien.

Braun—Blanquet, J. (1932). *Plant Sociology*.

Braun-Blanquet, J. (1951). *Pflanzensoziologie*, 2nd edn. Springer, Wien.

Bray, R. J. and Curtis, J. T. (1957). An ordination of the upland forest communities of southern Wisconsin. *Ecol. Monogr.,* **27**, 325–49.

Brereton, A. J. (1971). The structure of the species populations in the initial stages of salt-marsh succession. *J. Ecol.,* **59**, 321–38.

Brown, R. T. and Curtis, T. T. (1952). The upland conifer-hardwood forests of northern Wisconsin, *Ecol. Monogr.,* **22**, 217–34.

Burden, J. J. (1980*a*). Intra-specific diversity in a natural population of *Trifolium repens. J. Ecol.,* **68**, 717–35.

Burden, J. J. (1980*b*). Variation in disease-resistance within a population of *Trifolium repens. J. Ecol.,* **68**, 737–44.

Cain,S.A. (1932). Concerning certain phytosociological concepts. *Ecol. Mongor.*, **2**, 475.

Cain, S. A. (1934). Studies on virgin hardwood forest. II. – A comparison of quadrat sizes in a quantitative phytosociological study of Nash's Woods, Posey County, Indiana. *Amer. Midl. Nat.,* **15** 529.

Cain, S. A. (1938). The species-area curve. *Amer. Midl. Nat.*, **19**, 573.

Caldwell, P. A. (1957). The spatial development of *Spartina* colonies growing without competition. *Ann. Bot., Lond.,* **21**, 203–16.

Cavers, P. B. and Harper, J. L. (1966). Germination polymorphism in *Rumex crispus* and *Rumex obtusifolius. J. Ecol.,* **54**, 367–82.

Chadwick, M. J. (1960). *Nardus stricta*, a weed of hill grazings. *The Biology of Weeds. Blackwell Sci. Pub.,* Oxford.

Chou, C-H. and Muller, C. H. (1972). Allelopathic mechanisms of *Artostaphylos glandulosa* var. *zacaensis. Amer. Mid. Nat.,* **88**, 324–47.

Christian, C. S. and Perry, R. A. (1953). The systematic description of plant communities by the use of symbols. *J. Ecol.,* **41**, 100–5.

Clapham, A. R. (1936). Overdispersion in grassland communities and the use of statistical methods in plant ecology. *J. Ecol.,* **24**, 332–51.

Clarke, G. M. (1980). *Statistics and Experimental Design*. Edward Arnold, London.

Clements, F. E. (1904). Development and structure of vegetation. *Rep. Bot. Surv. Nebra.,* **7**.

Clements, F. E. (1916). Plant succession. An analysis of the development of vegetation. *Carnegie Inst. Washington*. No. 242.
Colinvaux, P. A. (1973). *Introduction to Ecology*. Wiley, New York.
Connell, J. H. and Slatyer, R. O. (1977). Mechanisms of succession in natural communities and their role in community stability and organization. *Amer. Natur.*, **111**, 1119–44.
Cooper, C. F. (1960). Changes in vegetation. Structure and growth of south-western pine forests since white settlement. *Ecol. Monogr.*, **30**, 129–64.
Cooper, C. F. (1961). Pattern in ponderosa pine forests. *Ecology*, **42**, 493–9.
Cooper, W. S. (1923). The recent ecological history of Glacier Bay, Alaska. *Ecology*, **6**, 197.
Cooper, W. S. (1926). The fundamentals of vegetational change. *Ecology*, **7**, 391–413.
Cooper, W. S. (1931). A third expedition to Glacier Bay, Alaska. *Ecology*, **12**, 61–95.
Cooper, W. S. (1937). The problem of Glacier Bay, Alaska. A study of glacier variations. *Geogr. Rev.*, **27**, 37–62.
Cooper, W. S. (1939). A fourth expedition to Glacier Bay, Alaska. *Ecology*, **20**, 130–59.
Cottam, G. (1949). The phytosociology of an oak wood in south-western Wisconsin. *Ecology*, **30**, 271–87.
Cottam, G. and Curtis, J. T. (1956). The use of distance measures in phytosociological sampling. *Ecology*, **37**, 451–60.
Cowles, H. C. (1899). The ecological relations of the vegetation on the sand dunes of Lake Michigan. *Bot. Gaz.*, **27**, 95–117, 167–202, 281–308, 361–91.
Cowles, H. C. (1901). The physiographic ecology of Chicago and vicinity. A study of the origin, development and classification of plant societies. *Bot. Gaz.*, **31**, 73.
Cowles, S. (1982). Preliminary results investigating the effect of lichen ground cover on the growth of black spruce. *Nat. Can.*, **109**, 573–81.
Crocker, R. L. and Major, J. (1955). Soil development in relation to vegetation and surface age at Glacier Bay, Alaska. *J. Ecol.*, **43**, 427–48.
Curtis, J. T. (1947). The paloverde forest type near Gonivaves, Haiti, and its relation to the surrounding vegetation. *Caribb. Forester*, **8**, 1–26.
Curtis, J. T. and McIntosh, R. P. (1950). The inter-relations of certain analytic and synthetic phytosociological characters. *Ecology*, **31**, 434–55.
Curtis, J. T. and McIntosh, R. P. (1951). An upland forest continuum in the prairies-forest border region of Wisconsin. *Ecology*, **32**, 476–96.
Dagnelie, P. (1960). Contribution à l'étude des communautés végétales par l'analyse factorielle. *Bull. Serv. Carte. Phytogeogr. Sér. B.*, **5**, 7–71, 93–195.
Dahl,E. and Hadaĉ,E. (1949). Homogeneity of plant communities. *Studia bot.Cechosl.*, **10**, 159–76.
Dale, M. B. (1975). On objectives of methods of ordination. *Vegetatio*, **30**, 15–32.
David, F. N. and Moore, P. G. (1954). Notes on contagious distributions in plant populations. *Ann. Bot., Lond.*, N.S. **18**, 47–53.
Davis, T. A. W. and Richards. P. W. (1933–4). The vegetation of Moraballi Creek, British Guiana, an ecological study of a limited area of Tropical Rain Forest. Parts I and II. *J. Ecol.*, **21**, 350–84; **22**, 106–55.
De Candolle, A. P. (1832). *Physiologie végétale*, T. **3**, 1474–5. Paris.
De Vries, D. M. (1953). Objective combinations of species *Acta bot. neerl.*, **1**, 497–9.
Dice, L. R. (1952). Measure of the spacing between individuals within a population. *Contr. Lab. vertebr. Biol. Univ. Mich.*, **55**, 1–23.
Dietvorst, P. Maarel, E. van der, and Pulten, H. van der. (1982). A new approach to the minimal area of a plant community. *Vegetatio*, **50**, 77–91.
Donald,C.M. (1951). *Aust.J.Agric.Res.*, **2**, 355.
Donald,C.M. (1961). *Symp.Soc.Exp.Biol.*, **51**, 282–313.
Draper, N. and Smith, H. (1981). *Applied Regression Analysis*. John Wiley, New York.
Drury, W. H. and Nisbet, I. C. T. (1973). Succession. *J. Arnold, Arboretum*, **54**, 331–68.
Du Rietz, G. E. (1921). Zur methodlogischen Grundlage der Modernen Pflanzensoziologie. *Akad. Abh. Uppsala.*
Egler,F.E. (1954). Vegetation science concepts. I. Initial floristic composition – a factor in old-field vegetation development. *Vegetatio*, **4**, 412–17.
Ellenberg, H. (1979). *Zeigerwerte der Gefässpflanzen Mitteleuropas* 2nd edn. Scripta Geobotanica, 9. Goltze, Göttingen.

Emmett, H. E. G. and Ashby, E. (1934). Some observations between H-ion concentration of the soil and plant succession. *Ann. Bot., Lond.*, **68**, 869.

Evans, F. C. (1952). The influence of size of quadrat on the distributional patterns of plant populations. *Contr. Lab. vertebr. Biol. Univ. Mich.*, **54**, 1–15.

Field, W. O. (1947). Glacier recession in Muir Inlet, Glacier Bay, Alaska. *Geogr. Rev.*, **37**, 369–99.

Fisher, R. A. (1940). The precision of discriminant functions. *Ann. Eugenics,* **10**, 422–9.

Fisher, R. A. and Yates, F. (1948). *Statistical tables for biological, agricultural and medical research.* Oliver and Boyd, London.

Fisher, R. A. and Yates, F. (1957). *Statistical Tables*, 5th edition. Oliver and Boyd, London.

Flenley, J. R. (1969). The vegetation of the Wabag region, New Guinea highlands: A numerical study. *J. Ecol.*, **57**, 465–90.

Friedman, J., Orshan, G. and Ziger-Cfir, Y. (1977). Suppression of annuals by *Artemisia herba-alba* in the Negev desert of Israel. *J.Ecol.*, **65**, 413–26.

Gates, M. A. and Hansell, R. I. C. (1983). On the distinctness of clusters. *J. Theor. Biol.*, **101**, 263–73.

Gauch,H.G. Jr. (1982). *Multivariate Analysis in Community Ecology.* Cambridge University Press, Cambridge.

Gauch,H.G. Jr. and Whittaker,R.H. (1972*a*). Coenocline simulation. *Ecology*, **53**,446–51.

Gauch,H.G. Jr. and Whittaker,R.H. (1972*b*). Comparison of ordination techniques. *Ecology*, **53**, 868–75.

Gauch,H.G. Jr. and Whittaker,R.H. (1976). Simulation of community patterns. *Vegetatio*, **33**, 13–6.

Gauch,H.G. Jr. and Whittaker,R.H. (1981). Hierarchical classification of community data. *J.Ecol.*, **69**, 537–58.

Gauch,H.G. Jr., Whittaker, R.H., and Singer,S.B. (1981). A comparative study of nonmetric ordinations. *J.Ecol.*, **69**, 135–52.

Gauch, H. G. Jr., Whittaker, R. H. and Wentworth, R. T. (1977). A comparative study of reciprocal averaging and other ordination techniques. *J. Ecol.*, **65**, 157–74.

Gimingham, C. H., Pritchard, N. M. and Cormack, R. M. (1966). Interpretation of a vegetational mosaic on limestone in the island of Gotland. *J. Ecol.*, **54**, 481–502.

Gittins, R. (1965). Multivariate approaches to a limestone grassland community. III. A comparative study of ordination and association – analysis. *J. Ecol.*, **53**, 411–25.

Gleason, H. A. (1917). The structure and development of the plant association. *Bull. Torrey Bot. Club*, **43**, 463–81.

Gleason,H.A. (1920). Some applications of the quadrat method. *Bull. Torrey Bot.Club*, **47**, 21–33.

Gleason, H. A. (1922). The vegetational history of the Middle West. *Ann. Ass. Amer. Geogr.*, **12**, 39–85.

Gleason,H.A. (1926). The individualistic concept of the plant association. *Bull. Torrey Bot. Club,* **53**, 7–26.

Godwin, H. and Conway, V. M. (1939). The ecology of a raised bog near Tregaron, Cardiganshire. *J. Ecol.*, **27**, 313–59.

Goodall, D. W. (1952*a*). Some considerations in the use of point quadrats for the analysis of vegetation. *Aust. J. sci. Res.*, **5**, 1–41.

Goodall, D. W. (1952*b*). Quantitative aspects of plant distribution. *Biol. Rev.*, **27**, 194–245.

Goodall, D. W. (1953). Objective methods for the classification of vegetation. I. – The use of positive interspecific correlation. *Aust. J. Bot.*, **1**, 39–63.

Goodall, D. W. (1954*a*). Vegetational classification and vegetational continua. *Angew. Pflanzensoziologie,* **1**, 168–82.

Goodall, D. W. (1954*b*). Objective methods for the classification of vegetation. 3 – An essay in the use of factor analysis. *Aust.J.Bot.*, **2**, 304–24.

Gorham, E. (1953). Chemical studies on the soils and vegetation of water-logged habitats in the Engish Lake District. *J. Ecol.*, **41**, 345–60.

Grant, R. E. and Clebsch, E. E. C. (1975). The allelopathic influences of *Sassafras albidum* in old-field succession in Tennessee. *Ecology,* **56**, 604–15.

Gray, R. and Bonner, J. (1948). Structure, determination and synthesis of a plant growth inhibitor, 3-acetyl-6-methoxybenzaldehyde found in the leaves of *Encelia farinosa. J. Amer. chem. Soc.*, **70**, 1249–53.

Greig-Smith, P. (1952*a*). The use of random and contiguous quadrats in the study of the structure of plant communities. *Ann. Bot., Lond.*, N.S. **16**, 293–316.

Greig-Smith, P. (1952*b*). Ecological observations on degraded and secondary forest in Trinidad, British West Indies. *J. Ecol.*, **40**, 316–30.

Greig-Smith, P. (1957). *Quantitative Plant Ecology*. Butterworths, London.

Greig-Smith, P. (1961*a*). Data on pattern within plant communities. I. – The analysis of pattern. *J. Ecol.*, **49**, 695–702.

Greig-Smith, P. (1961*b*). Data on pattern within plant communities. II. *Ammophila arenaria* (L.) Link. *J. Ecol.*, **49**, 703–8.

Greig-Smith, P. (1979). Pattern in vegetation. *J. Ecol.*, **67**, 755–79.

Greig-Smith, P. (1980). The development of numerical classification and ordination. *Vegetatio*, **42**, 1–9.

Greig-Smith, P. (1983). *Quantitative Plant Ecology, third edition*. Blackwell Scientific Publications, Oxford, London, Edinburgh.

Greig-Smith, P., Austin, M. P. and Whitmore, T. C. (1967). The application of quantitative methods to vegetation survey. I. Association – analysis and principal component ordination of rain-forest. *J. Ecol.*, **55**, 483–503.

Grime, J. P. (1974). Vegetation classification by reference to strategies. *Nature*, **250**, 26–31.

Grime, J.P. (1979). *Plant Strategies and Vegetation Processes*. John Wiley, Chichester.

Grime, J. P. (1984). Towards a functional description of vegetation. In: *Population structure of Vegetation*. Ed. J. White and J. Beeffink. Junk, The Hague.

Hall, J. B. (1967). Some aspects of the ecology of *Brachypodium pinnatum*. Ph.D. thesis, Imperial College, London.

Hall, J.B. (1971). Pattern in a chalk grassland community. *J. Ecol.*, **59**, 749–62.

Handley, W. R. C. (1963). Mycorrhizal associations and *Calluna* heathland afforestation. Bull. Forest Commun, London, 36.

Harper, J. L. (1960). Factors controlling plant numbers. *The Biology of Weeds*. (Br. Ecol. Soc. Symp.) Blackwell Sci. Pub., Oxford. pp. 119–32.

Harper, J. L. (1961). Approaches to the study of plant competition. *Symp. Soc. exp. Biol.*, **15**, 1–39.

Harper, J. L. (1977). *Population Biology of Plants*. Academic Press London, New York, San Francisco.

Harper, J. L. (1982). After description. In: *The Plant Community as a Working Mechanisms*. Ed. E. I. Newman, Blackwell Scientific Publications.

Harper, J. L. and Chancellor, A. P. (1959). The comparative biology of closely related species living in the same area. IV. *Rumex:* Interference between individuals in populations of one and two species. *J. Ecol.*, **47**, 679–95.

Harper, J. L. and Clatworthy, J. N. (1963). The comparative biology of closely related species. VI. Analysis of the growth of *Trifolium repens* and *T. fragiferum* in pure and mixed populations. *J. exp. Bot.*, **14**, 172–90.

Harper, J. L., Landragin, P. A. and Ludwig, J. W. (1955). The influence of environment on seed and seedling mortality. I. The influence of time of planting on the germination of maize. *New Phytol.*, **54**, 107–18.

Harper, J. L. and McNaughton, J. H. (1962). The comparative biology of closely related species living in the same area. VII. Interference between individuals in pure and mixed populations of *Papaver* species. *New Phytol.*, **61**, 175–88.

Harper, J.L. and Sagar, G.R. (1953). Some aspects of the ecology of buttercups in permanent grassland. *Proc. Br. Weed Control Conf. 1953*, pp. 256–65.

Harper, J. L., Williams, J. T. and Sagar, G. R. (1965). The behaviour of seeds in soil. I. The heterogeneity of soil surfaces and its role in determining the establishment of plants from seed. *J. Ecol.*, **53**, 273–86.

Harrison, C. M. (1970). The phytosociology of certain English heathland communities. *J. Ecol.*, **58**, 573–89.

Heslop-Harrison, J. (1964). Forty years of Genecology. In *Adv. Ecol. Res.*, **2**. Academic Press, London.

Hill, M. O. (1973). Reciprocal averaging: An eigenvector method of ordination. *J. Ecol.*, **61**, 237–49.

Hill, M. O. (1974). Correspondence analysis. A neglected multivariate method. *J. Roy. Stat. Soc. Series C.*, **23**, 340–54.

Hill, M. O. (1979*a*). *DECORANA – A FORTRAN program for detrended correspondence analysis and reciprocal averaging*. Ithaca, N.Y., Cornell University.

Hill,M.O. (1979*b*) *TWINSPAN – A FORTRAN program for arranging multi-variate data in an ordered two-way table by classification of the individuals and attributes*. Ithaca, N.Y., Cornell University.

Hill, M. O., and Gauch, H. G. (1980). Detrended correspondence analysis, an improved ordination technique. *Vegetatio*, **42**, 47–58.

Hirschfeld, H. O. (1935). A connection between correlation and contingency. *Proc. Camb. Phil. Soc.*, **31**, 520–4.

Hope-Simpson, J. F. (1940). On the errors in the ordinary use of subjective frequency estimations in grassland. *J. Ecol.*, **28**, 193–209.

Hopkins, B. (1955). The species-area relations of plant communities. *J. Ecol.*, **43**, 409–26.

Hopkins, B. (1957). The concept of the minimal area. *J. Ecol.*, **45**, 441–9.

Hopkins, B. (1968). Vegetation of the Olokemeji forest reserve, Nigeria. V. The vegetation on the savanna site with special reference to its seasonal changes. *J. Ecol.*, **56**, 97–115.

Horn, H. S. (1974). The ecology of secondary succession. *Ann. Rev. Ecol. Syst.*, **5**, 25–37.

Horn, H. S. (1975). Markovian properties of forest succession. In: *Ecology and Evolution of Communities*. Ed. M. Cody and J. Diamond. Harvard University Press. Cambridge, Mass.

Horn. H. S. (1981). Some causes of variety in patterns of secondary succession. In: *Forest Succession. Concepts and application*. Ed. D. C. West, H. H. Shugart, and D. B. Botkin. Springer-Verlag. New York, Heidelberg, Berlin.

Horsley,S.B. (1977). Allelopathic interference among plants II. Physiological modes of action. In *Proceedings of the fourth North American forest biology workshop*. H.E.Wilcox and A. F. Hamer, Eds. School of Continuing Education, College of Environmental Science and Forestry, Syracuse, New York.

Hulst,R.V. (1978). On the dynamics of vegetation: patterns of environmental and vegetational change. *Vegetatio*, **38**, 65–75.

Hunt,R. (1978). *Plant Growth Analysis*. Edward Arnold, London.

Hunt, R. (1982). *Plant Growth Curves*. Edward Arnold, London.

Ivimey-Cook, R. B., Proctor, M. C. F. and Wigston, D. L. (1969). On the problem of the 'R/Q' terminology in multivariate analyses of biological data. *J. Ecol.*, **57**, 673–5.

Jackson, J. R. and Willemson, R. W. (1976). Allelopathy in the first stages of secondary succession on the piedmont of New Jersey. *Amer. J. Bot.*, **63**, 1015–23.

Jobidon, R. and Thibault, J. R. (1982). Allelopathic growth inhibition of nodulated and unnodulated *Alnus crispa* seedlings by *Populus balsamifera*. *Amer. J. Bot.*, **69**, 1213–23.

Johnson, E. A. and Rowe, J. S. (1975). Fire in the wintering ground of the Beverley Caribou herd. *Amer. Midl. Nat.*, **94**, 1–14.

Jones, E. W. (1955–6). Ecological studies on the rain forest of southern Nigeria. IV. – The plateau forest of the Okomu Forest Reserve. *J. Ecol.*, **43**, 564–94; **44**, 83–117.

Kaminsky, R. (1981). The microbial origin of the allelopathic potential of *Adenostoma fasciculatum* H & A. *Ecol. Monogr.*, **51**, 365–82.

Kapustka, L. A. and Rice, E. L. (1976). Acetylene reduction (N_2-fixation) in soil and old field succession in central Oklahoma. *Soil Biol. Biochem*, **8**, 497–503.

Kays,S. and Harper,J.L. (1974). The regulation of plant and tiller density in a grass sward. *J. Ecol.*, **62**, 97–105.

Kendall, M. G. (1957). *A course in multivariate analysis*. London.

Kershaw, K. A. (1957). The use of cover and frequency in the detection of pattern in plant communities. *Ecology*, **38**, 291–9.

Kershaw, K. A. (1958, 1959). An investigation of the structure of a grassland community. I. The pattern of *Agrostis tenuis*.*J.Ecol.*, **46**, 571–92; II. The pattern of *Dactylis glomerata, Lolium perenne* and *Trifolium repens*; III. Discussion and conclusions. *J.Ecol.*, **47**, 31–53.

Kershaw, K. A. (1960*a*). The detection of pattern and association. *J. Ecol.*, **48**, 233–42.

Kershaw, K. A. (1960*b*). Cyclic and pattern phenomena as exhibited by *Alchemilla alpina*. *J. Ecol.*, **48**, 443–53.

Kershaw, K. A. (1961). Association and co-variance analysis of plant communities. *J. Ecol.*, **49**, 643–54.

Kershaw, K. A. (1962*a*). Quantitative ecological studies from Landmannahellir, Iceland. I. *Eriophorum angustifolium*. *J.Ecol*,, **50**, 163–9.

Kershaw, K. A. (1962*b*). Quantitative ecological studies from Landmannahellir, Iceland. II. The rhizome behaviour of *Carex bigelowii* and *Calamagrostis neglecta*. *J. Ecol.*, **50**, 171–9.

Kershaw, K. A. (1962*c*). Quantitative ecological studies from Landmannahellir, Iceland. III. The variation of performance of *Carex bigelowii*. *J. Ecol.*, **50**, 393–9.

Kershaw, K. A. (1963). Pattern in vegetation and its causality. *Ecology*, **44**, 377–88.

Kershaw, K. A. (1968). Classification and ordination of Nigerian savanna vegetation. *J. Ecol.*, **56**, 467–82.

Kershaw, K. A. (1976). The vegetational zonation of the East Pen Island salt marshes, Hudson Bay. *Can. J. Bot.*, **54**, 5–13.

Kershaw, K. A. (1977). Studies on lichen-dominated systems. XX. An examination of some aspects of lichen woodlands in Canada. *Can. J. Bot.*, **55**, 393–410.

Kershaw, K. A. (1978). The role of lichens in boreal tundra transition areas. *Bryologist*, **81**, 294–306.

Kershaw,K.A. and Rouse,W.R. (1971). Studies on lichen-dominated systems. II. The growth pattern of *Cladonia alpestris* and *Cladonia rangiferina*. *Can. J. Bot.*, **49**, 1401–10.

Kershaw, K. A. and Rouse, W. R. (1973). Studies on a lichen-dominated system. V. A primary survey of a raised beach system in N.W. Ontario. *Can. J. Bot.*, **51**, 1285–307.

Kershaw, K. A. and Rouse, W. R. (1976). The impact of fire on forest and tundra ecosystems. *INA Pub. No.* QS-8117-000-EE-A1.

Kershaw, K. A., Rouse, W. R. and Bunting, B. T. (1975). The impact of fire on forest and tundra ecosystems. *INA Pub. No.* QS-8038-000-EE-A1.

Kershaw, K. A. and Shepherd, R. (1972). Computer display graphics for principal component analysis and vegetation ordination studies. *Can. J. Bot.*, **50**, 2239–50.

Kershaw, K. A. and Smith, M. M. (1978). Studies on lichen-dominated systems. XXI. The control of seasonal rates of net photosynthesis by moisture, light and temperature in *Stereocaulon paschale* (L.) Hoffm. *Can. J. Bot.*, **56**, 2825–30.

Kershaw, K. A. and Tallis, J. H. (1958). Pattern in the high level *Juncus squarrosus* community. *J. Ecol.*, **46**, 739–48.

Larson, D. W. and Kershaw, K. A. (1974). Studies on lichen-dominated systems. VII. Interactions of the general lichen heath with edaphic factors. *Can. J. Bot.*, **52**, 1163–76.

Lawrence, D. B. (1953). Development of vegetation and soil in south-eastern Alaska, with special reference to the accumulation of nitrogen. Final Report ONR, Project NR, pp. 160–83.

Lawrence, D. B. (1958). Glaciers and vegetation in south-eastern Alaska. *Amer. Scient.*, **46**, 89–122.

Lazenby, A. (1955). Germination and establishment of *Juncus effusus* L. The effect of different companion species and of variation in soil and fertility conditions. *J. Ecol.*, **43**, 103–19.

Lodhi, M. A. K. (1979). Inhibition of nitrifying bacteria, nitrification and mineralization in spoil soils as related to their successional stages. *Bull. Torrey Bot. Club.*, **106**, 284–9.

Lodhi, M. A. K. and Killingbeck, K. T. (1980). Allelopathic inhibition of nitrification and nitrifying bacteria in a ponderosa pine. (*Pinus ponderosa* Dougl.) community. *Amer. J. Bot.*, **67**, 1423–9.

Looney, J. H. H. and Kershaw, K. A. (1985*a*). Ordination distortion and principal component analysis – some available alternatives. Analysis of a simple linear salt marsh gradient. *Vegetatio* (in submission).

Looney, J. H. H. and Kershaw, K. A. (1985*b*). Ordination distortion and principal component analysis – some available alternatives. Analysis of a beach ridge data set. *Ecology* (in submission).

Maarel, E. van der (1982). On the manipulation and editing of phytosociological and ecological data. *Vegetatio*, **50**, 71–6.

Maarel, E. van der, Janssen, J. G. M. and Louppen, J. M. W. (1978). TABORD, a program for structuring phyosociological tables. *Vegetatio*, **38**, 143–56.

MacArthur, R. H. and Wilson, E. O. (1967). *The Theory of Island Biography*. Princeton University Press, Princeton, N.J.

Maikawa, E. and Kershaw, K. A. (1976). Studies on lichen – dominated systems. XIX. The postfire recovery sequence of Black Spruce-lichen woodland in the Abitau Lake Region, N.W.T. *Can. J. Bot.*, **54**, 2679–87.

Marks, P. L. (1974). The role of pin cherry, (*Prunus pensylvanica* L) in the maintenance of stability in northern hardwood ecosystems. *Ecol. Monogr.*, **44,** 73–88.

McKell, C. M., Perrier, E. R. and Stebbins, G. L. (1960). Responses of two sub-species of orchard grass (*Dactylis glomerata* subspecies *lusitanica* and *judaica*) to increasing soil moisture stress. *Ecology,* **41**, 772–8.

McVean, D. N. and Ratcliffe, D. A. (1962). *Plant communities of the Scottish Highlands.* H.M.S.O., London.

Mead, R. (1974). A test for spatial pattern at several scales using data from a grid of contiguous quadrats. *Biometrics,* **30**, 295–307.

Michelmore, A. P. G. (1934). Vegetational succession and regional surveys with special reference to tropical Africa. *J. Ecol.,* **22**, 313–17.

Mooney, H. A. and Billing, W. D. (1961). Comparative physiological ecology of arctic and alpine populations of *Oxyria digyna. Ecol. Monogr.*, **31**, 1–29.

Moore, J. J. (1962). The Braun-Blanquet system. A reassessment. *J. Ecol.*, **50**, 761–9.

Moore. P. G. (1953). A test for non-randomness in plant populations. *Ann. Bot., Lond.*, N.S. **17**, 57–62.

Muller, C. H. (1969). Allelopathy as a factor in ecological process. *Vegetatio,* **18**, 348–57.

Neal, M. W. and Kershaw, K. A. (1973*a*). Studies on lichen-dominated systems III. Phytosociology of a raised-beach system near Cape Henrietta Maria, N. Ontario. *Can. J. Bot.*, **51**, 1115–25.

Neal, M. W. and Kershaw, K. A. (1973*b*). Studies on lichen-dominated systems. IV. The objective analysis of Cape Henrietta Maria raised beach systems. *Can. J. Bot.*, **51**, 1177–90.

Newnham, R. M. (1968). A classification of climate by principal component analysis and its relationship to tree species distribution. *Forest Sci.* **14**, 254–64.

Nicholson, L. A. and Robertson, R. A. (1958). Some observations on the ecology on an upland grazing in north-east Scotland with particular reference to Callunetum. *J. Ecol.*, **46**, 239–70.

Noble, I. R. and Slatyer, R. O. (1980). The use of vital attributes to predict successional changes in plant communities subject to recurrent disturbances. *Vegetatio*, **43**, 5–21.

Noy-Meir, I., and Austin, M. P. (1970). Principal component ordination and simulated vegetational data. *Ecology*, **51**, 551–2.

Noy-Meir, I., Walker D. and Williams, W. T. (1975). Data transformations in ecological ordination. II. On the meaning of data standardization. *J. Ecol.*, **63**, 779–800.

Odum, E. P. (1953). *Fundamentals of ecology*. W. B. Saunders, Philadephia.

Olson, J. S. (1958). Rates of succession and soil changes on Southern Lake, Michigan, sand dunes. *Bot. Gaz.*, **119**, 125–70.

Orloci, L. (1966). Geometric models in ecology. I. The theory and application of some ordination methods. *J. Ecol.*, **54**, 193–215.

Oswald, H. (1923). *Die vegetation des Hochmoores Komosse*. Akad. Abh. Uppsala.

Ovington, J. D. (1953). A study of invasion by *Holcus mollis* L. *J. Ecol.*, **41**, 35–52.

Parker, R. E. (1979). *Introductory Statistics for Biology*. Edward Arnold, London.

Pearson, G. A. (1942). Herbaceous vegetation a factor in natural regeneration of Ponderosa Pine in the south-west. *Ecol. Monogr.*, **12**, 315–38.

Peet, R. K. and Christensen, N. L. (1980). Succession: A population process. *Vegetatio,* **43**, 131–49.

Pénzes, A. (1958). A survival of stoloniferous plant colonies (polycormons) of a relict character. *Biologia, Sav.*, **13**, 253–64.

Pénzes, A. (1960). Über die Morphologie Dynamik und Zönologisch Volle der Sprosskolonien – bildenden Pflanzen (Polycormone). *Geobotanika,* **6**, 501–515

Persson, S. (1981); Ecological indicator values as an aid in the interpretation of ordination diagrams. *J. Ecol.*, **69**, 71–84.

Phillips, J. (1934–5). Succession, development, the climax and the complex organism. An analysis of concepts. Parts I and II. *J. Ecol.*, *22*, 559–71; **23**, 210–46; **23**, 488–508.

Phillips, M. E. (1954*a*). Studies in the quantitative morphology and ecology of *Eriophorum angustifolium* Roth. II. Competition and dispersion. *J.Ecol.*, **42**, 187–210.

Phillips, M. E. (1954*b*). Studies in the quantitative morphology and ecology of *Eriophorum angustifolium* Roth. III. The leafy shoot. *New Phytol.*, **53**, 312–43.

Pianka, E. R. (1983). *Evolutionary Ecology, third edition*. Harper and Row, New York.

Pickett, S. T. A. (1976). Succession: An evolutionary interpretation. *Amer. Natur.*, **110**, 107–19.

Pickett, S. T. A. (1982). Population patterns through twenty years of oldfield succession. *Vegetatio*, **49**, 45–59.

Pierce, W. G. and Kershaw, K. A. (1976). Studies on lichen-dominated systems. XVII. The colonization of young raised beaches in N.W. Ontario. *Can.J.Bot.*, **54**, 1672–83

Pigott, C. D. (1956). The vegetation of upper Teesdale in the north Pennines. *J. Ecol.*, **44**, 545–86.

Poore, M. E. D. (1955–6). The use of phytosociological methods in ecological investigations. I. The Braun-Blanquet System. *J. Ecol.*, **43**, 226–44; II. Practical issues involved in an attempt to apply the Braun-Blanquet system. *J. Ecol.*, **43**, 245–69; III. Practical applications. *J. Ecol.*, **43**, 606–51; IV. General discussion of phytosiological problems. *J. Ecol.*, **44**, 28–50.

Poore, M. E. D. (1968). Studies in Malaysian rain forest. I. The forests on triassic sediments in Jengka forest reserve. *J. Ecol.*, **56**, 143–96.

Popay, A. I. and Roberts, E. H. (1970*a*). Factors involved in the dormancy and germination of *Capsella bursa-pastoris* (L.) Medik. and *Senecio vulgaris* L. *J. Ecol.*, **58**, 103–22.

Popay, A. I. and Roberts, E. H. (1970*b*). Ecology of *Capsella bursa-pastoris* (L.) Medik. and *Senecio vulgaris* L. in relation to germination behaviour. *J. Ecol.*, **58**, 123–39.

Popma, J., Muncina, L., Tongeren, O., van; and Maarel, E. van der (1983); On the determination of optimal levels in phytosociological classification. *Vegetatio*, **52**, 66–75.

Proctor, J., Anderson, J. M., Chai, P. and Vallack, H. W. (1983). Ecological studies in four contrasting lowland rain forests in Gunung Mulu National Park, Sarawak. *J. Ecol.*, **71**, 237–60.

Raunkiaer, C. (1909). Formationsundersogelse og Formationsstatisk. *Bot. Tidsskr.*, **30**, 20–132.

Raunkiaer, C. (1928). Dominansareal, Artstaethed og Formationsdominanter. Kgl. Danske Videnskabernes Selskab. *Biol. Medd. Kbh.*, **7**, 1.

Raunkiaer, C. (1934). *The Life Forms of Plants and Statistical Plant Geography*. Translated by Carter, Fausboll and Tansley; Oxford Univ. Press.

Rice, E. L. (1974). *Allelopathy*. Academic Press, New York.

Rice, E. L. (1979). Allelopathy – an update. *Bot. Rev.*, **45**, 15–109.

Richards, P. W. (1936). Ecological observations on the rain forest of Mount Dulit, Sarawak. Parts I and II. *J. Ecol.*, **24**, 1–37; **24**, 340–60.

Richards, P. W. (1952). *The Tropical Rain Forest.* Univ. Press. Cambridge.

Richards, P. W., Tansley, A. G. and Watt, A. S. (1940). The recording of structure, life form and flora of tropical forest communities as a basis for their classification. *J. Ecol.*, **28**, 224–39.

Robinson, R. K. (1972). The production by roots of *Calluna vulgaris* of a factor inhibitory to growth of some mycorrhizal fungi. *J. Ecol.*, **60**, 219–24.

Rouse, W. R. (1976). Microclimate changes accompanying burning in subarctic lichen woodland. *Arctic and Alpine Res.*, **8**, 351–76.

Russell,E.W. (1961). *Soil Conditions and Plant Growth, ninth edition*. London.

Sarma, K. K. V. (1974). Allelopathic potential of *Echinops echinatus* and *Solanum swattense* on seed germination of *Argemone maxicana. Trop. Ecol.*, **15**, 156–7.

Sarukhán,J. and Harper,J.L. (1973). Studies on plant demography: *Ranunculus repens* L., *R. bulbosus* L., and *R. acris* L. Population flux and survivorship. *J. Ecol.*, **61**, 675–716.

Selleck, G. W. (1960). The climax concept. *Bot. Rev.*, **26**, 534–45.

Silvertown,J.W. (1982). *Introduction to Plant Population Ecology*. Longman, London and New York.

Skellam, J. G. (1952). Studies in statistical ecology. I. Spatial pattern. *Biometrica*, **39**, 346–62.

Snaydon, R. W. (1962). Micro-distribution of *Trifolium repens* and its relation to soil factors. *J. Ecol.*, **50**, 133–43.

Snedecor, G. W. and Cochran, W. G. (1980). *Statistical Methods*. (7th Ed.) Iowa State University press, Ames, Iowa.

Stokes,P. (1965). *Temperature and seed dormancy. Handb.Pfl.Phusoil.*, **15(2)**, 746–803.

Svedburg,T. (1922). Ettbidrag till de statiska metodernas anvandning inom voxtbiologien. *Svensk bot. Tidskr.*, **16**, 1–8.

Swan, J. M. A. (1970). An examination of some ordination problems by use of simulated vegetational data. *Ecology*, **51**, 89–102.

Tansley, A. G. (1916). The development of vegetation. *J. Ecol.*, **4**, 198–204.

Tansley, A. G. (1920). The classification of vegetation and the concept of development. *J. Ecol.*, **8**, 118–49.

Thibault,J-R., Fortin,J-A. and Smirnoff,W.A. (1982). In vitro allelopathic inhibition of nitrification by balsam poplar and balsam fir. *Amer. J. Bot.*, **69**, 676–9.

Thompson, H. R. (1958). The statistical study of plant distribution patterns using a grid of quadrats. *Aust. J. Bot.*, **6**, 322–42.

Thompson, G. W. (1952). Measures of plant aggregation based on contagious distributions. *Contr. Lab. Vertebr. Biol. Univ. Mich.*, **53**, 1–16.

Thurston, J. M. (1959). *The Biology of Weeds*. Blackwell. Oxford. pp. 69–82.

Toole, E. H., Hendricks, S. B. and Borthwick, H. A. and Toole, V. K. (1956). Physiology of seed germination. *A. Rev. Physiol.*, **7**, 299–324.

Turkington, R. A. (1975). Relationships between neighbours among species of permanent grassland. Ph.D. thesis, University of Wales.

Usher, M. B. (1969). The relation between mean square and block size in the analysis of similar patterns. *J. Ecol.*, **57**, 505–14.

Usher, M. B. (1975). Analysis of pattern in real and artificial plant populations. *J. Ecol.*, **63**, 569–85.

Vasilevich, V. I. (1961). Association between species and the structures of a phytocoenosis. *Dokl. obsch. Sobr. Ak. Nauk, S.S.S.R.*, **189**, 1001–4.

Vegis, A. (1963). Climatic control of germination, bud break, and dormancy. *Environmental Control of Plant Growth,* pp. 265–87. N.Y.

Walker, D. (1970). Direction and rate in some British postglacial hydroseres. In: *The vegetational history of the British Isles*. Ed. D. Walker and R. West. Cambridge University Press, Cambridge.

Ward, S. D. (1970). The phytosociology of *Calluna-Arctostaphylos* heaths in Scotland and Scandinavia. *J. Ecol.*, **58**, 847–63.

Warming, E. (1896). Lehrbuch der Oekologischen. *Pflanzengeographie*.

Warren Wilson, J. (1952). Vegetational patterns associated with soil movement on Jan Mayen Island. *J. Ecol.*, **40**, 249–64.

Watt,A.S. (1919). On the cause of failure of natural regeneration in British oak wood. *J. Ecol.*, **17**, 173–203.

Watt, A. S. (1925). On the ecology of British beechwoods with special reference to their regeneration. II. The development and structure of beech communities on the South Downs. *J. Ecol.*, **13**, 27–73.

Watt, A. S. (1940). Studies in the Ecology of Breckland. IV. The Grass Heath. *J. Ecol.*, **28**, 42–70.

Watt, A. S. (1947*a*). Pattern and process in the plant community. *J. Ecol.*, **35**, 1–22.

Watt, A. S. (1947*b*). Contributions to the ecology of bracken. IV. The structure of the community. *New Phytol.*, **46**, 97–121.

Watt, A. S. (1955). Bracken versus heather, a study in plant sociology. *J. Ecol.*, **43**, 490–506.

Watt, A. S. (1962). The effect of excluding rabbits from grassland. A (Xerobrometum) in Brackland. 1936–60. *J. Ecol.*, **50**, 181–98.

Watt, A. S. (1981). Further observations on the effects of excluding rabbits from grassland A in East Breckland: The pattern of change and factors affecting it (1936–73). *J. Ecol.*, **69**, 509–36.

Watt, A. S. and Fraser, G. K. (1933). Tree roots in the field layer. *J. Ecol.*, **21**, 404–14.

Weatherell,J. (1957). The use of nurse species in the afforestation of upland heaths. *Quat.J. For.*, LI, 1—7.

Webb, D. A. (1954). Is the classification of plant communities either possible or desirable? *Saer. bot. Tidssk.*, **51**, 362–70.

Welch, J. R. (1960). Observations on deciduous woodland in the eastern province of Tanganyika. *J. Ecol.*, **48**, 557–73.

Went, F. W. (1942). The dependence of certain annual plants on shrubs in South California deserts. *Bull. Torrey bot. Cl.*, **69**, 100–14.

Went, F. W. (1957). *The Experimental Control of Plant Growth.* Chronica Botanica Co., Waltham.

Wernham, C. C. (1951). Cold testing of corn. A chronological and critical review. *Pa. St. Coll. Agric. Exp. Sta. Progr. Report No. 47.*

Whitaker, R. A. (1953). A consideration of climax theory: The climax as a population and pattern. *Ecol. Monogr.*, **23**, 41–78.

White, J. and Harper, J. L. (1970). Correlated changes in plant size and number in plant populations. *J. Ecol.*, **58**, 467–85.

Whitford, P. B. (1949). Distribution of woodland plants in relation to succession and clonal growth. *Ecology*, **30**, 199–208.

Whittaker, R. H. (1967). Gradient analysis of vegetation. *Biol. Rev.*, **42**, 207–64.

Williams, W. T. and Dale, M. B. (1965). Fundamental problems in numerical taxonomy. *Adv. Bot. Res.*, **2**, 35–68.

Williams, W. T. and Lambert, J. M. (1959). Multivariate methods in plant ecology. I. Association analysis in plant communities. *J. Ecol.*, **47**, 83–101.

Williams, W. T. and Lambert, J. M. (1960). Multivariate methods in plant ecology. II. The use of an electronic digital computer for association-analysis. *J. Ecol.*, **48**, 689–710.

Williams, W. T. and Lambert, J. M. (1961). Multivariate methods in plant ecology. III. Inverse association-analysis. *J. Ecol.*, **49**, 717–29.

Woolhouse, H. W. (1981). Aspects of the carbon and energy requirements of photosynthesis considered in relation to environmental constraints. In *Physiological ecology: an Evolutionary Approach to Resource Use.* Ed. C. R. Townsend and P. Calow. Blackwell, London.

Yarranton, G. A. and Green, W. G. E. (1966). The distributional pattern of crustose lichens on limestone cliffs at Rattlesnake Point, Ontario. *Bryologist*, **69**, 450–61.

Yoda, K., Kira, T., Ogawa, H. and Hozumi, H. (1963). Self-thinning in over-crowded pure stands under cultivated and natural conditions. *J. Biol. Osaka Cy. Univ.*, **14**, 107–29.

Zinke, P. J. (1962). The pattern of individual forest trees on soil properties. *Ecology*, **43**, 130–3.

Index